Introduction to Optimization of Structures

N.V. Banichuk

Introduction to Optimization of Structures

With 66 Illustrations

Springer-Verlag
New York Berlin Heidelberg
London Paris Tokyo Hong Kong

N.V. Banichuk
Institute for Problems in Mechanics
USSR Academy of Sciences
117526 Moscow
Soviet Union

Translator
Vadim Komkov
Mathematics and Computer Science Department
Air Force Institute of Technology
Wright Patterson Air Force Base, OH 45433
USA

Original Russian edition *Vvedenie v Optimizatsiyu Konstruktsiy* published © 1986 by Nauka, Moscow.

Library of Congress Cataloging-in-Publication Data
Banichuk, Nikolaĭ Vladimirovich.
[Vvedenie v optimizat͡siiu konstrukt͡siĭ. English]
Introduction to optimization of structures/N.V. Banichuk.
p. cm.
Translation of: Vvedenie v optimizat͡siiu konstrukt͡siĭ.
Includes bibliographical references.
ISBN 0-387-97212-9 (alk. paper)
1. Structures, Theory of. I. Title.
TA645.B3513 1990
624.1′7—dc20 89-26165

Printed on acid-free paper.

Typeset by Asco Trade Typesetting Ltd., Hong Kong.
Printed and bound by Edwards Brothers, Inc., Ann Arbor, Michigan.
Printed in the United States of America.

9 8 7 6 5 4 3 2 1

ISBN 0-387-97212-9 Springer-Verlag New York Berlin Heidelberg
ISBN 3-540-97212-9 Springer-Verlag Berlin Heidelberg New York

Contents

Author's Introduction

Recently, substantial advances have been made toward the solution of certain important problems of mechanics related to volumetric reduction and the improvement of the mechanical properties of structures.

The concepts of the "best" structures, in one sense or another, and of criteria that are assigned to their performance have been generalized. Numerical optimization techniques have also been improved, thus permitting us to formulate effective estimates of sensitivity of some of the important structural properties to changes in design parameters, and thus to analyze techniques for obtaining optimal solutions. Such results permit, at least in part, one to make extensive use of optimization techniques for the eventual establishment of automated (i.e., computer-assisted) design of systems. However, there are many unsolved problems in optimal design theory that are now being intensively studied.

This volume contains an exposition of some of the fundamental concepts, as well as an attempt to present the "state of the art" in the theory of optimal design. It consists of two parts. The purpose of the first part is to introduce the reader to the theory and techniques of optimal design. Here we offer an exposition of problems of optimal design and techniques for transforming them, necessary conditions for optimality, analytic and numerical methods for optimization of structural systems with distributed parameters, and optimization techniques for discrete systems. We consider problems of designs with multiple objectives, designs with incomplete information, and also fundamental concepts for designs with multiple optimality criteria.

The second part of this book is primarily devoted to applying the individual criteria of strength, rigidity, stability, and weight to optimization processes. Here we examine optimal design solutions for beams, curved rods, trusses, plates, and shells, and for large (heavy) bodies that may be obtained by the use of such specific criteria.

This book also utilizes results of research from the Laboratory for Structural Optimization in the Institute for Problems in Mechanics in the USSR Academy of Sciences.

A substantial part of the material presented in this book originated with lectures given by the author to the students of the Physico-Technical Institute in Moscow.

The author offers his sincere thanks to A.Iu. Ishlinskii, who helped in formulating the overall theme of the present book and also supported subsequent work.

The author also wishes to acknowledge the assistance of B.D. Annin, N.H. Artunian, A.A. Barsuk, V.I. Biriuk, V.V. Bolotin, V.V. Vasiliev, S.Iu. Ivanov, V. Komkov, I.I. Koande, V.V. Kobelev, A.D. Larichev, V.P. Malkov, I.F. Obraztsov, A.G. Ugodchikov, V.M. Frolov, E.J. Haug, and A.V. Sharaniuk for their advice and comments on specific problems, and V.I. German, E.V. Makeev, M.V. Selishcheva, and, in particular, A.G. Mishina for their considerable efforts in the preparation of the manuscript.

Moscow, USSR *N.V. Banichuk*

Translator's Notes

A number of monographs dealing with the distributed parameter aspects of mechanical and structural optimization has appeared in the 1984 to 1988 time period. These include the Haug–Choi–Komkov monograph [55],[1] a translation of Banichuk's book [8], the Pironeau monograph [67], the Troitskii–Petukhov study [48], the Bogomolov–Simpson monograph [11], and Volume II of Komkov's monograph [57]. Expository works and long articles related to these topics include [53], [63], [65], [66], and [71]. They reinforce some older surveys, such as the excellent survey of Haug given in [23], the Olhoff survey of 1986 [63], the survey of Komkov [26] given at the meeting of the American Mathematical Society in 1983, the Niordson–Pedersen review [30] of 1973, Prager's work [37] of 1972, a shorter review by Venkayya [51], and the Brandt–Wasiutyński review [52] of 1963. A comparison of the recent and older monographs and surveys reveals an accelerated pace of ideas, and the steady incorporation of increasingly more sophisticated mathematical techniques and of new relevant physical data. The mathematical research of R.V. Kohn, Kohn and Vogelius, and Kohn and Strang established theoretical foundations for the "homogenization" of structural designs, an important aspect of shape optimization. Cea, Rousselet, Haug, Choi, Olhoff, and the author of the present study contributed significantly to a better understanding of the optimization of the stability of structural systems, which is absolutely essential in designs of slender aerospace-type structures and in earthquake-resistant designs.

The present monograph provides another and more recent presentation of the "state of the art" by a leading expert in this field. It concentrates on "difficult" aspects of systems with distributed parameters and on continuum mechanics, in general, as opposed to a discretized approach to structural design.

In some respects the distributed parameter theory is essential as a proper modeling background for discrete representation of many physical and engi-

[1] The references cited here are incorporated into the general bibliography given at the end of this monograph.

neering systems. Much of the theory offered in the present monograph is essential to efficient discretization schemes for structural designs and to the construction of numerical algorithms. Although this theory has outpaced the adaptation of relevant computing algorithms, it is the opinion of the translator that recent developments in rapid computational technology, particularly in the parallel computing field, have created a new environment for imaginative uses of this theory in computer-assisted design of very large and complex structural systems.

Wright Patterson Air Force Base, Ohio *Vadim Komkov*

Part 1
The Theory and Techniques of Structural Optimization

1 Fundamental Concepts and Problems of Optimal Design

1.1 The Choice of a Computational Scheme in the Theory of Optimal Design

The focus of the theory of optimal design is on the best possible choice of a system of forces, the shape and properties of materials, and the working environment for a structure. The study of general laws governing extremal solutions and the development of effective optimization techniques are also part of the theory of optimal design.

Research in design has made it possible to describe the maximal potential for of structural improvement, to estimate the quality of traditional (non-optimal) reinforcements, and to discover the most effective ways of improving these reinforcements. There is a great variety of problem statements in optimal design (cf. [5], [14], [24], and [44]), perhaps because the equations for the loads and deflections of a structure, and the constraints imposed on its expected mechanical properties substantially differ from one type of structure to the next (beams, columns, curved beams, plates, or shells), between the different rheological properties (elasticity, plasticity, creep), the different external loads (such as surface loads, body forces, static, or dynamic loads, "dead loads," loads that depend on the behavior of the structure itself, or the thermal loads), the different types of control variables (such as variables controlling the shape of a structure), different assumptions for measuring the completeness of information available on the working environment (i.e., problems with incomplete information concerning interaction between the structure and the external environment, or concerning the manner of supporting the structure). The accuracy of the model and of the relevant data also influences the formulation of such problems.

The choice of a computational scheme (i.e., model) is essential to both the analysis and the optimization of a structure. Therefore, optimization of design is impossible without first conducting a preliminary study of the assumptions made about both the real and imagined aspects of the structural response, and before preparing a scheme describing the working conditions, and before making various simplifications that would still preserve the adequacy of the computational scheme for a real-life structure. A mathematical model of the real behavior of an object stripped of all unrealistic properties or assumptions shall be called in this presentation "a computational scheme."

Such a description of computational schemes routinely takes place in typical strength of materials courses (e.g., [39]). But, in fact, the choice of computational scheme is not unique.

In some cases several differing schemes may be offered for the study of the same object. On the other hand, a single computational scheme may model several different real-life objects.

In design optimization one tries to apply computational schemes that would uniquely determine the relevant quantities of stress-deformed states, as well as specific values of the design variables.

However, this is not always possible to achieve, either because of the absence of exact data concerning the external loads, or incompleteness of the data concerning the manufacture of the product, the distribution of parameters that describe the materials of the structure, or other factors contributing to the incompleteness of information. To produce an adequate scheme in such a situation it makes sense to relax the demands of an "exact" description of a real-life object and to adopt either a computational scheme for the structure that assumes the worst possible scenario or else to adopt a scheme of stochastic description. These are the so-called "guaranteed" and "probabilistic" approaches to the adoption of a computational scheme.

With regard to the geometric aspects affecting the choice of a computational scheme, we shall discuss only the most commonly used schemes in the theory of optimal structural design. One may model a solid three-dimensional body whose characteristic dimensions are of the same order of magnitude in each spatial direction. Bodies having one of these dimensions "substantially larger" than the other two (such as rods, arches, or systems of beams) and bodies having one dimension "substantially smaller" that the other two dimensions (such as shells and plates) comprise the remaining two cases.

The general setting may vary in optimal design theory. It does depend substantially on the type of the designed structure; that is, is this a traditional structure, or is it an entirely new design? In the first case much useful information is available concerning the prototypes and much of the accumulated experience may be utilized in the form of a "support" solution, or perhaps as the initial approximation for the optimal design process. In that case only a few design parameters are regarded as unknown and optimization of the structure is accomplished with a relatively small number of basic design variations. In the second case the design is determined by a large number of design variables and the optimization process has to be carried out by considering a large number of admissible variations (i.e., of separate design projects). A degree to which the parameters are regarded as known, or conversely, the absence of any knowledge concerning the design parameters that would determine either the exterior shape or the internal properties of a structure is an important factor in determining the computational scheme.

The stage in the computational scheme in which the important unknown variables appear for the first time is critical, and in general, determines the effects of optimization procedures. The introduction of the design parameters frequently amounts to a stipulation of some additional requirements for the computational scheme. Let us clarify the last statement by offering an example of an optimal design of a thin shell with variable thickness. In the optimization we will use the classical theory of elastic thin shells. If, as an outcome of the optimization process, we obtain a distribution of the thicknesses having large gradients or having some singularities (such as discontinuities, a "zero," or an "infinite" thickness), the classical computational scheme is not valid and therefore one must introduce appropriate corrections.

Here the possible corrections are of two principal types:

1. An open introduction into the computational scheme of new constraints that are necessitated by the accepted theory, and the subsequent use of these constraints during the design process. This operation amounts to a "tightening" of the set of admissible design projects.
2. A "broadening" and generalizing of the classical computational schemes by taking into account trends that tend to alter the design projects, thus permitting an analysis of a larger class of admissible structures.

In choosing a computational scheme for optimal design problems an important role is played by the a priori knowledge of properties of the unknown solution. Information about the model, knowledge of some basic properties of the solution, and also a reexamination of the initial hypotheses all permit us to state some essential constraints in the formulation of the optimization problems and to exclude "second layer" considerations, making it possible to state the problem in proper form to apply some of the well-known numerical or even analytical techniques.

Therefore, a large body of results in the theory of optimal design is directly connected with some well-known computational schemes. However, it is frequently quite difficult to "guess" in advance the properties of the unknown optimal solution, and the optimization problem may turn out to be formulated so that the derived solutions violate the fundamental hypotheses essential to the model itself. For example, certain solutions of problems of plate design may have large gradients of thickness. This contradicts some of the assumptions that are fundamental to Kirchhoff's plate theory. Other well-known singularities arising in the use of classical models of plates and shells are caused by the presence of either zero or infinite thickness in the optimal solution. Thus, if one discovers a deviation from the model or a violation of hypotheses that are essential to the computational scheme, it becomes necessary to modify the system of relations used in a design procedure by the introduction of additional constraints, for example, on thickness in a problem concerning the bending of plates. Another method of escaping from such a situation may involve a generalization of the model, consisting in "softening" the hypotheses and constructing a computational scheme for a broader class of structures—one that includes some previously inadmissible designs. Thus, the choice of a model and its evolution constitute important aspects of the design process.

Let us summarize what has been said above. Generally, it is useful in any optimal design process to permit the possibility of making corrections in the computational schemes (that is in the mathematical model) thus sharpening or changing some of the initially assigned conditions.

1.2 Formulation of Problems in Optimal Structural Design

The optimal design problems considered here consist in determining the shapes, internal properties, and working conditions of a structure that obey assigned constraints, and produce an extremum (either a maximum, or a minimum) for a chosen quantity characterizing the design. In a rigorous postulation of a problem of optimal structural design one must include the statement of basic constitutive relations (the choice of a mathematical model) and define the functional to be optimized, asign all the necessary constraints to the state functions, and select the unknown control (design) variables. From a purely mathematical point of view, such problems may be classified by identifying the types of systems of differential equations and boundary conditions, the types of optimized functionals and initially given constraints, or, alternatively, by the dimension of the problem, the manner in which the unknown design variables enter into the fundamental relations (e.g., should we control the coefficients or the boundary of the region?), the completeness of information concerning the given initial data (i.e., are we given a problem with full information or one with incomplete information?), the nature of the extremum (e.g., are we given a single extremum or multiple extrema?) and the manner in which the extremum is defined (e.g., are we given a single criterion or multiple criteria?), and perhaps by some other properties of the system.

In this section we shall examine a classical formulation of optimal design problems. An exposition of some possible generalizations will be given in sections 1.5 to 1.8. As we have previously remarked, an essential feature in the formulation of such problems is the choice of a mechanical model. To begin with, we choose the state variables $\mathbf{u}$ and system of equations

$$L(\mathbf{x}, \mathbf{u}, \mathbf{h}, \mathbf{q}) = 0 \tag{1.1}$$

relating these variables to the design parameters and also to the external loads. Here the state variable $\mathbf{u} = \{u_1(x), \ldots, u_m(x)\}$ is a vector function that determines the state of the structure. The independent variable $\mathbf{x} = \{x_1, \ldots, x_l\}$ assumes all possible values in the region Ω. The symbol L in the eq. (1.1) denotes a differential operator acting on functions of the spatial coordinates x_i only.

Equation (1.1) can be regarded as a system of generally nonlinear differential equations. In this monograph we will be concerned with theories that assume the validity of both geometric and physical hypotheses of linearity. In accordance with such assumptions, the behavior of structures is described by operators that are linear with respect to the state variables.

The operator L in eq. (1.1) depends on the design vector function $\mathbf{h}$, where $\mathbf{h} = \{h_1(x), \ldots, h_n(x)\}$, and on the vector function of the external loads $\mathbf{q}$. The natural numbers m, n, and l are given. Here it is assumed that the boundary conditions determining the type of support and the loads applied to the structure are all specified in the definition of L.

Given the loads and the structural parameters, the system of differential equations must be "closed" (i.e., is well-posed), and thus should uniquely determine the state variables, which in turn characterize the stresses and deformations of the structure. Determining the state variables from the given design functions shall be referred to as the direct problem.

If the state equations reflect the laws of physics, the choice of the design variables appearing in the functionals considered in the design process, including the optimized functional (i.e., the quality, or cost functional) and the system of constraints, are all governed by the designated purpose of the structure, the working environment, and the technology available for its construction.

The functions $h_i(x)$ determine the shape and also the physical and mechanical properties of the construction materials. As our $h_i(x)$, we can choose the distribution of thickness and the cross-sectional area of the body, functions that determine the location of the median surface for curved beams or shells, the distribution of the density of the reinforcing material inside the structure, or perhaps the angles that define the orientation of the axes of anisotropy at each point inside the region occupied by an elastic body.

In an optimal design problem, besides the state functions and the design variables we also need to know the functionals that characterize the design. These functionals depend on the vectors $\mathbf{u}$, $\mathbf{h}$, and $\mathbf{q}$:

$$J_1 = J_1(\mathbf{u}, \mathbf{h}, \mathbf{q}), \ldots, J_r = J_r(\mathbf{u}, \mathbf{h}, \mathbf{q}).$$

Two types of such functionals are considered in optimal design problems: functionals of the first type are called integral functionals:

$$J_i = \int_\Omega f_i(\mathbf{x}, \mathbf{u}, \mathbf{h}, \mathbf{q})\,d\Omega, \qquad i = 1, \ldots, r_1, \tag{1.2}$$

and of the second type, local functionals, for example,

$$J_j = \max_x f_j(x, \mathbf{u}(x), \mathbf{h}(x), \mathbf{q}(x)), \tag{1.3}$$

with $j = r_1 + 1, \ldots, r_1 + r_2$.

Here f_i denotes given differential forms while r_1 and r_2, are assigned natural numbers with $r_1 + r_2 = r$.

Integral functionals or combinations of integrals of the type given in (1.2) may represent characteristic properties of the structure, such as the weight, the energy of the elastic deformation (compliance), natural frequencies of vibration, or a critical load that would cause the structure to lose its stability (cf. [1.11], [1.27], [1.39], and [1.40]). A local property could be the magnitude of the maximal deflection or the stress intensity (cf. [7], [8], and [28]).

If other demands are made concerning the properties of the structure, then appropriate constraints must be applied to the design and state variables. The given constraints may form a system of inequalities, which may be displayed in vector form:

$$\boldsymbol{\psi}(\mathbf{x}, \mathbf{u}, \mathbf{h}, \mathbf{q}, J_1, \ldots, J_r) \leq 0. \tag{1.4}$$

The components ψ_i of the vector $\{\boldsymbol{\psi}\} = \{\psi_1, \ldots, \psi_k\}$ are regarded as a priori known functions of their arguments. Various notations used for the constraints (1.4) are discussed in section 2.2. In specific cases inequalities (1.4) represent different types of bounds on the stresses, deformation, or displacements, the integrands in rigidity or compliance functionals, and also the natural frequencies of vibration and values of critical parameters that determine the loss of stability. One of these functionals, or perhaps a function of several functionals, $F(J_1, \ldots, J_r)$, is then chosen as the functional that is to be optimized.

The optimization problem now consists in finding a specific (vector) function that assigns a minimum (or a maximum) to the functional

$$J = F(J_1, \ldots, J_r) \tag{1.5}$$

and also satisfies (1.1) to (1.4).

Note that, generally speaking, there may be arbitrarily many functionals and assigned constraints, as long as they are not contradictory. However, there can be only a single optimized functional (or quality criterion) for the structure in each specific problem of the form of (1.1) to (1.5). Thus, in the bending of a beam with variable thickness it is possible to formulate a problem of weight minimization with a constraint on the maximal deflection, or the problem of minimizing the maximal deflection for a given total weight. However, simultaneous optimization of both these functionals does not make sense within the context of the classical concepts of optimality.

A correct formulation of optimization problems with a vector of assigned quality criteria becomes possible if one makes use of the concept of optimality in the sense of Pareto, or in some other sense appropriate to a multiple criteria optimization. A large number of publications (e.g., [46], [1.25], [1.37], and [1.38]) contain an exposition of the basic ideas of multi-criterion optimization theory. Nevertheless, approaches to structural design based on non-classical concepts of extrema are presently only in the initial stages of development.

It has been made clear in several recent research articles on optimal design theory that in many instances the formulation of optimization problems in the manner of (1.1) to (1.5) has some serious limitations, since the optimal solution $\mathbf{h}^*$ or $\mathbf{u}^*$ may not exist for the stated design limitations, even though there exists a minimal value J^* for the cost functional. Moreover, we may not be particularly interested in finding $\mathbf{h}^*$ or $\mathbf{u}^*$. It may suffice for our purposes to uncover the general trends while searching for an optimal solution and to find the sensitivity of functionals that are describing important characteristics of a structure with respect to changes in the values of the parameters. Therefore, it is of interest to broaden the formulation of the design problems, and to introduce mathematical rigor so as to derive the entire process of obtaining a solution by means of techniques of modern sensitivity analysis.

This generalized setting would include a construction of a minimizing sequence, where for each fixed superscript "i" the elements of that sequence satisfy (1.1) to (1.4) and the condition

$$\lim_{i\to\infty} (J^i) = J^*, \tag{1.6}$$

where $J^i = J(\mathbf{u}^i, \mathbf{h}^i, \mathbf{q})$.

In the case when a classical solution exists for the problem represented by (1.1) to (1.4), the limits: $\mathbf{h}^i \to \mathbf{h}^*$, and $\mathbf{u}^i \to \mathbf{u}^*$ (as $i \to \infty$) also exist.

1.3 Basic Functionals and Optimization Criteria

The choice of functionals in design optimization is the most important part of formulating optimization problems. Many factors enter into this choice, for example, the main purpose of the structure, working conditions, technology available for its construction, cost limitations, properties of the model used to describe the mechanical behaviour of the structure, and the a priori known properties of the optimal design problem.

In what follows we consider typical functionals that are most frequently considered in structural optimization:

1.3.1 One of the most important properties of a structure is its total weight, and consequently this functional is considered in most design optimization projects as either the quality criterion or as one of the assigned constraints. The weight of the structure determines both the quantity of materials used in the construction and some of its functional properties. For example, an increase in the weight of an aircraft results not only in the increase of the quantity of materials necessary for the construction, but will also produce higher fuel consumption and will no doubt worsen other important flight characteristics.

Weight is an integral property of the design. For homogeneous continua, weight is proportional to the volume occupied by the body:

$$J = \gamma \int_{\Omega} d\Omega, \tag{1.7}$$

where γ is the specific weight of the material. In this case, to change the weight of the structure it is necessary to vary the domain of integration Ω.

For thin-walled structures made of homogeneous materials, the weight of a structure is give by an integral whose integrand is a function of the distribution of "thicknesses" h:

$$J = \gamma \int_{\Omega} f(h) d\Omega. \tag{1.8}$$

For example, for a continuous plate supported along its boundary Γ in the xy plane (where x, y are Cartesian coordinates), with Ω the region enclosed by Γ, $h = h(x, y)$. In this case reducing the weight of the structure may be attained by either changing the function $h(x, y)$ (for a fixed domain Ω), or simultaneously varying both the thickness and the shape of the region.

However, in problems of design optimization the quantity of material may be classified in the "design dependent" category, with its quantity and its distribution to be found, or in the "design independent" category, with the quantity and manner of distribution already assigned.

Thus, in the design of a three-layered plate one most frequently considers the design of the two outer reinforcing layers only, with the middle layer regarded as fixed. Then the weight functional is given by the sum of the integrals $J_a + J_c$ (depending on the thickness of the reinforcing layers and the thickness of the middle layer of the plate, respectively), but minimization of J consists in minimizing the weight of the outer layers J_a only. This is a consequence of the additive property of the weight functional.

In cases involving optimization of inhomogeneous bodies, the weight functional depends on the internal composition of the material, for example, the densities of the binder and reinforcing additives for a composite material. If we denote by h_a, γ_a, h_c, and γ_c, the concentration of a material and its specific weight, respectively (reinforcing and binding material components, respectively),

$$J = \int_{\Omega} (\gamma_a h_a + \gamma_c h_c) d\Omega,$$

where Ω denotes the spatial region occupied by the structure.

1.3.2 In many research articles on optimal design the measure of rigidity of the structure is provided by the total work performed by the external forces in a quasi-static loading of an elastic body. This functional is known as the compliance of the structure. Let an elastic body be supported along a part of the boundary Γ_u, while another part of the boundary Γ_σ is subjected to the applied loads $\mathbf{q}$. Then the compliance is given by the integral along Γ_σ of the inner product of the vector of the elastic displacements $\mathbf{u}$ and the vector of the external loads:

$$J = \frac{1}{2}\int_{\Gamma_\sigma} (\mathbf{q}\cdot\mathbf{u})\mathrm{d}\Gamma. \tag{1.9}$$

One can generalize the definition of compliance by interpreting $\mathbf{q}$ and $\mathbf{u}$ as generalized forces and corresponding generalized displacements, respectively. For example, the moments applied to a beam may be regarded as the generalized forces, while the angles of rotation measured at the corresponding points the generalized displacements.

Note that, in general, the functional (1.9) is not really a strict measure of the rigidity of a structure. In fact, just because the value of the functional J is small, it does not follow that the displacements of some discrete points in the elastic body are also small. However, in some specific problems of optimal design the functional (1.9) does determine the maximal deflections and, therefore, may be regarded as the rigidity quality criterion. A study of some functionals that characterize the rigidity of structures is offerred in Section 7.1.

1.3.3 The theory of thin-walled structures subjected to compressive conservative forces may be concerned with Rayleigh-type functionals that determine critical values of the load parameters producing a loss of stability.

Let p denote the load parameter and pT the work performed in bringing the unit volume of the structure to the critical state, while Π is the potential energy density for the deformed structure after the loss of stability.

Recalling a classical theorem of elastic stability one can derive the following formula for the critical value of p:

$$p = \int_\Omega \Pi(u, h)\mathrm{d}\Omega \bigg/ \int_\Omega T(u, h)\mathrm{d}\Omega \tag{1.10}$$

A typical problem in the theory of optimal design of structures subjected to compressive loads is maximization of the critical value p_0 (where p_0 is the smallest eigenvalue) for a given weight of the structure, or minimizing the weight subject to the constraint $p_0 \geq \mu$, where μ is a given positive number. Solutions of some problems of this kind are offered in Chapter 8.

For the class of design problems in which stability constraints may be violated, in the absence of other spectral problems, computations generally involve a search only for the smallest eigenvalue, which may involve constraints on the higher eigenvalues (i.e., constraints on higher natural frequencies in optimal control of the spectrum of natural frequencies of vibrations).

However, there exist cases in an optimal design procedure when it becomes necessary to study higher eigenvalues in order to estimate a bound on the smallest eigenvalue. This situation turns up specifically in optimization problems in which multiple critical loads may occur (cf. [32], [34], and [1.5]).

1.3.4 We discussed above integral–type functionals that arise in optimization problems, applying techniques from the classical calculus of variations. However, several important problems lead to computations of functionals that depend on the values of the state functions at points whose locations are not initially known (cf. [7], [24], and [1.7]). All basic strength and deformation properties are local. The maximal deflection at some point in the structure is an example of a property of structural rigidity that is described by a local functional. Applying these ideas to the bending of beams and plates, we may estimate the magnitude of the maximal deflection,

$$J = \max_{x \in \Omega} |u(x)|, \tag{1.11}$$

and thus the problem of optimizing the rigidity of a structure by varying the design function $h(x)$ can be restated as one of minimizing the maximum deflection.

Many strength properties of structures are determined by local functionals. A typical such local functional may be written as

$$J = (f)_{\Omega_0} = \max_{x} f. \tag{1.12}$$

Here $f = f(\sigma_{ij}(x))$ denotes the intensity of the stress σ_{ij}, or perhaps some other measure of the stressed state at an arbitrary point of the structure, and Ω_0 is the set of points on which f attains its maximum. Note that the demand to extend the class of admissible loads to objects subjected to distributed loads only leads at once to the problem of minimizing J.

A more detailed discussion of strength and rigidity functionals as well as specific solutions to the associated optimization problems are given in Chapters 6 and 7.

1.4 Design Parameters

The main purpose of many optimal design research projects is to find the most effective optimization technique. In the design of optimal structures it is customary to have a wide choice of design variables whose variations may influence the magnitude of the quality criterion. For example, a reduction in the weight of a structure may be accomplished by altering the distribution of the thicknesses of the members, controlling the anisotropy, reinforcing, prestressing, etc.... It is important to discover which specific methodology of optimization or which combination of techniques results in the largest gain for a given functional.

But even in the cases when prohibitive cost or technological difficulties make the optimal outcome impossible, or at least most unlikely, it is still important to conduct a study of optimal design, as such a study yields a theoretical estimate of the quality of a non-optimal traditional structure. Indeed, it is possible that an actually constructed structure may turn out to be close to optimal and any further design improvements would not be economically justified. However, this can be made clear only by conducting an optimal design study. In the following section we shall study problems arising in the introduction of the design variables.

1.4.1 *The Dimension.* The dimension of a vector function will not be defined by the usual count of the number of components of this vector, but rather by the number of independent variables which influence the scalar state functions. The dimension of a state function depends on several factors: the shape of the structure, the material properties, the nature of internal forces, the manner of support, etc. In some specific problems (e.g., in thin-walled structures) our ability to reduce the dimension of a state function from three to two, or even to one could depend on the presence of symmetry, and on averaging of basic properties along one or two of the spatial coordinates.

Unlike the state function, the design variables may have quite arbitrary dimension, and this fact could be utilized in the formulation of a problem. Thus, in an optimal design of a noncircular plate with variable thickness, the distribution of deflections (i.e., the state function) is a function of two independent coordinates (x and y) which are drawn in the plane of the plate. On the other hand, thickness of the plate (the design variable) may be regarded either as a function of two variables $h = h(x, y)$, or as a function of only a single independent variable, for example $h = h(x)$.

We observe that the study of "controls" in a lower dimension is of definite interest because it is much easier to implement it, even if in general it is not as effective with respect to the possible gain in the value of the quality functional.

Another example of the use of a one-dimensional control in a two-dimensional problem is an optimal design of rods and meshes used for the purpose of reinforcing of slabs or shells.

1.4.2 How should the cross-sectional area of a beam vary along its length is a most common problem considered in the design optimization of beams. There are various possibilities for varying the type of cross-section and different ways of changing the cross-sectional parameters along the length of the beam.

For example, we may consider only beams having a circular cross-section with the radius r varying along its length, that is $r = r(x)$, or we may decide to select beams with a rectangular cross-section, having a variable height $h_1(x)$ and width $h_2(x)$. Thus, we may choose as the design variable (let us denote it by h) one, or several functions, so that all of them uniquely define the shape of the beam, that is the shape of its cross-section, and also the rule for changing the geometric sizes (the parameters) of the cross-section along its entire length. Thus, for a continuous beam with a circular cross-section specifying a function $r = r(x)$ completely determines its shape, while for continuously varying beams, having rectangular cross-sections, two functions must be assigned to determine the shape of each beam.

The quantity EI (where E is Young's modulus and I the moment of inertia of the cross-sectional area about a line perpendicular to the bending plane and passing through the neutral axis) is of fundamental importance in characterizing the rigidity of a beam. It appears in the differential equations of equilibrium, and in the boundary conditions. The use of this quantity reveals how a distribution of "thickness" affects the deflection function.

We shall limit our study to relations of the type:

$$EI(x) = A_\alpha h^\alpha(x), \tag{1.13}$$

where A_α is a constant depending on the type of the beam's cross-section and on Young's modulus. For beams that have everywhere a constant rectangular cross-section we have $\alpha = 0$ and $A_0 = Eh_2h_1^3/12$ (here h_2 is the width, and h_1 the height of the cross-section). For rectangular beams with variable height, but constant width $\alpha = 3$, $h = h_1$, and $A_3 = Eh_2/12$. In the case of variable width, but of constant height $\alpha = 1$, $h = h_2$, and we have in this case $A_1 = Eh_1^3/12$. If the ratio of height to width of the cross-section is constant (i.e., $h_1/h_2 = \mu$) along the entire length, and if all cross-sectional areas consist of similar rectangles, then $\alpha = 4$, $h = h_1$, and $A_4 = E/12\mu$.

For three-layered beams with variable thickness of the outer reinforcing layers equal to $1/2\ h_1$, and having a constant width H of the middle layer, with H obeying $H \gg \max_x h(x)$, we have $\alpha = 1$, $h = h_1$, and $A_1 = Eh_2H^2/4$, where h_2 is the variable width of the cross-section.

In the process of optimal design of beams it is frequently convenient to choose the variation of the cross-sectional area $S = S(x)$ as the control function. The relation between S and EI differs for different types of beams, but can be described by the general formula

$$EI(x) = C_\alpha S^\alpha(x), \tag{1.14}$$

that is analogous to (1.13). For rectangular beams having variable width but constant height this relation becomes

$$EI = Eh_1^2 S(x)/12.$$

In the case of constant width $h_1(x)$ but variable height $h_2(x)$,

$$EI = (E/(12h_2^2))S^3(x).$$

For beams with a circular cross-section but with a variable radius,

$$EI = (E/(4\pi))S^2(x).$$

Equation (1.13) is also used in optimal designs of elastic plates for finding the distribution of thickness. It relates the cylindrical rigidity EI to the thickness of the plate.

The values $\alpha = 1$ and $\alpha = 3$ correspond to the cases of the tri-layered and the uniform plates, respectively. When $\alpha = 3$ then h in (1.13) denotes the thickness of the plate, while $A_3 = E/(12(1 - \nu^2))$, where the constant ν denotes Poisson's ratio for the plate's material. In the case of $\alpha = 1$, the design variable $(\frac{1}{2}h)$ denotes the thickness of either one of the plate's outer reinforcing layers, while $A_1 = EH^2[4(1 - \nu^2)]$, where H denotes the constant thickness of the middle layer.

1.4.3 The coefficients of the equations, the boundary conditions, the constraints assigned to the problem, and even the boundary of the domain in which all these relations are defined (for problems with unknown boundaries) may be regarded as design variables in a structural optimization process. Problems that involve the optimization of coefficients and boundaries are generally considered separately and entirely different techniques are used in solving this class of problems. However, all such classifications are quite arbitrary. In some instances it is possible to use the properties of a mapping of the unknown domain into some canonical region. Then the unknown functions defining this transformation may be regarded as unknown coefficients both in the relevant equations and in the functionals, which are now defined in the canonical domain.

More detailed studies of this problem are contained in [1], [6], and [22].

1.5 Dynamic Problems of Structural Optimization

Formulating the optimization problems in Section 1.2 we have postulated that the laws of statics are applicable, and therefore that the state equations and constraints are time-independent. The unknown state functions u and design functions h depend only on the spatial coordinates; that is, $u = u(x)$ and $h = h(x)$. In the more general case, when structures are subjected to dynamic loads, all basic relations include dependence on time t, and the response of a structure is given by a state function $u = u(x, t)$. In this case the definition of the control function must include the dependence on both t and x, that is: $h = h(x, t)$, and we have to deal with an optimal control problem. We remark that the problems of determining h, which is a function of both x and t, may also arise in the planning of constructions, where both the geometry and certain structural properties may change with the passage of time. Research techniques for such problems are given in several expository works dealing with the theory of optimal control for systems with distributed parameters.

A formulation and solution of that kind of problem is of considerable interest. However, the main thrust of this monograph is directed towards a somewhat different class of dynamic optimization problems, whereby the design variable is considered to be a function of the spatial coordinates only, i.e., $h = h(x)$. At the present time a development of answers to corresponding problems in structural design and computation of the corresponding dynamic responses is of major theoretical and practical importance (cf. [13], [20], [24], [44], [1.13], and [1.31]).

We shall now introduce a formulation of a dynamic optimization problem. We assume that the structure shall perform its functions in some interval of time $[0, t_k]$, where t_k is a fixed final instant of time. A well-known scheme of loading is assigned to that time interval, that is, a vector function $q(x, t)$ of the assigned loads is presumed to be known. The equations which determine the behavior of a structure, and relate to each other the state function $u(x, t) = \{u_1(x, t), \ldots, u_m(x, t)\}$ and the design function $h = \{h_1(x), \ldots, h_n(x)\}$ shall be written symbolically as follows:

$$L(x, t, u, h, q) = 0. \tag{1.15}$$

The symbol L in (1.15) denotes a differential operator that contains differentiation with respect to the time variable t, and with respect to the spatial coordinates x_i. It is assumed that the initial and boundary conditions which determine the state of the structure at time $t = 0$, and the manner in which it is supported are all included in the definition of L.

To define the operator L it is essential to have the information concerning the time-dependence of the applied loads, and of the expected response of the structure, and also of changes (with time) of structural parameters, and of parameters determining the composition of the material. One should also have sufficient data concerning changes in the field of applied forces, changes in the manner of support, of possible reinforcements, and of deterioration of some properties due to wear.

The following inequalities must be obeyed through the entire expected life of the structure:

$$\psi(x, t, u, h, J_1, \ldots, J_r) \leq 0, \tag{1.16}$$

since these inequalities describe the strength, rigidity, or other constraints that were initially assigned to the structure. The components ψ_i of the vector $\psi = \{\psi_1, \ldots, \psi_k\}$ are known functions of their arguments, and J_i are functionals that depend on u, h, and q.

Integral functionals are of the form

$$J_j = \int_0^{t_k} \int_\Omega f_j(x, t, u, h, q)\,\mathrm{d}\Omega\,\mathrm{d}t, \tag{1.17}$$

where f_j are given functions of the above arguments.

Various integral functionals may be represented in the form (1.17). In particular, the weight functional for a thin-walled structure is given by

$$\gamma/t_k \left\{ \int_0^{t_k} \int_\Omega h\,\mathrm{d}\Omega\,\mathrm{d}t \right\}. \tag{1.18}$$

The function $h(x)$ describes the distribution of thickness for the structural member while γ denotes the specific weight of the material. Local functionals, such as

$$J_j = \max_t \max_x f_j(x, t, u(x, t), h(x), q(x, t)) \tag{1.19}$$

differ from the corresponding functionals (1.3) previously introduced in section 1.2, because the function f_j given in (1.19) is time-dependent, and an additional requirement of finding a maximum with respect to time has been assigned to it.

The problem of optimizing a dynamically loaded structure consists in finding a function $h(x)$ minimizing (or maximizing) the functional (1.5) and satisfying the condition (1.16) for any $t \in [0, t_k]$, and for any $x \in \Omega$. In checking the conditions (1.16) functionals J_j are computed according to (1.17) and (1.19), while the function $u(x, t)$, which depends on both the design $h(x)$ and the load $q(x, t)$, is found by solving the boundary-value problem (1.15).

Typical dynamic optimization problems take place, for example, in designing aircraft wings, when one has to take into account variable gusting loads (cf. [1.7]). One deals in a computational scheme with loads similar to the gusts of wind by choosing some functions that are determined experimentally, and then approximated by some analytical relations. Optimal design consists of finding distribution of the material providing the strength, and also the distribution of other nonstructural mass along the wing, in order to have the smallest possible weight of the wing and at the same time to make certain that the gusts cause amplitudes of vibration and stresses that do not exceed some prescribed bounds. Somewhat similar problems occur in designs of shells when impact loads are considered in the computations.

We should mention some cases in which it is possible to exclude the time variable t, and to reduce a dynamic optimization problem to a much simpler static problem, such as the one postulated in Section 1.2.

An important class of dynamic problems is related to computations of steady-state forced vibrations caused by periodic loads. If we assume here that the amplitudes of vibrations are small, and that the structural response is linear elastic, then the corresponding initial-boundary-value problem for the displacement vector function can be written in operator notation:

$$L(x, t, u, h, q) = Au_{tt} + Cu - q = 0. \tag{1.20}$$

Here A is the inertia matrix operator and C is the rigidity matrix containing operators of differentiation with respect to the spatial variables, while the subscript t denotes differentiation with respect to time. The coefficients in the operators A and C depend on the design variables h, that is, $A = A(h)$, $C = C(h)$. The modelling techniques which formulate the relationships in a dynamic problem in the form (1.20), and the properties of the operators A and C are described for example in [1.10]. It is assumed that the external loads are given by a harmonic function of time and the resulting vibrations are also harmonic. This assumption allows us to represent both the load and the displacement of the structure in the following manner:

$$u = U(x)e^{i\omega t}, \qquad q = Q(x)e^{i\omega t}. \tag{1.21}$$

Here ω denotes the frequency of the forced vibrations, i is the usual imaginary unit, and $U(x)$ and $Q(x)$ are the amplitude functions for the displacement and the load, respectively. The representation (1.21) permits us to exclude the time variable t from all subsequent considerations and to rewrite the defining relations (1.20) as a system of equations and boundary conditions for the unknown displacement amplitude function

$$CU - \omega^2 AU = Q. \tag{1.22}$$

Using eq. (1.22) and regarding $U(x)$ as a function that describes the response of the structure, it is possible to formulate the optimal design problem in a manner similar to that of Section 1.2.

For example, it is possible to use eq. (1.22) to study the problem of minimizing the weight $J(h)$ for a thin-walled structure, while constraining its dynamic compliance

$$\int_{\Omega} QU \mathrm{d}\Omega - c \le 0 \tag{1.23}$$

and imposing additional constraints on the distribution of the thickness variables h. It is assumed in formulation of the constraints (1.23) that the vectors Q and U belong to the same vector space, c is a given constant, and that the integrand in (1.23) represents an inner product. A similar exposition of problems arising in optimization of structures subjected to forced harmonic oscillations, may be found in [1.14 to 1.19], [1.28], and [1.35 to 1.36].

One of the intensively researched problems in the theory of optimal structural design is the reduction of weight of a structure with bounds assigned to natural frequencies of vibrations, and also the dual problem of optimizing the natural frequencies of vibration for a given weight of the structure. Spectral problems of optimization arising in studies of free vibrations are considered in several books and in expository articles (e.g., [3] to [5], [11], [14] to [16], [19] to [24], [28], [30], [32] to [35], [40], [44], [48], and [70]).

The corresponding problems of optimal design are not given a detailed exposition in this book, and in this paragraph we shall limit our discussion to explanations concerning formulation of such problems.

The natural frequencies and eigenfunctions of free vibrations may be determined from the system of equations (1.22) if we substitute $Q = 0$. The variables ω and U that solve the equation of free vibrations (1.22) may be also represented by a Rayleigh quotient-type relation:

$$\omega^2 = \int_{\Omega} \Pi(U, h) \mathrm{d}\Omega \Big/ \int_{\Omega} T(U, h) \mathrm{d}\Omega. \tag{1.24}$$

Here U denotes the amplitude of elastic displacements, h the control function, Π and T are, respectively, the amplitude values of the potential and kinetic energy per unit volume:

$$\Pi = (CU, U), \qquad T = (AU, U). \tag{1.25}$$

The parentheses in (1.25) denote the scalar (or inner) product.

In the case of longitudinal vibrations of a rectangular elastic rod having a variable cross-sectional area $h = h(x)$, and

$$\Pi = hU_x^2, \qquad T = hU.$$

Here and below we shall make use of dimensionless variables.

In transverse vibrations of thin continuous plates with variable thickness $h = h(x)$,

$$\Pi = h^3[(U_{xx} + U_{yy})^2 + 2(1 - \nu)(U_{xx}U_{yy} - U_{xy}^2)], \qquad T = hU^2,$$

where $U = U(x, y)$ is the amplitude function of plate displacements.

The eigenfunctions ω_i $(i = 1, 2, \ldots)$ corresponding to different eifenvectors U^i constitute a spectrum of vibrations $0 \leq \omega_0 \leq \omega_1 \leq \omega_2 \leq \ldots$ If the frequency of an external load applied to the structure coincides with the natural frequency of vibrations, then some undesirable resonance phenomena may occur. Thus, if we design a structure and wish to avoid resonances it is necessary to put some distance between the natural frequencies and the frequencies of the applied loads. The design of a structure is such that if the load frequency ω satisfies the condition $0 \leq \omega < \omega_0$, or a two-sided inequality $\omega_k < \omega < \omega_{k+1}$, the structure will not be subject to resonance.

Attempts to enlarge the resonance-free strip $0 \leq \omega < \omega_0$ leads to the equivalent design problem of maximizing the first (or principal) natural frequency ω_0 by choosing an appropriate design variable $h(x)$. It is a common practice to assign an isoperimetric constraint to $h(x)$ by requiring that the volume (or the weight) of the structure remains constant.

For a frequency strip $\omega_k < \omega < \omega_{k+1}$, it is common practice in structural design to maximize ω_{k+1}, keeping ω_k fixed, or else minimize ω_k, keeping ω_{k+1} fixed. It is also possible to formulate the problem of maximizing the difference $\Delta\omega = \omega_{k+1} - \omega_k$ with a constraint that requires the optimal resonance-free strip to include a corresponding pair of natural frequencies in all subsequent redesign improvements of the nonoptimal structure.

1.6 Multipurpose Structural Design

Classical single-purpose problems of structural design optimization, that is, problems such as those considered in Section 1.2, are characterized by the assumption that the loads, the manner of support, and various demands concerning the stressed-deformed state are stated in definite, unique fashion, and that the purpose of the structure is also unique. Therefore, the design process may be accomplished within the framework of a single computational scheme. For example, in beam designs we consider only the bending loads as the external forces, and in the design of columns only the buckling loads. It is possible to consider the simultaneous application of various types of loads within the traditional framework within which such problems are formulated. However, designs of single-purpose structures (single-objective designs) turn out to be severely restrictive, and to approximate the realistic working conditions of the structure one must enlarge the formulation of the problem and take into account multiple objectives.

It is particularly important to study such problems of multiobjective optimization when it is necessary to compute a structure for different working conditions. Some of the simplest cases comprise optimal design problems for beams and plates, which may be subjected to bending loads only, and, in particular, to concentrated loads that may be applied to various locations within the structure. Within the framework of multiobjective design we may also consider problems of structural optimization for moving loads, where it is assumed that the movement and variation in the load intensity is quasi-static, and therefore dynamic effects may be ignored. Among the research works dealing with optimization of structures subjected to moving loads, or to multiple types of loads we refer the reader to [5], [1.20 to 1.26], [1.29], [1.30], [1.32], and [1.34].

In a study of optimal design for structures subjected to moving loads one may introduce a sequence of loads chosen to represent the loading program. These may comprise transverse loads that do not exceed a given magnitude. This type of problem is distinguished by the property that only bending loads are admitted. Consequently, in the optimization process one needs to consider only the same basic equations of state (i.e., the equations of pure bending) and exactly the same boundary conditions throughout. The only computational variants of applied loads consist in different applications of the loads. Of equal theoretical and practical interest is the structural design in which computations must include applications of different types of loads. In that case one should apply a corresponding computational scheme to each loading variant and then consider the appropriate state equations, boundary conditions, and the characteristic properties of the structure in the design process.

The formulation of problems in multiobjective optimization consists in a description of the applied computational schemes and the choice of the unknown design parameters, the optimized quality criterion, and the constraints imposed on the response of the structure under various types of loads. Let us denote by j the index of a loading variant ($j = 1, 2, \ldots, s$), and by $u^j = \{u_1^j(x), \ldots, u_m^j(x)\}$ and $J^j = \{J_1^j(u^j, h), \ldots, J_r^j(h^j, h)\}$, the state vector function of the structure and the functional characteristics, respectively. We also introduce equations of state, constraints, and the quality functional:

$$L^j(x, u^j, h, q^j) = 0, \tag{1.26}$$

$$\Psi^j(x, u^j, h, q^j, J^j) \le 0, \tag{1.27}$$

$$J = F(J^1, \ldots, J^j) \to \min_h . \tag{1.28}$$

Relations (1.26) to (1.27) represent a system of differential equations and inequalities that connect the design variables, state function, and functional characteristics.

From a mathematical point of view, the problems posed by multiobjective optimization have the following properties. The state function labeled by the index j occurs only in those equations and inequalities in (1.26) to (1.27) that are labeled by the same index. It also occurs in the expression for the optimized functional. In other words, the functions u^j ($j = 1, 2, \ldots, s$) turn out to be local variables in the optimization problem. The manner in which the design variable h occurs in the equations and inequalities indexed by different values of j makes these relations interdependent. Unlike the local variables u^j, the function h is a global design variable.

Another property of multiobjective design problems is the following. In the optimization of a structure it is not essential to study the sequence of different variants of applied loads. To find an optimal solution it suffices to have the information concerning the set of all admissible parameters and the load characteristics (variants of structural support, set of all realizable loads, and set of all requirements assigned to the structure for different types of loads).

If the distribution of h is given, then (1.26) represent a closed system of differential equations and boundary conditions (i.e., one having a unique solution) that may be solved for u^j, while their integrals may be found independently. This is an essential property of the principle of decomposition applied to the theory of optimal design. In finding the state function this principle permits us to simplify the solution due to a reduction in the dimension of the corresponding "direct" analysis problems.

Problems of multiobjective optimization may be formulated in many different ways. One possible formulation involves the optimization of integral-type functionals with inequality–type constraints for both maximin and minimax optimization problems. The formulation of minimax problems in multiobjective optimization is discussed in [1.2] and [8], and may be described as follows.

Let every jth state of a structure be characterized by some functional J^j and let every functional J^j $(j = 1, 2, \ldots, s)$ have the same physical dimension.

Then the problem of multiobjective optimization may be reduced to finding a maximin or minimax:

$$J^* = \min_h \max \{J^1, \ldots, J^s\}, \tag{1.29}$$

$$J^* = \max_h \min \{J^1, \ldots, J^s\}. \tag{1.30}$$

The second occurrence of the symbols max and min, respectively, in (1.29) and (1.30) denote finding a maximum or minimum of the quantities enclosed in the braces. Utilizing the criterion (1.29) or (1.30), the function F introduced in (1.28) becomes

$$F = \max\{J^1, \ldots, J^s\} \qquad \text{or} \qquad F = \min\{J^1, \ldots, J^s\}, \quad \text{respectively.}$$

As an example of a multiobjective optimization consisting in the minimization of a maximum over a certain set of quantities (cf. [1.22]), we could offer the case of finding the thickness h for an elastic beam subjected to a sequential application of different transverse loads discussed above.

Thus, the design may consist in attaining the criterion of minimizing the greatest deflection. In this case the functional J characterizes the rigidity of the beam when it is subjected to the jth type loading, and its value is equal to the magnitude of the maximal deflection, that is $J^j = \max_x u^j(x)$. The greatest of the quantities J^j becomes the minimized (cost) functional.

As an example of a maximization of a minimum we can use the design of an elastic beam having variable height h and a quality criterion consisting in the optimization of the fundamental frequencies of different types of free vibrations (cf. [1.23]). A sequence of states of the beam represents three different types of natural vibrations ($j = 1, 2, 3$): longitudinal (l), torsional (t), and bending (b). The function J^j denote in this case the minimal values of the natural frequencies that correspond to the lower harmonics for each of these types of vibrations (i.e., $J^1 = \omega_1^l$, $J^2 = \omega_1^t$, and $J^3 = \omega_1^b$). As the quality criterion to be optimized, or rather maximized, we choose in this case the smallest of these three frequencies of vibrations, that is, $J = \min\{J^1, J^2, J^3\}$.

We comment that both the minimax and maximin problems of multiobjective optimization may be reformulated as classical problems of minimization, or maximization, respectively, for a single quantity with a finite number of constraining inequalities. Thus, the multiobjective design problem given by (1.26), (1.27), (1.29) is equivalent to minimizing the functional J_U,

$$J^* = \min_h J_U, \tag{1.31}$$

under the conditions (1.26) and (1.27) and the inequality constraints

$$J^j \leq J_U, \qquad j = 1, 2, \ldots, s. \tag{1.32}$$

Similarly, the problem expressed in (1.26) and (1.30) is equivalent to finding the maximum of J_L:

$$J_* = \max_h J_L, \tag{1.33}$$

under the conditions (1.26) and (1.27) and inequalities

$$J^j \geq J_L, \qquad j = 1, 2, \ldots, s. \tag{1.34}$$

1.7 Designing with Incomplete Information: "Guaranteed" Versus Probabilistic Approach

A majority of problems in the theory of optimal design of structures, and, in particular, problems discussed in the preceding section, are considered within a framework of the deterministic approach, that is, one assumes that complete information is provided with regard to applied loads, properties of the structural materials, and boundary conditions.

To solve such problems we may apply techniques from the calculus of variations and the methodology of design optimization for systems with distributed parameters. Problems in the design of optimal structural systems with incomplete information are essentially different, both in their formulation and in research techniques. We shall now consider various approaches to optimization problems with incomplete information and, trying to be specific, consider the problem of finding the shapes of elastic bodies having the smallest possible weight, subject to appropriate constraints on strength and rigidity. The formulation and solution of such optimization problems using deterministic techniques result in shapes that commonly exhibit the property that even the most insignificant changes in external conditions (such as changing the point of application of a load) will cause the structure of the given shape to violate some strength or geometric constraints. However, in many cases we do not have complete information regarding loads, or else it is known that different types of loads may be applied at different times. Therefore, it is of interest to consider not only a deterministic setting, but also more general problems of structural optimization in which the optimization process is carried out for entirely different classes of applied forces (cf. [5], [1.3], [1.4]).

The minimax approach is one of the feasible approaches to the formulation and solution of these types of problems with "incomplete information". In this minimax approach (otherwise known as the "guaranteed approach") one must assume that a set containing all possible realizable external loads is known and that we need only determine the shape of a structure having minimal weight and satisfying all strength and geometric constraints for all possible loads contained in the realizable set.

Such a structural shape will be called *optimal* if for any structure with a smaller weight, it is possible to select a system of loads belonging to the admissible set such that either some assigned strength or geometric constraint has been violated. One of the following two alternatives has to be realized when one solves a problem in this approach.

In the first case it turns out that there exists, in fact, a "worst load" such that a structure having the smallest possible weight can be found by computing it for the particular load and that this structure satisfies all assigned strength and geometric constraints for all other loads belonging to the given admissible set.

Then this structural shape solves the design optimization problem for an entire class of admissible loads and, therefore, it solves the originally posed problem. Or else, there is no "worst load," and the solution that optimizes the entire class of admissible loads fails to be optimal for any specific load chosen from the admissible set. Examples illustrating both cases are given in [5] and [8]. We remark that the minimax approach may be applied to problems with incomplete information concerning either the actual boundary conditions or the properties of the structural materials.

Let a complete system of equations and boundary conditions given in the operator notation

$$L(x, u, h, q) = 0, \tag{1.35}$$

describe the equilibrium of a structure and related various states, depending on the designs and applied external forces.

The type of load applied to the structure is not given a priori. Instead we assume that the admissible set R_q of all realizable loads is known, that is,

$$q \in R_q. \tag{1.36}$$

In our considerations of design problems below, we shall assume that only loads that satisfy (1.36) will be considered. If, for example, the optimized object is a plate, and the external forces comprise only one-sided transverse loads, whose magnitudes do not exceed P, then the set R_q is given by

$$R_q\colon q \geq 0, \qquad \int_\Omega q \, d\Omega \leq P,$$

where Ω denotes the region bounded by the contour of the plate.

If h and q are given, the boundary-value problem (1.35) is assumed to be uniquely solvable with respect to the state variable u. The optimization problem consists in finding a function $h(x)$ that minimizes the functional $J(h)$ (i.e., the weight), and satisfies for any choice of q in (1.36) all strength and rigidity constraints

$$\psi(x, u, h, q, J_1, \ldots, J_r) \leq 0, \tag{1.37}$$

where ψ is a given vector function. Conditions (1.37) form a system of scalar inequalities.

It is possible to consider the probabilistic approach to problems with incomplete information on equal footing with the minimax (or "guaranteed") approach. Let the external forces depend on a random variable ξ, that is: $q = q(x, \xi)$. The density distribution function for the random variable ξ is assumed to be known.

The behavior of the designed structure, as described by (1.35), depends implicitly (through the dependence on q) on the random variable ξ. Therefore a random character is inherited by the state variable u and the functions ψ appearing on the left sides of (1.37).

The choice of probability density distribution for the variable ξ permits us, at least in principle, to determine the moments of the random quantities, and in particular, their mathematical expectation and dispersion. This permits us, in turn, to control the probability of violating the assigned constraints (on the strength, rigidity, etc.) and to study the following problem of optimal design, consisting in finding a function h that minimizes the functional $J(h)$ (the weight of the object) and satisfies the system of inequalities

$$\mathbf{E}\psi(x, h, u) \leq 0, \tag{1.38}$$

$$\mathbf{D}\psi(x, h, u) \leq \varepsilon, \tag{1.39}$$

while taking care of (1.35).

Here **E** and **D** denote the mathematical expectation and dispersion, respectively, for a random variable, and ε denotes a sufficiently small number. Conditions (1.38) and (1.39) imply that some conditions must accompany a choice of the control variable h. The strength and rigidity conditions given in (1.37) must be satisfied in "the average" case, and the deviation of the quantity ψ must not exceed some chosen sufficiently small value.

1.8 Basic Ideas of Multicriteria Optimization

The calculus of variations or optimal control theories are generally concerned with minimizing or maximizing some scalar functional under a given set of constraints. The problem of optimizing simultaneously several functionals, or a single vector functional, within the framework of traditional definitions of optimality, is, in general, ill-posed. A well-posed formulation of optimization problems for vector functionals has been called multicriteria optimization. Such formulation is possible only if the classical definition of an extremum is generalized. The Pareto concept of an optimal solution is a genuine generalization and extension of the classical ideas. Application of the optimality concept in the sense of Pareto permits, in particular, to extend the class of problems that are usually considered in the classical theory of optimal design, and to study some basic ideas partaining to designs of structures that are optimal with respect to several quality criteria. By using the multicriteria theory it is also possible to attain a better understanding of some rules used in determining "real-life" structural shapes (cf. [1.38]).

Let us introduce the concept of optimality in the sense of Pareto. Let a system (a structure) be characterized by the r-quality criteria $J^i(h)$, $i = 1, \ldots, r$, that depend on the design variable $h \in \mathbf{H}$, where $\mathbf{H}$ denotes the set of admissible controls. Let $J(h)$ denote the vector functional: $J(h) = \{J_1(h), \ldots, J_r(h)\}$. We shall consider the problem of minimizing this functional. The design variable $h_* \in \mathbf{H}$ is optimal in the Pareto sense if for any $h \in \mathbf{H}$ the validity of inequalities: $J(h_*) \leq J(h)$, (respectively $J_i(h) \leq J_i(h_*)$ for every $i = 1, \ldots, r$) implies that $J(h) = J(h_*)$.

In other words, if the design variable h_* is Pareto optimal then it is impossible to decrease any further the value of any criterion J_i without simultaneously increasing the value of at least one other functional J_j, $(j \neq i)$.

For solutions $h_* \in \mathbf{H}$ that are optimal in the Pareto sense, the following assertion is true. For an arbitrary fixed index j the functional J_j is minimizes by h_*, which is chosen from the admissible set of solutions $h \in \mathbf{H}$, if we assign the condition that the remaining functionals J_i, $(i \neq j)$ must not have values greater than those attained when $h = h_*$, that is

$$J_j(h_*) = \min_{h \in \mathbf{H}} J_j(h), \tag{1.40}$$

$$J_i(h) \leq J_i(h_*), \qquad i = 1, 2, \ldots, r, \qquad i \neq j. \tag{1.41}$$

The properties (1.40) and (1.41) can be applied effectively to obtain necessary conditions for optimality of the design variables in the sense of Pareto.

Solutions of multicriteria optimization problems and finding of the optimal design variables may be reduced to the minimization of a scalar function defined by the equation

$$G(h) = c_1 J_1(h) + c_2 J_2(h) + \cdots + c_r J_r(h). \tag{1.42}$$

By means of the function $G(h)$ it is possible to restate sufficient optimality conditions in the sense of Pareto in the following manner.

Suppose that $\mathbf{c} \geq 0$, and $h_* \in H$ minimizes $G(h)$ over the set $h \in H$. Then $h_* \in \mathbf{H}$ is a minimal solution in the sense of Pareto. A proof of this assertion may be found in [40].

Application of the Pareto concept of optimality allows us to formulate the following general design problem. Let the response of a structure be described by the boundary-value system (1.1), and by a system of r-characteristic (quality) functionals $J_k(u, h, q)$, $k = 1, 2, \ldots, r$. Constraints (1.4) are assigned to the design variables $\mathbf{h} = \{h_1, \ldots, h_n\}$ and to the state variables $\mathbf{u} = \{u_1, \ldots, u_m\}$, reflecting original demands that were made concerning the structural design. The design quality criterion vector F contains s-scalar functionals F_i as components, that is $F = \{F_1, \ldots, F_s\}$. The functionals F_i represent known functions of the characteristic functionals J_k:

$$F_i = F_i(J_1, J_2, \ldots, J_r), \qquad i = 1, 2, \ldots, s. \tag{1.43}$$

The multicriteria optimization problems have the following common properties. In general, there is no unique solution, instead there exists a minimizing set of solutions that are optimal in the above given sense, the *Pareto set*. The construction of the Pareto set provides important information concerning different creative possibilities for the structure's design and reveals hidden reserves for optimization.

The choice of a single design function from the Pareto set may be resolved by using appropriate model of multilevel optimization. For example, in a two-level optimization, at the first level we solve the multicriterion problem and find the appropriate set of design functions (which are optimal in the Pareto sense). Then on the second level, after introducing an additional scalar criterion, we find a unique optimal solution over the previously obtained set of solutions.

We note that the problems where some of the functionals $F_1, \ldots, F_l$, are minimized, while others $F_{l+1}, \ldots, F_s$ are maximized, may also be represented formally by (1.1), (1.4), (1.43), and (1.44) with a vector quality functional, all of whose components are to be minimized. It suffices to define the quality criterion vector in the following way in terms of the scalar components $F_1, \ldots, F_s$:

$$F' = \{F_1, \ldots, F_l, -F_{l+1}, \ldots, -F_s\}. \tag{1.44}$$

It is clear that minimization of the functional (1.45) with indicated additional conditions is equivalent to minimization of the functionals $F_1, \ldots, F_l$, and maximization of the functionals $F_{l+1}, \ldots, F_s$.

References for Chapter 1

1.1. Adali, S., The design of beams on Vinkler-Pasternak foundations for minimum dynamic response and maximum eigenfrequency, *J. Méchanique Théor. Appl.*, 1982, *1*, No. 6, pp. 975–993.

1.2. Albul, A.V. Banichuk, N.V., and Koande, I.I., On a dynamic problem of optimal design, in: *Numerical Techniques of Mechanics*, Kishinev, Shtiintsa, 1981, *64*, pp. 3–7.

1.3. Banichuk, N.V., Game problems in the theory of optimal design, In: *Proceedings of IUTAM Symposium on Optimization in Structural Design*, Warsaw, 1973, Springer Verlag, Berlin, 1975, pp. 111–121.

1.4. Banichuk, N.V., Minimax approach to structural optimization problems, *J. Optimiz. Theory Appl.*, 1976, *20*, No. 1, pp. 111–127.

1.5. Banichuk, N.V. and Barsuk, A.A., On a technique for optimizing the elastic stability in the case of multiple critical loads, in: *Applied Problems of Strength and Plasticity. Statics and Dynamics of deformable Bodies.* Gorkii, Gorkii University, 1983, pp. 85–89.

1.6. Banichuk, N.V. and Ivanova, S.Iu., On optimal design of structures with dynamic loads, in: *Applied Problems of Strength and Plasticity*. All-Union Higher Institute of Learning Collection. Gorkii, Gorkii University, 1984, pp. 88–96.

1.7. Banichuk, N.V., Ivanova, S.Iu, and Sharaniuk, A.V., Sensitivity analysis and optimal design of structures computed for dynamic loading, *Izv. Akad. Nauk S.S.S.R., M.T.T.*, 1985, No. 4, pp. 166–172.

1.8. Bogomolov, S.I., Simpson, E.A., and Sukiasova, N.G., Optimization of rods according to deformation characteristics, in: *Dynamics and Strength of Machines*, Kharkov, Vish. Shkola, 1980, *30*, pp. 70–76.

1.9. Bogomolov, S.I., Simpson, E.A., and Sukiasova, N.G., On an approach to the problem of isolating vibrations, in: *Dynamics and Strength of Machines*, Kharkov, Vish. Shkola, 1980, *32*, pp. 69–75.

1.10. Bolotina, V.V. (ed.), *Vibrations in Technology*, Reference Guide, Vol. 6, part 1, *Vibrations of Linear Systems*, Moscow, Mashinostroenie, 1978, 352 pp.

1.11. Collatz, L. Eigenvalue Problems, Nauka, Moscow, 1968, 504 pages.

1.12. Feodosiev, V.I., Strength of Materials, Moscow, Nauka, 1970, 544 pp.

1.13. Huang, H.C., Minimum weight design of vibrating elastic structures with dynamic deflection constraints. *J. Appl. Mech.*, 1976, *43*, No. 1, pp. 178–180.

1.14. Icerman, L.J., Optimal structural design for given deflection, *Intern. J. Solids Struct.*, 1969, *5*, pp. 473–490.

1.15. Ivanova, S.U., Some problems of weight reduction for structures working in the forced harmonic vibrations mode, *Izv. Akad. Nauk S.S.S.R., M.T.T.*, 1984, No. 5, pp. 147–154.

1.16. Ivanova, S.U., On optimal designs of plates made of composite material subjected to forced harmonic vibrations, in: *Second All–Union Science and Technology Conference, (Strength, Rigidity and Technology of Manufacture from Composite Materials*), Erevan, University of Erevan, 1984, Vol. 2, pp. 31–36.

1.17. Ivanova, S.U., Optimal design of Plates performing a forced harmonic oscillation, in: *Problems of Structural Mechanics and Strength of Aircraft, Trudy MAI*, Moscow, 1985, pp. 19–24.

1.18. Johnson, E.H., Disjoint design spaces in the optimization of harmonically excited structures, *A.I.A.A. J.*, 1976, *14*, No. 2, pp. 259–262.

1.19. Johnson, E.H., Rizzi, P. Ashley, H., and Segenreich, S.A., Optimization of continuous one-dimensional structures under steady harmonic excitation, *A.I.A.A. J.*, 1976, *14*, No. 12, pp. 1690–1698.
1.20. Karihaloo, B.L., Optimal design of multi-purpose structures. *J. Optimiz. Theory Appl.*, 1979, *27*, No. 3, pp. 449–461.
1.21. Karihaloo, B.L., Optimal design of multi-purpose tie beams, *J. Optimiz. Theory Appl.*, 1979, *27*, No. 3, pp. 427–438.
1.22. Karihaloo, B.L., Optimal design of multi-purpose tie column of solid construction, *Intern. J. Solids Struct.* 1979, *15*, pp. 103–109.
1.23. Karihaloo, B.L. and Parberry, R.D., The optimal design of beam columns, *Intern. J. Solids Struct.*, 1979, *15*, pp. 855–859.
1.24. Karihaloo, B.L. and Parberry, R.D., Minimum weight multi-constraint vibrating cantilevers. *Intern. J. Solids Struct.*, 1982, *18*, No. 5, pp. 419–430.
1.25. Koski, J., Multicriterion optimization in structural design, in: *New Direction in Optimum Structural Design*, Atrek, E. et al. (ed.), Chichester, Wiley, 1984, pp. 483–503.
1.26. Martin, J.B., Optimal design of elastic structures for multipurpose loading, *J. Optimiz. Theory Appl.*, 1970, *6*, No. 1, pp. 22–40.
1.27. Mikhlin, S.G., Variational methods of Mathematical Physics, Nauka, Moscow, 1970, 512 pp.
1.28. Mróz, Z, Optimal design of elastic structures subjected to dynamic harmonically-varying loads, *ZAMM*, 1970, *50*, pp. 303–309.
1.29. Olhoff, N., Optimization of columns against buckling, in: *Optimization of Distributed Parameter Structures*, Iowa City, Iowa, 1980, Haug, E.J. and Cea, J. (eds.), Sijthoff-Noordhoff Publ., Alphen aan den Rijn, 1981, vol. 1, pp. 333–361.
1.30. Prager, W. and Shield, R.T., Optimal design of multi-purpose structures, *Intern. J. Solids Struct.*, 1968, *4*, No. 4, pp. 469–475.
1.31. Rabinovich, I.M., On computation of trusses and beams having minimal volume subjected to the effects of dynamic loads and their own weight. in: *Studies of Theory of Reinforcement*, Stroiizdat., Moscow, 1967, *15*, pp. 151–158.
1.32. Save, M. and Prager, W. Minimum weight design of beams subjected to fixed and moving loads, *J. Mech. Phys. Solids*, 1963, *11*, No. 4, pp. 255–267.
1.33. Seyranian, A.P., Optimal design of a beam with constraints on the natural frequency of vibrations and the loss of stability, *Izv. Akad. Nauk S.S.S.R., M.T.T.*, No. 1, 1976, pp. 147–152.
1.34. Shield, R.T., Optimum design methods for multiple loading, *ZAMP*, 1963, *14*, pp. 38–45.
1.35. Sofronov, Iu.D., Computation of a beam having minimal weight and subjected to cyclic loading, *Stroit. Mekhanika, Raschet Sooruzh.*, 1969, No. 6, pp. 40–43.
1.36. Sofronov, Iu.D., Beam of minimal volume subjected to cyclic loads, *Trans. KAI*, 1970, *116*, pp. 10–17.
1.37. Stadler, W., Preference optimality in multi-criteria control and programming problems, *Nonlinear Analysis Theory, Methods and Applications*, 1980, *4*, No. 1, pp. 51–65.
1.38. Stadler, W., Natural structural shapes (the static case), *Quart. J. Mech. Appl. Math.*, *31*, part 2, 1978, pp. 169–217.
1.39. Timoshenko, S.P., Stability of Elastic Systems, Moscow–Leningrad, Gostekhteorizdat., 1946, 532 pp.
1.40. Timoshenko, S.P., *Vibrations in Engineering*, Moscow, Nauka, 1967, 444 pp.

2 Reformulation of Optimal Design Problems

2.1 Introduction of Auxiliary Design Parameters

Frequently, in structural optimization one may assign to the design parameters constraints that are independent of either the state function or the basic functional. For example, in the search for an optimal distribution of thickness h in a beam, plate, or shell we assign the constraint $h \geq 0$, which is a direct consequence of the physical meaning of the specific design variable. In optimization problems in which the optimal distribution of thickness turns out to be 0 at some interior points in the domain of optimization Ω, this constraint may cause some difficulties. For this reason it is convenient to use a device based on an idea of Valentine, namely to introduce auxiliary functions that "automatically" take care of the assigned conditions. For example, if we introduce a new design variable φ related to h by the equality $h = \varphi^2$, then obviously, the inequality $h \geq 0$ is satisfied for arbitrary real values of φ.

Several choices of techniques for introducing auxiliary variables have been suggested in specific optimization problems resulting in a substantial simplification of the procedures for construction of optimal solutions. Several methods of introducing auxiliary functions $h = \Phi(\varphi)$ to impose the constraint

$$0 \leq h \leq h_{\max} \tag{2.1}$$

are introduced in the monograph [2.16]. The symbol $h_{\max}$ in (2.1) denotes the maximal admissible value that may be assigned to the design variable $h(x)$.

These techniques all have the same remarkable property that, for any variation of the auxiliary variable in an infinite interval $(-\infty, +\infty)$, the variation of the variable h inside the interval is given by (2.1).

The functions $h = \Phi(\varphi)$, given by the formulas

$$\Phi = h_{\max}|\cos\varphi| \quad \text{or} \quad \Phi = h_{\max}|\sin\varphi|,$$

$$\Phi = \frac{1}{2}h_{\max}\,(1 + (2/\pi)\arctan\varphi) \quad \text{or} \quad \Phi = \frac{2}{\pi}h_{\max}|\arctan\varphi|,$$

$$\Phi = h_{\max} e^{-\varphi^2}, \qquad \text{or} \quad \Phi = h_{\max} e^{-|\varphi|};$$

$$\Phi = h_{\max}(1 - e^{-\varphi^2}), \qquad \text{or} \quad \Phi = h_{\max}(1 - e^{-|\varphi|}), \tag{2.2}$$

have all the required properties.

We also note that auxiliary control variables have been introduced in [2.12] with the help of inverse trigonometric functions.

Somewhat more general bounds assigned to the control function (i.e., the distribution of thickness) assume the form

$$h_{\min} \le h(x, y) \le h_{\max}. \tag{2.3}$$

Several articles on optimal design elaborate on the use of techniques dealing with these inequalities. Here we only wish to note a simple trigonometric formula replacing the inequalities (2.3) (cf. [5] and [2.3]), $h = \alpha + \beta \sin \varphi$, where

$$\alpha = \tfrac{1}{2}(h_{\max} + h_{\min}) \quad \text{and} \quad \beta = \tfrac{1}{2}(h_{\max} - h_{\min}), \tag{2.4}$$

which is widely used in solutions of applied problems. It is easy to check that if h is computed from (2.4) and substituted in (2.3), these inequalities are satisfied for any arbitrary value of φ.

In addition to inequalities (2.4), in problems of optimal design frequently one has to consider isoperimetric conditions of the type

$$\int_\Omega h d\Omega = 1. \tag{2.5}$$

This condition can be also be excluded from consideration in such problems if we introduce the auxiliary function φ, related to h by the formula

$$h = \varphi \bigg/ \int_\Omega \varphi d\Omega. \tag{2.6}$$

Substituting h as given by (2.6) into eq. (2.5) conforms that this equality is satisfied for an arbitrary choice of (an integrable function) φ.

We also observe that a more general integral constraint

$$\int_\Omega \chi h d\Omega = 1, \tag{2.7}$$

where χ is some given function, may be excluded from future consideration by applying the substitution

$$h = \varphi \bigg/ \int_\Omega \chi \varphi d\Omega. \tag{2.8}$$

In concluding this short exposition of various techniques used for eliminating constraints, we observe that the introduction of auxiliary design variables may be accomplished in various (and nonunique) ways. This feature may be utilized in numerical solutions of such problems by improving the convergence of the algorithms.

2.2 Constraint Formulation Techniques

In the theory of optimal design both inequality–equality-type constraints are assigned to the response of a structure. Modern techniques of optimal design permit us to deal with both these types of constraints.

However, equality-type constraints can be realized in a particularly simple manner in several techniques, and for this reason in most research articles dealing with the optimization of structures inequalty-type constraints are converted into equality-type constraints.

Bearing this in mind, we shall apply certain transformations and introduce auxiliary variables. We shall also pay attention to certain techniques that are very commonly used in the theory of optimal design.

The operation of computing the absolute value of a quantity permits to replace an inequality of the form (1.4) by an equivalent relation of the form

$$\psi_j + |\psi_j| = 0. \tag{2.9}$$

The subscript j in the above expression and in the remainder of this section may assume all values from 1 to k.

Indeed, if inequality (1.4) is satisfied (or if it is not satisfied), condition (2.9) is also true (or also false, respectively).

An analogous substitution device may utilize the sign operator. The system of equalities

$$1 + \operatorname{sign} \psi_j = 0 \tag{2.10}$$

is equivalent to conditions (1.4).

It is easy to see that if the constraints on the response of a structure are given by two-sided inequalities $0 \le \psi_j \le c_j$ (where c_j are given constants), instead of the canonical form (1.4), the idea used in obtaining (2.9) may be employed in an analogous manner to derive the simultaneous equalities

$$\begin{gathered} \psi_j - |\psi_j| = 0, \\ \psi_j - c_j + |\psi_j - c_j| = 0, \end{gathered} \tag{2.11}$$

while the technique used in obtaining (2.10) results in the equalities

$$\begin{gathered} 1 - \operatorname{sign} \psi_j = 0, \\ 1 + \operatorname{sign}(\psi_j - c_j) = 0. \end{gathered} \tag{2.12}$$

A different set of techniques employed for reducing inequality-type constraints to equality constraints utilizes the concept of auxiliary variables. Introducing auxiliary variables χ into the inequality constraints (1.4) may result in relations of the form

$$\psi_j + \chi_j^2 = 0, \tag{2.13}$$

where $\chi_j(x)$ are arbitrary unknown real functions $(-\infty \le \chi \le \infty)$. For a system of two-sided inequality constraints

$$\psi_{j\min} \leq \psi_j \leq \psi_{j\max}, \tag{2.14}$$

the following technique for introducing auxiliary functions χ_j permits us to transform these constraints into equality constraints valid for $-\infty \leq \chi_j \leq \infty$:

$$(\psi_j - \psi_{j\min})(\psi_{j\max} - \psi_j) - \chi_j^2 = 0. \tag{2.15}$$

We could point out some other techniques for converting inequality constraints into equality-type constraints that are utilized in articles on optimal control and optimization of structures. However, in the particular techniques for converting inequalities into equalities discussed above in this section, a well-known methodology of the classical calculus of variations may be directly applied to our research.

2.3 Replacement of Local Properties by Integral Functions

Let us agree to classify problems of optimal structural design into two classes, depending on the types of optimized functionals and constraints. In the first class we include problems of optimization in which the quality criteria and the constraints are given by integrals of unknown functions. In this case the constraints have the general form of "isoperimetric" equalities and inequalities. Most frequently, in studies on optimal design one encounters such integral characteristics as weight, the total energy of deformation (compliance), forces causing the loss of stability, or the frequency of natural vibrations. The optimization problems included in the second class have quality criteria and constraints of a local (nonintegral) type. In this class we also include mixed problems in which one has to consider both the integral and the local characteristics of a structure.

Typical local functionals that may be minimized in an optimal structural design include, for example, the maximum displacement in a deformed object or the maximum value of stress intensity.

The first class of problems comprises a considerable body of research derived in the theory of optimal design. To start with, consider the following explanation. Well-known methods of the classical calculus of variations and linear programming are available for solving problems with integral functionals. Such techniques permitted a successful completion of analytical and numerical research of many problems and helped to reveal several interesting general laws.

For some types of problems with integral functionals the research was made easier because of the fact that the equations of equilibrium are "natural" for these functionals and permit one to ignore the corresponding differential relations (cf. [5], [2.2], and [3.17]).

Fewer studies have been devoted to the investigation of two-dimensional problems with local functions, mainly because of the absence of sufficiently general effective techniques for solving problems that belong to this class. Several typical difficulties arise in solving problems of this class.

For example, in minimizing the maximum deflection of a plate and finding the best distribution of its thickness, the derivation of the necessary conditions for optimality and the numerical application of such conditions are complicated by the fact that we do not have a prior knowledge of the position of the point of maximum deflection. Moreover, the location of that point depends not only on the nature of the external loads, but also on the unknown distribution of thickness (i.e., the design variable).

If we consider optimization problems with inequality constraints applied to local-type characteristics, analogous difficulties arise in locating points or regions in which these inequalities become, in fact, strict equalities. A real simplification of problems in the second class may be achieved if the points at which the local characteristics of the structure are to be computed or at which the functional attains its extremum are known beforehand, possibly because of some symmetry conditions present in the problem.

We remark that in comparing integral and local functionals, we have in mind primarily problems of optimal design in dimensions higher than 1.

Following on what has been said above, it would be of interest to develop techniques that would reduce more complex optimization problems with local functionals to corresponding problems with integral quality criteria. We shall indicate some methods that can be used for such reduction using as our example the problem given by relations (1.1) to (1.5), assuming that the local constraints occurring in the system of inequalities (1.4) and the minimized functional have the form

$$\max_x |f_j(x, u(x), u_x(x), h(x)| \leq 1, \qquad j = 1, \ldots, r, \tag{2.16}$$

$$J_* = \min_h J = \min_h \max_x f(x, u(x), u_x(x), h(x)), \tag{2.17}$$

where f and f_j are known functions of x, $u(x)$, $u_x(x)$, and $h(x)$. Hence, $u = \{u_1, \ldots, u_m\}$ and $h = \{h_1, \ldots, h_n\}$ are vector functions; $x = \{x_1, \ldots, x_e\}$ is a vector of independent variables, $u_x = \{u_{1_{x_1}}, \ldots, u_{m_{x_e}}\}$ is a vector of partial derivatives of the state function; and m, n, e, and r are given integers.

The minimized functional and the expressions given on the left side of (2.16) represent a norm assigned to the functions f and f_j in the space C of continuous functions, that is,

$$J = \|f\|_C, \qquad J_j = \|f_j\|_C. \tag{2.18}$$

Besides the norms assigned in C we can consider norms assigned in the space L_p of functions whose pth power is integrable. It is well known in functional analysis that for arbitrary $p \geq 1$ the following inequality holds:

$$\|f\|_L \leq \|f\|_C, \tag{2.19}$$

and that

$$\lim_{p\to\infty} \|f\|_{L_p} = \|f\|_C. \tag{2.20}$$

Relations analogous to (2.19) and (2.20) are valid for the functions f_j ($j = 1, 2, \ldots, r$). Since the difference between the C and L_p norms is small for sufficiently large p, it is possible to replace the functionals (2.18) by the functions

$$\left.\begin{aligned} J_p &= \left\{(1/\mu(\Omega)) \int_\Omega |f|^{p_j} d\Omega\right\}^{1/p}, \\ (J_j)_{p_j} &= \left\{(1/\mu(\Omega)) \int_\Omega |f_j|^{p_j} d\Omega\right\}^{1/p}, \end{aligned}\right\} \tag{2.21}$$

where $\mu(\Omega)$ denotes the measure of the set Ω. We shall make use of this property in converting problems with local functionals into problems with integral functionals.

We shall consider as approximate solutions of the optimization problem functions u and h that satisfy instead of the relations (2.16) and (2.17), the conditions

$$(J_{p_j}) \le 1, \qquad j = 1, 2, \ldots, r_l \tag{2.22}$$

$$J = \min_h J_p = \min_h \left\{\left(\frac{1}{\mu(\Omega)}\right) \int_\Omega |f|^p d\Omega\right\}^{1/p} \tag{2.23}$$

In this manner the original optimization problem (2.16), (2.17) in local functionals is replaced by a problem (2.22), (2.23) in integral functionals. In investigating the transformed problem it is possible to use well-established techniques from the calculus of variations applicable to integral functionals.

For a large number of constraints of the form (2.16), (i.e., for large r_l) the presence in the transformed problem of the $r_l + 1$ parameters p, p_j, may lead to some difficulties in estimating the values of p, p_j that would ensure the necessary accuracy. In that case it is possible to employ a more "rigid" scheme of analysis with only a single parameter p, assuming that $p_1 = p_2 = \cdots p_{r_l} = p$. Again, the system of constraints (2.22) can be simplified by regarding the functions f_j ($j = 1, \ldots, r_l$) as components of a vector $f^v = \{f_1, \ldots, f_{r_l}\}$ and defining the quantity

$$\|f^v\|_C = \max_j \max_x |f_j(x, u(x), u_x(x), h(x)|. \tag{2.24}$$

Thus, the system of inequalities (2.16) may be reduced to the single inequality

$$\|f^v\|_C \le 1. \tag{2.25}$$

Further, replacing the C-norm in (2.25) by the L_p-norm we arrive at the final inequality

$$\left\{\frac{1}{\mu(\Omega)} \int_\Omega \left[\frac{1}{r_l} \sum_{j=1}^{r_l} |f_j|^p\right] d\Omega\right\}^{1/p} \le 1. \tag{2.26}$$

Note that the inequality $\|f^v\|_C \le \|f^v\|_{L_p}$ and the corresponding condition (2.25) imply that the original system of inequalities (2.16) is only approximately satisfied. To make certain that condition (2.25) is not violated, we may replace the derived inequality (2.26) by the condition

$$\left\{\left(\frac{1}{\mu(\Omega)}\right)\int_{\Omega}\left[\frac{1}{r_l}\sum_{j=1}^{r_l}|f_j|^p\right]d\Omega\right\}^{1/p}\leq 1-\varepsilon(p), \tag{2.27}$$

where $\varepsilon(p)$ is some function of the parameter p. These arguments are also fully applicable to the system of inequalities (2.22). The function $\varepsilon(p)$ may be found using the estimates for the L_p norm.

We shall now estimate the error arising in conversion of problems with local functionals to problems with integral functionals. Let J and J_p denote, respectively, the functionals

$$J = \|f(x, u(x), h(x)\|_C \quad \text{and} \quad J_p = \|f(x), u(x), h(x)\|_{L_p},$$

and h_* and h_{p^*} the functions assigning a corresponding minima to the functionals J and J_p. We introduce the error estimate obtained in [2.3] (cf. [5] and [8]):

$$\left.\begin{aligned} &0\leq[J(h_*)-J_p(h_{p^*})]/J(h_*)\leq 1-\chi(p),\\ &\chi(p)=1+\frac{\delta}{p}\ln\{\delta/(p+\ln H)\}+\frac{1}{p}[\ln H-\delta].\end{aligned}\right\} \tag{2.28}$$

Note that the error estimate (2.28) is based on the a priori information $f\in W_{H,\delta}$. We shall say that the function f belongs to the space $W_{H,\delta}$ if there exists $\varepsilon_0>0$ such that for all $0\leq\varepsilon\leq\varepsilon_0$, $\mu(\Omega_\varepsilon)\geq\mu(\Omega)H\varepsilon^\delta$, where $H\geq 0$ and $\delta\geq 0$ are given constants. Ω_ε $(0\leq\varepsilon\leq 1)$ denotes the set of points such that $|f(x)|\geq\|f(x)\|_C\cdot(1-\varepsilon)$.

Let us indicate another technique for converting local conditions to integral constraints widely applied in the theory of optimal design (cf. [21] and [24]). In this optimization problem, let the local constraints be of the form

$$f_j(x, u(x), u_x(x), h(x)) - 1 \leq 0. \tag{2.29}$$

It can be easily shown that inequalities (2.29) can be replaced by the system of integral equations

$$\int_{\Omega}(|f_j-1|+f_j-1)d\Omega = 0. \tag{2.30}$$

Note that there is strict equivalence between the systems of inequalities (2.29) and relations (2.30). This equivalence relation can be checked by direct substitution.

2.4 Elimination of Differential Relations by Means of Variational Principles

Variational principles play an extremely important role in the mechanics of deformable solids. First, they permit us to formulate the boundary-value problems of mechanics in a very compact manner and in greater generality.

The equilibrium equations and some of the boundary conditions (called the "natural conditions") can be deduced from variational principles as necessary conditions for the existence of an extremum. Second, a variational formulation of problems of the loads and deformations of structures permits one to apply effectively direct techniques of the calculus of variations to obtain solutions of such problems. All this applies not only to boundary-value problems of mechanics of deformable bodies, but also to many other problems of mathematical physics.

The application of variational principles to the problems of optimal design permit us to disregard differential couplings and avoid the necessity of introducing adjoint equations. Thus, the order of the general boundary–value optimization problem is reduced, and the derivation of the optimality conditions is simplified. In addition, the variational principles and variational inequalities that are derivable from these principles turn out to be most useful in analytic investigations of optimization problems and in validiting optimal solutions.

2.4.1 Let us assume that an elastic body occupies a region Ω bounded by the surface $\Gamma = \Gamma_u + \Gamma_\sigma$. The body is supported along the surface Γ_u while some external tractions q are applied to the body along the remaining part of the boundary Γ_σ. We shall consider the problem of minimizing the compliance

$$J = \frac{1}{2}\int_\Omega qu\,d\Gamma_\sigma \to \min_{\Gamma_\sigma} \tag{2.31}$$

by finding the shape of Γ_σ.

In "closed" formulation of such problems the basic equations (i.e., the equations of equilibrium) need to be postulated and additional constraints need to be stated. Then in solving the optimization problem, we can introduce an adjoint variable that obeys a complementary system of equations.

It is possible to take care of differential coupling relations in this manner. However by using the general form of the functional (2.31) by means of variational principles of the theory of elasticity (such as the principle of minimum potential energy), we may be able to restate the optimization problem in a suitable form, which makes the introduction of adjoint variables unnecessary.

Let us offer a demonstration. According to a variational principle, the actual distribution of the displacement vector $u(x)$ in a continuous medium minimizes the functional (cf. [2.1] and [2.8] to [2.9])

$$\prod(u, \Gamma_\sigma) = \frac{1}{2}\int_\Omega (\sigma_{ij}\varepsilon_{ij})\,d\Omega - \int_\Gamma (uq)\,d\Gamma_\sigma \to \min_u \tag{2.32}$$

under the constraint $(u)_\Gamma = 0$.

It is assumed in (2.32) that the stresses σ_{ij} and strains ε_{ij} are related to the displacements by Hooke's law and certain kinematic relations. For a physically realistic function $u(x)$, that is, one that minimizes the functional (2.32), $J = -\Pi$. Using this equality and the variational principle (2.32), we can connect the compliance J and functional Π:

$$J = -\min_{u} \Pi, \tag{2.33}$$

and consequently, the problem of minimizing the compliance may be reduced to assigning a minimum and maximum to the functional

$$J_* = \min_{\Gamma_\sigma} J = \min_{\Gamma_\sigma}\left(-\min_{u} \Pi\right) = -\max_{\Gamma_\sigma} \min_{u} (\Pi). \tag{2.34}$$

The minimum with respect to u occurring in the middle term of (2.34) is to be computed with additional constraint $(u)_\Gamma = 0$ and the maximum with respect to Γ_σ in the same term may be found with certain other conditions assigned to the permissible variations of the boundary Γ_σ, such as, for example, the isoperimetric condition, which states that the total volume of the region Ω occupied by a continuous medium remains constant.

It is unnecessary to consider the equations of equilibrium for each change of the boundary, since they were already incorporated in the formulation of problem (2.34) as necessary conditions for the existence of an extremum of Π with respect to u. A similar remark applies to the conditions assigned to the boundary Γ_σ.

Note that we can choose as control quantities not only functions that determine the shape of Ω, but also variables h that determine the interior structure of the medium (e.g., the distribution of the moduli of rigidity and coefficients of anisotropy). Even in that case all the above arguments remain valid, but the maximum in the middle term of (2.34) must be taken over h as well as over Γ_σ.

2.4.2 Let us assume that $q = 0$ on Γ_σ, but that on the boundary Γ_u is assigned a nonzero distribution of displacements (the body is in a deformed state).

As a specific example, consider a strip (see Figure 2.1), in which the displacements parallel to the x-axis are $(u_1)_{B_1B_4} = 0$ and $(u_1)_{B_3B_2} = U$, (where $U \geq 0$ is a constant), and the vertical displacements u_2 are identically equal to 0. We assume that $q = 0$ along curves B_1B_4 and B_3B_2 that comprise the boundary of Γ_σ. We introduce an integral J that will be used to obtain an estimate on the tensile rigidity of the strip:

$$J = \frac{1}{2}\int_{\Gamma_u} qu\,d\Gamma_u = \frac{1}{2}U\int_{B_2B_3} q\,d\Gamma_u = \frac{1}{2}(UP). \tag{2.35}$$

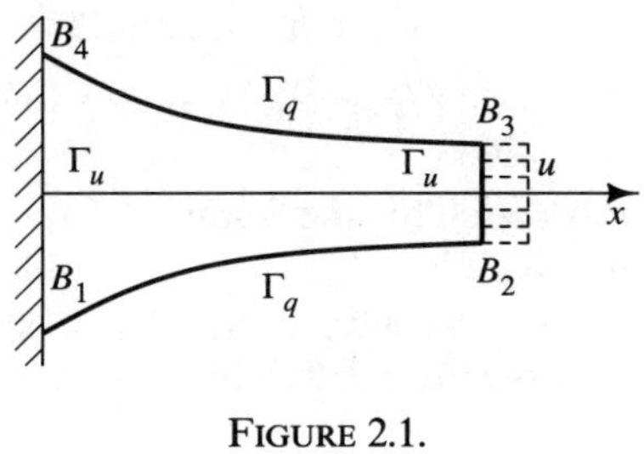

FIGURE 2.1.

Assuming that the constant U is given, we maximize the resultant force P, and therefore the functional J, by finding an appropriate shape of the boundary Γ_σ. Since

$$J = \min_u \Pi,$$

$$\Pi = 1/2 \int_\Omega \sigma_{ij}\varepsilon_{ij} d\Omega = 1/2 \int_\Omega \{a(u_{1x}^2 + u_{2y}^2) + 2bu_{1x}u_{2y} + (c/2)(u_{1y} + u_{2x})^2\} d\Omega, \tag{2.36}$$

we thus derive the formulation

$$J_* = \max_{\Gamma_\sigma} (J) = \max_{\Gamma_\sigma} \left(\min_u \Pi \right), \tag{2.37}$$

which is analogous to (2.34). The constants a, b, and c in (2.36) may be expressed in terms of Poisson's ratio ν and Young's modulus E. The relevant formulas are $a = E(1 - \nu)/(1 + \nu)(1 - 2\nu)$, $b = \nu E/[(1 + \nu)(1 - 2\nu)]$, and $c = E/(1 + \nu)$ in the case of a two-dimensional distribution of strains, and $a = E/(1 - \nu^2)$, $b = \nu E/(1 - \nu^2)$, and $c = E/(1 + \nu)$ for a two-dimensional stress distribution.

2.4.3 Let the response of a structure be described in a region Ω with boundary Γ by the linear differential equation $L(h)u = q$ with a self-adjoint operator L. Here h denotes the control function that enters into the coefficients of the operator L and u is the state function. The shape of the contour Γ is not fixed and is to be found as a result of the optimization procedure. The energy of the elastic body is to be determined from the formula

$$\Pi = \frac{1}{2} \int_\Omega (uL(h)u - 2qu) d\Omega. \tag{2.38}$$

If we recall that the function u assigns a minimum to the functional Π in (2.38), it is clear that the following equality holds:

$$J \equiv \frac{1}{2} \int_\Omega qu d\Omega = -\Pi, \tag{2.39}$$

and we arrive at the same conclusion as we did in Section 2.4.1, that for the given differential relation, the minimization of the compliance J defined by (2.39) is equivalent to the evaluation of a minimum with respect to u and of maxima with respect to Γ and h of the functional Π:

$$J_* = \min_\Gamma \min_h (J) = -\max_\Gamma \max_h \min_u (\Pi). \tag{2.40}$$

In a particular case, the search for the optimal distribution of the thickness and an optimal shape of a plate is reduced to the problem described by (2.40). Then Ω denotes the region bounded by the plate's contour, where u is the distribution of displacements and q the transverse loads.

As a specific example we could consider the torsion of an elastic anisotropic cylindrical rod (cf. [2.10] and [2.11]) whose axis coincides with the direction of the z-axis in a Cartesian coordinate system $\{x, y, z\}$. The function $\varphi(x, y)$ is defined in the region Ω having Γ as its boundary, where Ω occupies the transverse cross-section of the rod. φ is related to the components of the stress tensor by $\tau_{xz} = \theta\varphi_y$, $\tau_{yz} = -\theta\varphi_x$, where θ denotes the angle of twist per unit length of the rod. It satisfies the differential equation

$$L\varphi \equiv -(a\varphi_x - c\varphi_y)_x - (b\varphi_y - c\varphi_x)_y = 2, \tag{2.41}$$

$$(\varphi)_\Gamma = 0, \tag{2.42}$$

where a, b, and c denote the elastic constants of an anisotropic body, which, as is well known, satisfy the conditions $a > 0$ and $ab - c^2 > 0$ (cf. [2.11]). We pose the problem of finding the shape of the rod (i.e., the shape of its transverse cross-section Ω) for which the torsional rigidity of the rod

$$K = 2\int_\Omega \varphi \mathrm{d}\Omega \tag{2.43}$$

attains its maximum.

Note that if the coefficients a, b, and c satisfy these inequalities, for a given contour Γ a solution of the boundary-value problem (2.41) and (2.42) assigns a minimum value to the functional

$$\Pi = \int_\Omega (\varphi L\varphi - 4\varphi)\mathrm{d}\Omega = \int_\Omega (a\varphi_x^2 + b\varphi_y^2 - 2c\varphi_x\varphi_y - 4\varphi)\mathrm{d}\Omega, \tag{2.44}$$

in the appropriate class of functions satisfying condition (2.42). Moreover, when the functional (2.44) attains its minimum, the equality $K = -\Pi$ becomes valid.

In this fashion the initial problem of maximizing the torsional rigidity (2.41), (2.43) is transformed into the form

$$J_* = \max_\Gamma K = -\min_\Gamma \min_\varphi \Pi. \tag{2.45}$$

The (inner) minimum with respect to φ in (2.45) is computed for a fixed shape of the contour Γ under the boundary conditions (2.42). The (outer) minimum with respect to Γ must be found among the contours satisfying certain additional geometric conditions assigned to the admissible boundary shapes in specific problems. As such a geometric constraint, we could consider an isoperimetric condition stating that the area of the cross-section of the rod (i.e., the area Ω) must remain constant.

2.4.4 A different kind of an optimal design problem which permits one to eliminate differential relations and use variational principles occurs in the optimization of eigenvalues in self-adjoint boundary-value problems.

We have already mentioned in Sections 1.3 and 1.5 that problems of minimizing (or maximizing) eigenvalues are found in optimizing the stability or natural frequency of vibrations of conservative elastic systems.

Let an elastic system occupy a region Ω with boundary Γ. Its response is described by the differential equation

$$C(h)u - \lambda A(h)u = 0 \tag{2.46}$$

with homogeneous boundary conditions $N(h)u = 0$.

We shall assume (after taking into account the boundary conditions) that the operators $C(h)$ and $A(h)$ are self-adjoint and positive having coefficients that depend on the control variables h. Then it is possible to apply Rayleigh's variational principle to compute the minimal eigenvalue and the corresponding eigenfunction(s).

$$\left.\begin{aligned} &\lambda_0 = \min_u (J_1/J_2), \\ &J_1 = \int_\Omega (u \cdot C(h)u)\mathrm{d}\Omega, \qquad J_2 = \int_\Omega (u \cdot A(h)u)\mathrm{d}\Omega. \end{aligned}\right\} \tag{2.47}$$

Equation (2.46) is the Euler equation for the functional (2.47), and consequently the function u that yields the minimum of this functional "automatically" satisfies eq. (2.46).

Thus, the problem of maximizing the minimal eigenvalue by varying h and Γ has been reduced by the application of Rayleigh's principle to a sequential search, first for a minimum with respect to u and then for a maximum with respect to h and Γ:

$$\lambda_{0*} = \max_\Gamma \max_h \lambda_0 = \max_\Gamma \max_h \min_u (J_1/J_2). \tag{2.48}$$

Also note that if we consider the optimization of the nth eigenvalue instead of the minimal eigenvalue (according to our ordering), the minimum with respect to u in (2.47) and (2.48) is to be found under additional orthogonality constraints applied to the first $n - 1$ eigenfunctions. (cf. [1.27] and [2.7]).

2.5 Mutually Equivalent Design Problems

In the theory of optimal design it is frequently necessary to consider "mutually equivalent problems" whose solutions differ from each other only by multiplicative scale factors. In many cases an indirect solution of a problem may be bypassed by pointing out that a similar problem has been already investigated and that its solution may be obtained by an extension of an already known technique.

Therefore, a separate study of such mutually equivalent problems is useful in optimal design research, since it permits us to reduce the number of problems that need to be solved.

This mutual equivalence property is used in several research articles to solve specific problems of optimal design. A rigorous investigation of this topic that was carried out in [2.15] was specifically applied to problems with homogeneous functionals.

Let us consider an extremizing problem with homogeneous functionals. Let A denote a linear vector space and the set K a cone in that space. (This means that if an element $h \in K$ then $\lambda h \in K$ for any $\lambda =$ constant > 0.)

Let J_1 and J_2 be homogeneous functionals acting on A with degrees of homogeneity α and β, respectively. That is, $J_1(\lambda h) = \lambda^\alpha J_1(h)$ and $J_2(\lambda h) = \lambda^\beta J_2(h)$ for any $\lambda > 0$. We assume that these functionals are positive for any $h \in K$ and consider the extremal problem

$$\left.\begin{aligned} &\min J_1(h), \qquad h \in K, \\ &J_2(h) \geq c_2 > 0, \end{aligned}\right\} \tag{2.49}$$

where c_2 is a constant.

It is obvious that this problem makes sense if α and β have the same sign. Otherwise, no solution exists to this problem. To be specific, let us assume that both α and β are positive numbers.

We will demonstrate that if h^* solves problem (2.49) then $J_2(h) = c_2$, in other words, the minimum is attained on the boundary.

Indeed, if we assume that $J_2 > c_2$, we can choose a multiplier $\lambda = (c_2/J_2(h^*))^{1/\beta} < 1$. The element λh^* is admissible in the postulated problem, since $J_2(h^*) = \lambda^\beta J_2(h^*) = c_2$. Moreover, $J_1(\lambda h^*) < J_1(h^*)$, since $\lambda < 1$ and $\alpha > 0$, thus contradicting the optimality of h^*.

Now, let us examine a different problem with identical functionals (here c_1 is a given constant).

$$\left.\begin{aligned} &\max J_2(h), \quad h \in K, \\ &J_1(h) \leq c_1, \quad c_1 > 0. \end{aligned}\right\} \tag{2.50}$$

In this problem the extremum is also attained on the boundary, that is, $J_1(h^{**}) = c_1$. Here h^{**} denotes a solution of problem (2.50).

We shall prove the following proposition:

1. If problem (2.49) has a solution h^*, then the element $h^{**} = \gamma h^*$, where $\gamma = (c_1/J_1(h^*))^{1/\alpha}$, solves problem (2.50).

To prove this proposition let us take an arbitrary element $h \in K$ such that $J_1(h) \leq c_1$ and show that $J_2(h) \leq J_2(h^{**})$. Let us choose a multiplier $\kappa = (c_2/J_2(h))^{1/\beta}$. Then κh is an admissible element for problem (2.49) because $\kappa h \in K$ and $J_2(\kappa h) = c_1$. As a consequence of the optimality of h^*, we must have $J_1(\kappa h) \geq J_1(h^*)$, which is equal to $(c_2/J_2(h))^{\alpha/\beta}$. Therefore, $J_1(h) \geq J_1(h^*)$. We have already shown that $J_2(h^*) = c_2$. Now, it follows from this fact and the inequality stated above that

$$\left.\begin{aligned} J_2(h) &\leq J_2(h^*)\left[\frac{J_1(h)}{J_2(h^*)}\right]^{\beta/\alpha} \leq J_2(h^*)\left[\frac{c_1}{J_1(h^*)}\right]^{\beta/\alpha} \\ &= J_2\left[\left[\frac{c_1}{J_1(h^*)}\right]^{1/\alpha}_{h^*}\right] = J_2(\gamma h^*) = J_2(h^{**}). \end{aligned}\right\} \tag{2.51}$$

Now we apply the inequality $J_1(h) \leq c_1$. Thus, the proof of proposition 1 is complete.

The converse of this proposition is also true.

2. If h^{**} is a solution of problem (2.50), then the element $h' = \mu h^{**}$ with $\mu = (c_2/J_2(h^{**}))^{1/\beta}$ is a solution of problem (2.49). The proof of this proposition is identical to the one given above.

Now let us show that $\gamma\mu = 1$. Indeed, we use the equalities $c_2 = J_2(h^*)$, $c_1 = J_1(h^{**})$ to obtain

$$\begin{aligned} \gamma\mu &= [c_1/(J_1(h^*))]^{1/\alpha}[c_2/(J_2(h^{**}))]^{1/\beta} \\ &= [(J_1(h^{**}))/(J_1(h^*)]^{1/\alpha}[(J_2(h^*))/(J_2(h^{**}))]^{1/\beta} \\ &= \mu^{-1}[(J_1(h'))/(J_1(h^*))]^{1/\alpha}\mu[(J_2(h^*))/(J_2(h'))]^{1/\beta} = 1. \end{aligned} \tag{2.52}$$

In deriving the last estimate in (2.52), we used the optimal property of both h^* and h', concluding that $J_1(h^*) = J_1(h')$ and $J_2(h^*) = J_2(h')$.

Problems (2.49) and (2.50) have been called mutually equivalent with respect to each other. For such mutually equivalent problems the following assertion is true:

3. If either problem (2.49) or problem (2.50) has a unique solution, the mutually equivalent problem also has a unique solution, and these solutions are related by the equalities

$$h^{**} = \gamma h^*, \qquad \gamma = (c_1/J_1(h^*))^{1/\alpha} = (c_2/J_2(h^{**}))^{1/\beta}. \tag{2.53}$$

To prove this assertion we only need apply Propositions 1 and 2 and (2.53).

Also, because of the assumption concerning the existence and uniqueness of solutions, we can replace the "less than or equal" signs in the first two relations of (2.51) by strict inequalities.

Remark 1. Because the functionals in problems (2.49) and (2.50) are homogeneous, the solutions h^* and h^{**} may be represented in the form: $h^* = c_2^{1/\beta}h_1$, $h^{**} = c_1^{1/\alpha}h_2$, where h_1 and h_2 are the respective solutions of these problems for the values $c_2 = 1$ and $c_1 = 1$.

Remark 2. Even if the degrees of homogeneity of the basic functionals are of different sign, it would still make sense to consider the following problems:

$$\left.\begin{aligned} &\min J_1(h), \\ &J_2(h) \leq c_2 > 0, \\ &h \in \mathbf{K}. \end{aligned}\right\} \qquad \left.\begin{aligned} &\min J_2(h), \\ &J_1(h) \leq c_1 > 0, \\ &h \in \mathbf{K}. \end{aligned}\right\}$$

These problems are mutually equivalent with respect to each other.

It is not hard to check that the substitution $J_2'(h) = 1/J_2(h)$ reduces this problem to the equivalent mutual pair of problems given by (2.49) and (2.50) for the functionals J_1, and J_2'.

Let us discuss some examples of mutually equivalent problems that arise in designs of elastic beams. We shall make use of dimensionless variables, and assume that the beam has a unit length, that its axis coincides with the x-coordinate axis, and that it is simply supported at the endpoints $x = 0$ and $x = 1$.

Let $S(x)$ denote the distribution of the area of the transverse cross-section as a function of the length, and V, P, ω^2, and w, denote, respectively, the volume of the beam, the critical load causing loss of stability due to compressive forces applied at both ends, the square of the natural frequency of the transverse vibration, and the magnitude of deflection at the midpoint $x = \frac{1}{2}$ caused by a concentrated force applied at that point. Some formulas for these quantities are as follows:

$$\left.\begin{aligned} &V=\int_0^1 S\,dx, \qquad P=\min_u\left[\int_0^1 u_x^2\,dx\Big/\int_0^1 S^{-2}u^2\,dx\right],\\ &\omega^2=\min_u\left[\int_0^1 S^2u_{xx}^2\,dx\Big/\int_0^1 Su^2\,dx\right],\quad w=\int_0^1 S^{-2}\chi(x)\,dx,\\ &\chi(x)=\begin{cases}3x^2/8 & \text{if } 0\le x\le\frac{1}{2}\\ 3(1-x^2)/8 & \text{if } \frac{1}{2}\le x\le 1.\end{cases}\end{aligned}\right\} \tag{2.54}$$

We have assumed here that the shape of the cross-section remains basically the same as the area varies along the length of the beam (see Section 1.4).

Let V_0, P_0, ω_0, and w_0 be given positive constants. We choose the function $S(x)$ as the unknown control function. In addition to other constraints assigned to the problem, we also require that $S(x) \geq 0$.

Example 1. We consider the problem of minimizing the volume of a beam whose load consists of a single concentrated transverse force applied at the point $x = 1/2$, with a constraint assigned to the magnitude of the deflection at the point of application of the load:

$$V \to \min, \qquad w^{-1} \geq w_0^{-1}. \tag{2.55}$$

The functionals V and w^{-1} defined in (2.54) are homogeneous with respect to S with degrees of homogeneity $\alpha = 1$ and $\beta = 2$, respectively. A mutually equivalent problem with respect to (2.55) is the problem of minimizing w (i.e., maximizing the inverse quantity w^{-1}) with a constraint assigned to the volume of the beam:

$$w^{-1} \to \max_S, \qquad V \leq V_0. \tag{2.56}$$

If a function S^* assigns a minimum to the volume functional with a constraint on the magnitude of deflection, the function S^{**} that minimizes the magnitude of the deflection as long as the volume of the beam does not exceed a certain prescribed magnitude, must satisfy, according to (2.53), the relation

$$S^{**} = V_0 S^* \Big/ \int_0^1 S^*\,dx. \tag{2.57}$$

But if the solution S^{**} of the mutually equivalent problem (2.56) is known, the solution of the problem (2.57) can be related to S^{**} by the formula

$$S^* = (S^*/\sqrt{w_0})\left[\int_0^1 (S^{**})^{-2}\chi \mathrm{d}x\right]^{1/2}. \tag{2.58}$$

Example 2. By choosing a distribution of cross-sectional areas we wish to minimize the volume of a compressed column with a constraint, which stops the loss of stability:

$$V \to \min_S, \qquad P \geq P_0. \tag{2.59}$$

The degrees of homogeneity of the functionals V and P are, respectively, $\alpha = 1$ and $\beta = 2$. Dual to (2.59) is the problem of maximizing the critical load P causing a loss of stability of the elastic column with a bound assigned to the volume:

$$P \to \max_S, \qquad V \leq V_0. \tag{2.60}$$

The solution S^* of problem (2.59) and S^{**} of the dual problem (2.60) are related by the formulas

$$\left.\begin{aligned} S^{**} &= \gamma S^*, & \gamma &= V_0 \Big/ \int_0^1 S^* \mathrm{d}x. \\ S^* &= \mu S^{**} & \mu &= \left[P_0 \int_0^1 (S^{**})^{-2} u^2 \mathrm{d}x \Big/ \int_0^1 u^2 \mathrm{d}x\right]^{1/2}. \end{aligned}\right\} \tag{2.61}$$

which follow directly from eq. (2.53).

Example 3. The problem mutually equivalent to minimization of the volume of a beam with a constraint on the natural frequency of the transverse vibrations,

$$V \to \min_S, \qquad \omega^2 \geq \omega_0^2, \tag{2.62}$$

is that of maximizing the natural frequency with a bound on the volume of the beam, which must not exceed a given magnitude:

$$\omega_0^2 \to \max_S, \qquad V \leq V_0. \tag{2.63}$$

The degrees of homogeneity of the functionals V and ω^2 are $\alpha = 1$ and $\beta = 1$. By substituting these values of α and β into (2.53), we may derive formulas relating the solutions S^* and S^{**} of problems (2.62) and (2.63)

$$\left.\begin{aligned} S^{**} &= \gamma S^*, & \gamma &= V_0 \Big/ \int_0^1 S^* \mathrm{d}x. \\ S^* &= \mu S^{**} & \mu &= \omega_0 \int_0^1 S^{**} u^2 \mathrm{d}x \Big/ \int_0^1 (S^{**})^2 u_{xx}^2 \mathrm{d}x. \end{aligned}\right\} \tag{2.64}$$

2.6 Dual Optimization Problems

Dual transformations and studies of dual problems have assumed an increasingly important role in the theory of optimal design. Research results in the theory of duality permit the introduction of new techniques for estimating the magnitude of a global extremum and evaluation of certain bounds on the optimization processes. In many cases it is possible to establish that the values of the quality functionals are close for some designs, or perhaps are equal to the global extremum. Following the ideas of [2.13], we shall consider here the optimization of freely vibrating thin plates. We remark that techniques based on the theory of dual optimization that were developed in [2.13] are also applicable to problems of optimization of three-dimensional continua (e.g., minimization of the mass of material with bounds on the stresses, or minimization of the energy in an elastic deformation with a bound assigned to the total mass of the material).

Let the median surface of the plate occupy a region Ω having a boundary Γ in the xy plane. To determine the fundamental natural frequency of free vibrations, we apply the relations

$$\begin{aligned}
\omega^2 &= \min_{w \in V} J(h, w), \qquad J(h, w) = \int_\Omega \Pi(h, w)d\Omega \Big/ \int_\Omega T(h, w)d\Omega \\
\Pi &= [Eh^3/12(1-\nu^2)]\psi \\
&= [Eh^3/12(1-\nu^2)][(\Delta w)^2 - 2(1-\nu)(w_{xx}w_{yy} - w_{xy}^2)], \\
T &= \rho h \varphi = \rho h w^2, \\
V &= \{v \in W_2^2(\Omega), \qquad v = \partial v/\partial n = 0 \text{ on } \Gamma_1, v \equiv 0 \text{ on } \Gamma_2\}, \qquad (2.65)
\end{aligned}$$

where E is Young's modulus, ν Poisson's ratio, ρ the density, and w the amplitude of the deflection and $\partial w/\partial n$ its derivative in the direction normal to the boundary. The optimization problem consists in finding h_* and w_* such that

$$J(h_*, w_*) = \sup_{h \in \mathbf{H}} \inf_{w \in V} J(h, w),$$

where

$$\left.\begin{aligned}
\mathbf{H} &= \left\{h \in L_\infty(\Omega): \int_\Omega h d\Omega = h_m \operatorname{meas}(\Omega), h_{\min} \le h \le h_{\max}\right\}, \\
&0 \le h_{\min} < h_m < h_{\max}.
\end{aligned}\right\} \qquad (2.66)$$

In deriving a solution one must realize that the problem posed by (2.66) is not convex. In problems of this kind one must expect that along with piece-wise-continuous solutions there may exist generalized solutions (cf. [2.5] to [2.6]). This may cause certain well-known difficulties in constructing an exact solution. Therefore, we may be interested in finding some function $h \in \mathbf{H}$

that assigns to the quality functional a value only slightly smaller than the supremum. To do so we must first estimate the value of the supremum. This task may be accomplished by simplifying the dual problem.

By the definition of duality (cf. [2.14], in the dual problem we wish to find h_* and w_* such that

$$J(h_*, w_*) = \inf_{w \in V} \sup_{h \in \mathbf{H}} J(h, w). \tag{2.67}$$

Obviously, the following inequality is true:

$$\sup_{h \in \mathbf{H}} \inf_{w \in V} J(h, w) \leq \inf_{w \in V} \sup_{h \in \mathbf{H}} J(h, w). \tag{2.68}$$

This inequality can be used to produce an upper bound on the magnitude of the supremum for problem (2.66). Let us define the functional

$$J_0 = \sup_{h \in \mathbf{H}} J(h, w_0), \tag{2.69}$$

where w_0 denotes some arbitrary function belonging to the set V.

It follows easily from (2.67) to (2.69) that J_0 is an upper bound on the quality functional:

$$\sup_{h \in \mathbf{H}} \inf_{w \in V} J(h, w) \leq \inf_{w \in V} \sup_{h \in \mathbf{H}} J(h, w) \leq \sup_{h \in \mathbf{H}} J(h, w_0) \equiv J_0.$$

To find J_0 we need to construct a function h_0 such that $J(h_0, w_0) = J_0$.

It was proved in [2.13] that no intermediate values $h_{\min} < h < h_{\max}$ solve problem (2.69). We shall prove a corollary to this proposition.

We introduce a "control" function $\chi = (h_{\max} - h)/(h_{\max} - h_{\min})$. Then

$$J_0 = \sup_{\chi \in \mathbf{M}} J(\chi, w_0), \qquad J = \int_\Omega (A\chi + B)\psi(w_0)\,d\Omega \Big/ \int_\Omega (C\chi + D)\varphi(w_0)\,d\Omega,$$

$$A = h_{\min}^3 - h_{\max}^3, \qquad B = h_{\max}^3, \qquad C = h_{\min} - h_{\max}, \qquad D = h_{\max}. \tag{2.70}$$

Here $\mathbf{M}$ stands for a set of functions that take value 0 or 1 at each point of the region Ω and satisfy the isoperimetric condition

$$\int_\Omega (\chi h_{\min} + (1 - \chi)h_{\max})\,d\Omega = h_m \operatorname{meas}(\Omega).$$

Let N denote the closure of the set $\mathbf{M}$ in the weak* topology of the space $L_\infty(\Omega)$:

$$N = \{\chi \in L_\infty(\Omega) : 0 \leq \chi \leq 1, \qquad \int_\Omega (\chi h_{\min} + (1 - \chi)h_{\max})\,d\Omega = h_m \operatorname{meas}(\Omega).$$

Clearly,

$$J_0 = \sup_{x \in N} J(\chi, w_0). \tag{2.71}$$

It is not hard to show that a solution to problem (2.71) exists, that is, there exists a function $\chi_0 \in N$ for which the functional J attains its upper bound

[2.13]. It is also shown in [2.13] that the function χ_0 that assigns a maximum value to the functional $J(\chi, w_0)$ assumes one of the two values (0 or 1), and also, if Ω_0 and Ω_1 are proper subdomains on which $\chi_0 = 0$ or $\chi_0 = 1$, respectively, then

$$\left.\begin{aligned} &R(x, y) \geq R(\xi, \eta), \forall(x, y) \in \Omega_1, \qquad \forall(\xi, \eta) \in \Omega, \\ &R(u, v) = \psi(u, v) - J_0\varphi(u, v)/(h_{\min}^2 + h_{\min}h_{\max} + h_{\max}^2). \end{aligned}\right\} \tag{2.72}$$

Inequality (2.72) permits us to construct the function χ. First we choose some value J. This value, together with the corresponding value of J_0, is substituted in (2.72) and the corresponding function χ constructed.

Now we compute the functional $J^* = J(\chi, w_0)$. J_0 corresponds to the value of J for which $J = J^*$ and may be computed by standard techniques used for finding the roots of transcendental equations.

As a specific example, consider the project of designing a rectangular plate with sides a and b and thickness $h = h(x, y)$. The sides $y = 0$ and $y = b$ are simply supported while the edges $x = 0$ and $x = a$ are free. Problem (2.65) reduces to

$$\omega^2 = \inf J(h, F), \qquad J(h, F) = \int_0^a \Pi(h, F)\mathrm{d}x \bigg/ \int_0^a T(h, F)\mathrm{d}x,$$

$$\Pi = \{Eh^3/12(1 - v^2)\}[(F'' - \alpha^2 vF)F'' - \alpha^2(vF'' - \alpha^2 F) + 2\alpha(1 - v)(F')^2],$$

$$T = hF^2, \quad F \in U, \quad U = \{u \in W_2^2[0, a], \quad u(0) = u(a) = 0\}, \quad \alpha = n\pi/b.$$

It is necessary to find a function $h_* \in \mathbf{H}_2$ such that

$$w^2(h_*) = \sup_{h \in \mathbf{H}} w^2(h), \tag{2.73}$$

$$H_2 = \{h \in L_\infty[0, a]: h_{\min} \leq h \leq h_{\max}, \qquad \int_0^a h\mathrm{d}x = h_m a\}.$$

Using (2.73) we can obtain the following estimate for the supremum ($a = b = 1$):

$$J_0 = \pi^4(h_{\min}^2 + h_{\max}^2 + h_{\min}h_{\max} - h_{\min}^2 h_{\max} - h_{\max}h_{\min}^2).$$

It is shown in [2.13] that by changing the formulation of the problem (2.73), it is possible to determine solutions for which the value of the optimized functional is close to J_0.

We should add that the estimates obtained may depend rather strongly on either an unfortunate choice of w_0 or on the unavoidable existence of large differences between the values of quantities appearing in inequalities (2.68).

References for Chapter 2

2.1. Abovskii, N.P., Andreev, N.P., and Deruga, A.P., *Variational Principles in the Theory of Elasticity and in the Theory of Shells*. Moscow, Nauka, 1978, 288 pp.

2.2. Banichuk, N.V., On a variational problem with an unknown boundary, and determination of the optimal shapes of elastic bodies, *P.M.M.* 1975, *39*, No. 6, pp. 1082–1092.
2.3. Banichuk, N.V., Kartvelishvili, V.M., and Mironov, A.A., On a numerical method for solving two-dimensional optimization problems in the theory of elasticity. In: *Fifth All-Union Conference on numerical Methods for Solving Problems of the Theory of Elasticity and Plasticity. Part* 2. Novosibirsk, V.Ts.S.O., Akad. Nauk. S.S.S.R., 1978, pp. 3–14.
2.4. Cea, J., Optimization, Theory and algorithms; (Russian translation) Moscow, Mir, 1973, 244 pp. Originally published in French by Dunod Co., Paris, 1971.
2.5. Filippov, A.F., On some problems in the theory of optimal control, *Vestnik Moscow University, Ser. 1, Mathematics and Mechanics*, 1959, No. 2, pp. 25–32.
2.6. Gamkrelidze, R.V., On sliding optimal states, *D.A.N. Akad. Nauk S.S.S.R.*, 1962, *143*, No. 6, pp. 1243–1245.
2.7. Gould, S., *Variational Techniques in Eigenvalue Problems*, (Russian translation), Moscow, Mir, 1970, 328 pp.
2.8. Hill, R., *Mathematical Theory of Plasticity*, Clarendon Press, Oxford 1956, Russian translation: Moscow, Gostekhizdat., 1956, 407 pp.
2.9. Leibenzon, L.S., *Variational Methods for Solving Problems in the Theory of Elasticity*, Moscow and Leningrad, Gostekhizdat, 1943, 287 pp.
2.10. Lekhnitskii, S.G., *Theory of Elasticity of an Anisotropic Body*, Moscow, Nauka, 1977, 416 pp.
2.11. Lekhnitskii, S.G., *The Torsion of Anisotropic Inhomogeneous Rods*, Moscow, Nauka, 1971, 240 pp.
2.12. Miele, A., The calculus of variations in applied aerodynamics and flight mechanics, in *Optimization techniques*, pp. 99–170, Academic Press, New York, 1962, Russian translation: Moscow, Nauka, 1965, pp. 130–208.
2.13. Petukhov, L.V., Application of duality techniques in shape optimization problems for elastic bodies. *P.M.M.*, 1984, *48*, No. 5, pp. 830–838.
2.14. Rozin, L.A., *Variational Formulation of Problems for Elastic Systems*, Leningrad, Leningrad State University, 1971, 168 pp.
2.15. Seyranian, A.P., Homogeneous functionals and structural optimization problems, *Intern. J. Solids Structures*, 1979, *15*, No. 10, pp. 749–759.
2.16. Trukhaev, R.I., and Khomeniuk, V.V., *Theory of Nonclassical Variational Problems*, Leningrad, Leningrad State University, 1971, 168 pp.

3
Optimality Conditions

3.1 Necessary Conditions for Optimality in the Simplest Problems with Linear Equations Modeling the Behavior of Structures

We shall describe a variational approach used to obtain optimality criteria and reduce the optimization problem to a "closed" boundary–value problem for the corresponding differential equations.

Let $\mathbf{u}(x) = \{u_1(x), \ldots, u_n(x)\}$ be a vector function that satisfies in a region Ω the system of differential equations

$$L(h)\mathbf{u} = \mathbf{q} \tag{3.1}$$

and on the boundary Γ of Ω the boundary conditions

$$(N(\mathbf{h})\mathbf{u})_\Gamma = 0, \tag{3.2}$$

where $\mathbf{h} = \{h_1(x), \ldots, h_n(x)\}$ is a vector of design variables; $\mathbf{x} = \{x_1, \ldots, x_e\}$ is a vector of independent variables and q a vector function of the external loads. Systems of equations of the form (3.1) with boundary conditions (3.2) occur in the designs of linear-elastic, thin-walled structures. If q is independent of u and h, we are dealing with "dead loads" and external forces that do not depend on the geometry or the structural characteristics of the structure. These remarks are also applicable in the case when the loads depend linearly on deformation or structural displacements represented by the vector $\mathbf{u}$. Therefore any terms of eqn. (3.1) that reflect a linear dependence of $\mathbf{q}$ on $\mathbf{u}$ can be included formally in the left-hand side of that equation.

Let $J(u, h, q)$ and $J_i(u, h, q)$, $(i = 1, 2, \ldots, r)$, denote the integral functionals

$$J = \int_\Omega f(x, u, h, q)\mathrm{d}\Omega,$$

$$J_i = \int_\Omega f_i(x, u, h, q)\mathrm{d}\Omega, \tag{3.3}$$

where f and f_i are given functions of the variables $\mathbf{x}$, $\mathbf{u}$, $\mathbf{h}$, and $\mathbf{q}$.

We shall consider the problem of minimizing J:

$$J_* = \min_h J(u, h, q), \tag{3.4}$$

with integral constraints assigned to the design variables and the state function

$$J_i(u, h, q) - c_i \leq 0, \qquad i = 1, 2, \ldots, r. \tag{3.5}$$

Here c_i designate some given constants.

We shall derive optimization criteria for problem expressed by (3.1) to (3.5). With this in mind, we write the first variations of the integrals (3.3) and variational forms corresponding to eqs. (3.1) and (3.2):

$$\delta J = \int_\Omega [(\partial f/\partial u)\delta u + (\partial f/\partial h)\delta h] \mathrm{d}\Omega,$$

$$\delta J_i = \int_\Omega [(\partial f_i/\partial u)\delta u + (\partial f_i/\partial h)\delta h] \mathrm{d}\Omega, \tag{3.6}$$

$$L(h)\delta u + M(u, h)\delta h = 0, \tag{3.7}$$

$$N(h)\delta u + T(u, h)\delta h = 0, \tag{3.8}$$

where $(\partial f/\partial u) = \{\partial f/\partial u_1, \ldots, \partial f/\partial u_m\}$ and $(\partial f_i/\partial u) = \{\partial f_i/\partial u_1, \ldots, \partial f_i/\partial u_m\}$. The variations δJ and δJ_i depend on both the variation of the design variables and the variation of the state function. The latter are related linearly to δu and δh by (3.7) and (3.8). The variational relations (3.7) and the boundary relations (3.8) are obtained by substituting $u + \delta u$ and $h + \delta h$ for u and h, respectively in (3.1) and (3.2) and retaining only terms linear in δu and δh. The symbols $M(u, h)$ and $T(u, h)$ denote operators acting on the vector δh.

Let us express the first variation of the minimized functional in terms of the variation δh. To do so we follow a classical scheme of the calculus of variations and introduce an auxiliary vector function (the adjoint variable) $v(x) = \{v_1(x), \ldots, v_m(x)\}$ that will be determined so that the expression for the variation of the minimized functional will not contain any variation of the state function δu. Consequently v must obey a system of differential equations and boundary conditions, that is, v is a solution of a boundary–value problem.

We form the scalar product of the left-hand side of eq. (3.7) and the vector $v(x)$, and integrate this product over the domain Ω:

$$\int_\Omega v[L(h)\delta u + M(u, h)\delta h] \mathrm{d}\Omega = 0$$

(in effect, redefining the inner product).

After integration by parts, this integral is transformed into

$$\int_\Omega v[L(h)\delta u + M(u, h)\delta h] \mathrm{d}\Omega = \int_\Omega [\delta u L^*(h)v + \delta h M^*(u, h)v] \mathrm{d}\Omega. \tag{3.9}$$

Equation (3.8) connects the boundary values of the variations δu and δh

because of condition (3.2), which implies that the contour integrals occuring in the integration by parts are equal to 0. The remaining terms can also be equated to 0 because of the boundary condition, which may be assigned to the function v: $N^*(h)v = 0$. The coefficients of L^* are functions of h, but do not depend on u, though the coefficients of M^* may be functions of both h and u. The operators L^*, N^*, and M^* are adjoints of the operators L, N, and M, respectively.

For the sake of convenience, in the derivation of necessary optimality conditions, we shall replace the inequality constraints (3.5) by equality constraints. Following Miele, we shall use auxiliary unknowns (or slack variables) μ_i:

$$J_i - c_i + \mu_i = 0. \tag{3.10}$$

If strict inequality is enforced in (3.5) for the optimal solution, then $\mu_i \neq 0$. However if this constraint is "active" and the equal sign occurs in (3.5), then $\mu_i = 0$.

Let us consider (3.6) to (3.8) and extend our integral, making use of Lagrangian multipliers. Varying u, h, and μ_i, we represent the variation δJ in the form

$$\left.\begin{aligned} \partial J = \int \{[\partial f^\lambda/\partial h + M^*(u, h)v]\delta h \\ + [L^*(h)v + \partial f^\lambda/\partial u]\delta u\} d\Omega + 2 \sum_{i=1}^{r} \lambda_i \mu_i \delta\mu_i, \\ \partial f^\lambda/\partial h = \partial f/\partial h + \Sigma\lambda_i(\partial f_i/\partial h), \quad \partial f^\lambda/\partial u = \partial f/\partial u + \sum_{i=1}^{r} \lambda_i(\partial f_i/\partial u). \end{aligned}\right\} \tag{3.11}$$

To simplify the notation in (3.11), in all subsequent formulas we shall write

$$f^\lambda = f + \sum_{i=1}^{r} \lambda_i f_i, \tag{3.12}$$

where $\lambda = \{\lambda_1, \ldots, \lambda_r\}$ is a vector of Lagrangian multipliers, which can be determined from condition (3.5). By our remarks, we insist that the terms that multiply δu and $\delta\mu_i$ in (3.11) must vanish, that is, we set $L^*(h)v + (\partial f^\lambda/\partial u) = 0$ and $\lambda_i \mu_i = 0$ $(i = 1, 2, \ldots, r)$.

Thus, we derive the desired expression for the first variation of the optimized functional relating δJ to δh:

$$\delta J = \int [M^*(u, h)v + (\partial f^\lambda/\partial u)]\delta h d\Omega. \tag{3.13}$$

In deriving (3.13) we assumed that the function v satisfies the differential equation and boundary condition

$$\left.\begin{aligned} L^*(h)v + \partial f^\lambda/\partial u = 0, \quad (3.14) \\ (N^*(h)v)_\Gamma = 0, \quad (3.15) \end{aligned}\right\}$$

and that the quantities λ_i, μ_i satisfy the relation

$$\lambda_i \mu_i = 0, \qquad i = 1, 2, \ldots, r. \tag{3.16}$$

Using a common term of optimal control theory we shall call the function v "the adjoint variable." The quantities λ_i and μ_i can be determined by solving eq. (3.10) and (3.16). Condition (3.16) implies that for inactive constraints, $\mu_i \neq 0$. The corresponding Lagrangian multipliers λ_i must be then equal to 0, and consequently the corresponding constraints can be ignored in all subsequent developments. However, if $\lambda_i \neq 0$, $\mu_i \doteq 0$ and the constraint indexed by i turns out to be "active."

A necessary condition for minimizing the functional J assumes the form $\delta J = 0$. As a consequence we can derive a necessary optimality condition:

$$M^*(u, h)v + \partial f^\lambda / \partial h = 0. \tag{3.17}$$

In this manner, the optimization problem (3.1) to (3.5) has been reduced to the problem of solving the boundary-value problems given by (3.1) and (3.2) and by (3.14) and (3.15) and determining the vector function h (i.e., the design variable) from the optimality condition 3.17. The constants λ_i, μ_i are determined from conditions (3.10) and (3.16).

Let us suppose that f and f_i are linear functions of u, that is, $f = ug(h)$ and $f_i = ug_i(h)$, where g and g_i do not depend on u. In that case in the boundary-value relations (3.14) and (3.15) v is not a function of u, and the boundary-value problem for v can be stated independently of u:

$$L^*(h)v + g + \sum_{i=1}^{r} \lambda_i g_i = 0,$$

$$(N^*(h)v)_\Gamma = 0, \tag{3.18}$$

and the boundary-value problems (3.1), (3.2), and (3.18) are connected only by the choice of the design function h.

3.2 Conditions for an Extremum in Problems with Nonadditive Functionals

In addition to finding extrema of some integrals (subject to the assigned constraints), more general problems may arise in the theory of optimal design, such as the minimization or maximization of nonadditive functionals $J_1, \ldots, J_r$ that are given functions of several integrals. Such nonadditive functionals occur, for example, in problems of optimizing the natural frequencies of the vibrations of elastic systems and maximizing the critical loads that would cause a loss of stability in an elastic member of a structure.

Such nonadditive functionals also arise in problems involving the local properties of structures and in attempts to replace such properties by integrals, as described above in Section 2.3. Constraints on phase functions and control variables are also frequently represented by nonadditive functionals.

We shall now introduce some necessary conditions for optimality in problems with nonadditive functionals in which the minimized (or maximized) functional representing the quality criterion, and possibly the constraints, are given in the form of nonadditive functionals of integral characteristics:

$$J = \mathbf{F}(J_1, \ldots, J_r), \tag{3.19}$$

$$\mathbf{F}_i(J_1, \ldots, J_r) \leq 0, \qquad i = 1, 2, \ldots, k, \tag{3.20}$$

where

$$J_i = \int_\Omega f_i(x, u, h, q) d\Omega. \tag{3.21}$$

Here $u = \{u_1(x), \ldots, u_m(x)\}$, $h = \{h_1(x), \ldots, h_n(x)\}$, and $x = \{x_1, \ldots, x_e\}$.

In obtaining the necessary conditions for a minimum (maximum) of the functional J with appropriate constraints (3.20) and differential relations (3.1) and (3.2) assigned to the state function u and the design variables h, we make use of the formulas given in Section 3.1 and proceed to decompose the function

$$\mathbf{F}^\lambda = \mathbf{F} + \sum_{i=1}^{k} \lambda_i \mathbf{F}_i \tag{3.22}$$

into a sum of δJ_i, while retaining only the first-order terms:

$$\sum_{i=1}^{r} (\partial \mathbf{F}^\lambda / \partial J_i) \delta J_i = \int_\Omega \left\{ \sum_{i=1}^{r} (\partial \mathbf{F}^\lambda / \partial J_i) [(\partial f_i / \partial u) \delta u + (\partial f_i / \partial h) \delta h] \right\} d\Omega. \tag{3.23}$$

$$\mathbf{F}_i(J_1, \ldots, J_r) + \mu_i^2 = 0. \tag{3.24}$$

The symbols λ_i in (3.22) denote Lagrangian multipliers.

Using the auxiliary variables μ_i ($i = 1, 2, \ldots, k$) we transform inequality-type constraints, such as (3.20), into equality type constraints, in a manner similar to the preceding section. We shall display only the final formulas.

Taking into account the differential relations (3.1) and (3.2) and the boundary conditions assigned to the adjoint variable, that is,

$$(N^*(h))_\Gamma = 0, \tag{3.25}$$

as well as conditions (3.24) and formula (3.23), we present a representation of the variation of the functional J:

$$\begin{aligned} \delta J = \int_\Omega \Bigg\{ &\left[\sum_{l=1}^{r} (\partial \mathbf{F}^\lambda / \partial J_l)(\partial f_l / \partial u) + L^*(h) v \right] \delta u \\ &+ \left[M^*(u, h) v + \sum_{l=1}^{r} (\partial \mathbf{F}^\lambda / \partial J_l)(\partial f_l / \partial h) \delta h \right] \Bigg\} d\Omega + 2 \sum_{i=1}^{k} (\lambda_i \mu_i) \delta \mu_i, \end{aligned} \tag{3.26}$$

where

$$\begin{cases} \partial f_l / \partial u = \{\partial f_l / \partial u_1, \ldots, \partial f_l / \partial u_m\}, \\ (\partial f_l / \partial h) \delta h = \{(\partial f_l / \partial h_1) \delta h_1 + \cdots + (\partial f_l / \partial h_n) \delta h_n\}. \end{cases}$$

We equate the expressions that multiply (the arbitrary variations) δu and $\delta \mu_i$ to 0. Thus, we obtain a (vector) equation for the adjoint variable and a relation between the quantities λ_i and μ_i:

$$L^*(h)v + \sum_{i=1}^{r} \left(\frac{\partial \mathbf{F}^\lambda}{\partial J_l}\right)\left(\frac{\partial f_l}{\partial u}\right) = 0, \tag{3.27}$$

$$\lambda_i \mu_i = 0, \qquad i = 1, 2, \ldots, k. \tag{3.28}$$

For an "inactive" ith constraint, $\mu_i \neq 0$, and, clearly, the corresponding Lagrangian multiplier λ_i must be equal to 0. However, if the ith condition is "actice," $\lambda_i \neq 0$ and $\mu_i = 0$. Thus, if the state function and the design variable satisfy the boundary–value problem (3.1), (3.2) and (3.25), (3.27), while the quantities μ_i and λ_i obey conditions (3.24) and (3.28), the variation δJ is a function of δh:

$$\delta J = \int_\Omega \left\{\left[M^*(h, u)v + \sum_{i=1}^{r} (\partial \mathbf{F}^\lambda/\partial J_i)(\partial f_i/\partial h)\right]\delta h\right\} d\Omega. \tag{3.29}$$

In solving optimization problems formula (3.29) can be effectively used to construct "improvement variations" or to conduct sensitivity analysis. Also, this relation between δJ and δh implies a necessary optimality condition:

$$M^*(u, h)v + \sum_{i=1}^{r} (\partial \mathbf{F}^\lambda/\partial J_i)(\partial f_i/\partial h) = 0. \tag{3.30}$$

We observe that the coefficients appearing in the necessary conditions for an extremum and in the equations for the adjoint variables are computed for values of $J_1, \ldots, J_r$ that correspond to the extremal value (i.e., to the solution) of the optimization problem. Thus, (3.27) and (3.30) are integrodifferential equations. A derivation of extremality conditions for nonadditive functionals may be found in [3.8] and [3.14].

3.3 Generalization of Extremality Conditions to the Case of Constraints that Contain Derivatives

In Sections 3.1 and 3.2, we assumed that the integrands in the defining expressions (3.3) depend only on x, u, h, and q, and are independent of the derivatives of the state function or the design variable. However, functionals do depend on the derivatives of the functions u and h in the problems of structural design optimization. Therefore, the case when $f = f(x, u, u_x, h, h_x, q)$ $f_i = f_i(x, u, u_x, h, h_x, q)$ is of considerable practical interest. In that case, we need to modify the integrand in (3.6) by including in the expression for the variation of δJ the sums

$$(\partial f/\partial u_x)\delta u_x = (\partial f/\partial u_{1_{x_1}})\delta u_{1_{x_1}} + \cdots + (\partial f/\partial u_{m_{x_l}})\delta u_{m_{x_l}},$$
$$(\partial f/\partial h_x)\delta h_x = (\partial f/\partial h_{1_{x_1}})\delta h_{1_{x_1}} + \cdots + (\partial f/\partial h_{n_{x_l}})\delta h_{n_{x_l}}.$$

Identical expressions including sums representing $(\partial f_i/\partial u_x)\delta u_x$ and $(\partial f_i/\partial h_x)\delta h_x$ occur in the formulas for δJ_i. For the sake of simplicity, let us assume that the boundary conditions (3.2) and the side conditions governing the behavior of h on the boundary of Ω are such that all non-integral terms obtained by integration by parts of such additional sums in the formulas for J and J_i vanish identically. Therefore,

$$\delta J_i = \int_\Omega [(\partial f/\partial h - \partial/\partial x(\partial f_i/\partial h_x)\delta h + (\partial f_i/\partial u - \partial/\partial x(\partial f_i/\partial u_x))\delta u)]\,\mathrm{d}\Omega \tag{3.31}$$

and analogous expressions for δJ. Utilizing the definition (3.12) and performing routine algebra, we can repeat word for word the arguments given in the section 3.1, and derive the following expression for δJ:

$$\delta J = \int_\Omega \{[\partial f^\lambda/\partial h - \partial/\partial x(\partial f^\lambda/\partial h_x) + M^*(u, h)v]\delta h$$

$$+ [L^*(h)v + \partial f^\lambda/\partial u - \partial/\partial x(\partial f^\lambda/\partial u_x)]\delta u\}\,\mathrm{d}\Omega + 2\sum_{i=1}^{r}(\lambda_i\mu_i)\delta\mu_i, \tag{3.32}$$

where, as before,

$$\partial/\partial x(\partial f^\lambda/\partial u_x) = \sum_{i,j=1}^{m,l} \partial/\partial x_j(\partial f^\lambda/\partial u_{i_{x_j}}).$$

The boundary terms (i.e., the non-integral terms) in (3.9) are made to vanish by assigning to the vector function v a system of boundary conditions $N^*(h)v = 0$ and applying (3.8), which connects the boundary conditions for the variations of the state function and the permissible variations of the design variable. Now we use (3.32) and conclude that in this case the boundary–value problem for the adjoint variable and the expression relating the variation of the quality functional to the variation of the design variable is of the form

$$L^*(h)v + \partial f^\lambda/\partial u - \partial/\partial x(\partial f^\lambda/\partial u_x) = 0, \tag{3.33}$$

$$N^*(h)v = 0, \tag{3.34}$$

$$\delta J = \int_\Omega [M^*(u, h)v + \partial f^\lambda/\partial h - \partial/\partial x(\partial f^\lambda/\partial h_x)]\cdot\delta h\,\mathrm{d}\Omega. \tag{3.35}$$

Here the quantities λ_i and μ_i satisfy (3.16).

A necessary optimality condition that may be used to determine h may be obtained by setting the bracketed expression in (3.35) equal to 0:

$$M^*(u, h)v + \delta f^\lambda/\partial h - \partial/\partial x(\partial f^\lambda/\partial h_x) = 0. \tag{3.36}$$

If second-order derivatives with respect to u occur in the formulas for f and f_i, they can be handled in an identical fashion. Without going into details, we shall only state that also in this case the optimality condition assumes the general form of (3.36), though some additional terms appear in the boundary–value conditions for the adjoint variable:

$$L^*(h)v + \partial f^\lambda/\partial u - \partial/\partial x(\partial f^\lambda/\partial u_x) + 1/2 \sum_{j,i=1}^{e} \partial^2/\partial x_j \partial x_i(\partial^2 f_i/\partial u_{x_j} \partial u_{x_i}) = 0. \tag{3.37}$$

Let us now begin our discussion of problems with nonadditive functionals. We assume that the integrands in J and J_i are functions of u, u_x, h, h_x, and q. We retain all our previous assumptions concerning the behavior of the state function and the design variables on the boundary of Ω, and we will make use of (3.31) and (3.9). To obtain extremality conditions the discussion should be along the lines of that in Section 3.2, with the slight difference that formulas (3.6) for δJ, and δJ_i must be replaced by (3.31). As a result, we derive a necessary condition for optimality

$$M^*(u, h)v + \sum_{i=1}^{r} \partial F^\lambda/\partial J_i(\partial f_i/\partial h - \partial/\partial x(\partial f_i/\partial h_x)) = 0 \tag{3.38}$$

and the equation

$$L^*(h)v + \sum_{i=1}^{r} \partial F^\lambda/\partial J_i(\partial f_i/\partial u - \partial/\partial x(\partial f_i/\partial u_x)) = 0, \tag{3.39}$$

defining a boundary–value problem for an adjoint variable obeying the boundary conditions (3.34). Here F^λ denotes the quantity defined by (3.22).

If the functions f_i depend on higher-order derivatives of the state function, we obtain necessary optimality conditions in an analogous manner. Moreover the optimality condition and the relation between δJ and δh preserve their general form as given in (3.29) and (3.30), and the only changes take place in the equations and boundary conditions applied to the adjoint variable.

For example, if the integrands in (3.21) contain second–order derivatives of the state function $\delta^2 u_l/\partial x_i \partial x_j$, $(l = 1, \ldots, m;\ i, j = 1, \ldots, e)$, eqs. (3.27) for v assume the form:

$$L^*(h)v + \sum_{i=1}^{r} \partial F^\lambda/\partial J_i \Bigg[\partial f_i/\partial u - \sum_{j=1}^{e} \partial/\partial x_j(\partial f_i/\partial u_{x_j}) + 1/2 \sum_{j,d=1}^{e} \partial^2/\partial x_j \partial x_d(\partial^2 f_i/\partial u_{x_j} \partial u_{x_d}) \Bigg] = 0. \tag{3.40}$$

3.4 Self-Adjoint Optimization Problems

In the theory of optimal structural design an important class of stress-deformation problems described by equations of the form (3.1) and (3.2) consists of the "self-adjoint problem" in which it is possible to ignore the adjoint variables.

Let us return to the problem (3.1) to (3.5), assuming that the operators L and N are self-adjoint, that is,

$$L = L^*, \qquad N = N^*. \tag{3.41}$$

We choose the problem (3.1) and (3.2) with inhomogeneous boundary conditions, that is, with $q \neq 0$, and assume that

$$f = uq, \qquad f_i = g_i(h), \qquad i = 1, \ldots, r, \tag{3.42}$$

where g_i are known functions of the design variable. In this case the boundary-value problem (3.14) and (3.15) becomes

$$L(h)v = -q, \tag{3.43}$$

$$(N(h)v)_\Gamma = 0. \tag{3.44}$$

The boundary-value problem (3.43) and (3.44) is the same as the boundary-value problem (3.1) and (3.2) except for the sign of the right-hand side of eq. (3.43). Therefore, we can set

$$v = -u. \tag{3.45}$$

The function v can be eliminated from the optimality condition (3.18). Now

$$M^*(u, h)u + \sum_{i=1}^{r} \lambda_i(\partial g_i/\partial h) = 0. \tag{3.46}$$

In this manner, by using (3.45) we need no longer consider the adjoint variable in any self-adjoint problem satisfying the above additional assumptions. Thus, the order of the boundary–value problem has been halved.

Similarly, in the case $f = g(h)$, $f_i = g_i(h)$, $(i \neq k)$, and $f_k = uq$,

$$L(h)v + \lambda_k q = 0, \tag{3.47}$$

with boundary conditions (3.44). Then $v = -\lambda_k u$, and the optimality condition can be written in the form

$$-\lambda_k M^*(u, h)u + \partial g/\partial h + \sum_{i=1, i\neq k}^{r} (\lambda_i \partial g_i/\partial h) = 0. \tag{3.48}$$

The relations derived above can be illustrated by the following example involving optimization of the weight of a beam with a bound assigned to its compliance.

Let the beam be simply supported at its "left" end (i.e., at $x = 0$), and be rigidly built-in at the "right" end ($x = l$). We denote by q the transverse loads acting on the beam and by w the deflection function, and write down the equation of state, the boundary conditions, the bounds on compliance and the minimized functional:

$$L(h)w = (h^\alpha w_{xx})_{xx} = q,$$

$$w(0) = (h^\alpha w_{xx})_{x=0} = 0, \qquad w(l) = w_{xx}(l) = 0,$$

$$\int_0^l (qw)dx = c, \qquad \int_0^l h dx \rightarrow \min. \tag{3.43a}$$

The variational equation

$$(h^\alpha \delta w_{xx})_{xx} + \alpha(h^{\alpha-1} w_{xx} \delta h)_{xx} = 0 \tag{3.43b}$$

(which is another version of the state equation for beams in bending) can be multiplied by v and integrated by parts:

$$\begin{aligned}
&\int_0^l v[(h^\alpha \delta w_{xx})_{xx} + \alpha(h^{\alpha-1} w_{xx} \delta h)_{xx}]\mathrm{d}x \\
&\quad = \int_0^l [(h^\alpha v_{xx})_{xx} \delta w + \alpha(h^{\alpha-1} w_{xx} v_{xx}) \delta h]\mathrm{d}x \\
&\qquad + [v((h^\alpha \delta w_{xx})_x + \alpha(h^{\alpha-1} w_{xx} v_{xx}) \delta h]\mathrm{d}x \\
&\qquad + [v((h^\alpha \delta w_{xx})_x + \alpha(h^{\alpha-1} w_{xx} \delta h)_x) \\
&\qquad - v_x(h^\alpha \delta w_{xx} + \alpha h^{\alpha-1} w_{xx} \delta h) + \delta w_x(h^\alpha v_{xx}) \\
&\qquad - \delta w(h^\alpha v_{xx})_x]_{x=0}^{x=l}.
\end{aligned} \tag{3.43c}$$

To make certain that the nonintegral terms vanish, it suffices to make use of suitable boundary conditions by setting $v(0) = (h^\alpha v_{xx})_{x=0} = 0$, and $v(l) = v_x(l) = 0$. These relations represent boundary conditions for the adjoint variable v. At the same time v must satisfy the following differential equation and necessary optimality condition:

$$(h^\alpha v_{xx})_{xx} + \lambda q = 0, \qquad \alpha h^{\alpha-1}(w_{xx} v_{xx} + 1) = 0, \tag{3.43d}$$

where λ is a Lagrangian multiplier that corresponds to the isoperimetric condition. Comparing the boundary–value problems for w and for v, we conclude that $w = \lambda v$. Therefore, the adjoint variable may now be eliminated, and using (3.43d) we obtain

$$h^{\alpha-1}(w_{xx})^2 = \text{const.}$$

The optimality condition just derived, together with the basic equation of bending and the boundary conditions, constitute a "closed" boundary-value system for determining the distribution of thicknesses $h(x)$ and the distribution of (bending) deflections $w(x)$.

A different class of homogeneous self-adjoint problems may be characterized by conditions (3.41) and the equalities

$$q = 0, \qquad \partial f^\lambda / \partial u = cL(h)u, \tag{3.49}$$

where c is some constant. If (3.41) and (3.49) are true, the boundary-value problem (3.14) and (3.15) for the adjoint variable may be written in the form

$$L(h)v = 0, \qquad (N(h)v)_\Gamma = 0. \tag{3.50}$$

If u solves the boundary-value problem (3.1) and (3.2) with $q = 0$, $v = \gamma u$, (where γ is some constant) must satisfy (3.50), and the necessary optimality condition assumes the form

$$\gamma M^*(u, h)u + \partial f^\lambda / \partial h = 0. \tag{3.51}$$

Now let us consider the variational problems (3.1), (3.2) and (3.19) to (3.21) with nonadditive functionals, which include, in particular, such problems as optimizing the natural frequencies of vibrations or optimizing critical loads that cause a loss of stability. For such problems to become self-adjoint, conditions (3.41) must hold, though the following equalities must be satisfied as well:

$$q = 0, \qquad \sum_{i=1}^{r} \partial F^{\lambda}/\partial J_i(\partial f_i/\partial u - \partial/\partial x(\partial f_i/\partial u_x)) \equiv L(h)u. \tag{3.52}$$

By comparing eqs. (3.1), (3.2), (3.27), (3.41), and (3.52), one can see that the boundary-value problem for the adjoint variable v coincides with the original boundary-value problem for the state function u. By our previous discussion, this permits us to express the adjoint variable as a function of the state variable and to halve the order of the associated system of differential equations.

3.5 Optimality Conditions for Problems with Unknown Boundaries

Until now we have considered only the cases when the design variables are to be determined in a (fixed) domain Ω. The values of the design variables determine the values of functionals studied in the problem and also determine the coefficients of the differential equations and the boundary conditions, so that the state function $u(x)$ is also determined for given values of $h(x)$.

The boundary of Ω is perceived as fixed, and cannot be varied.

However, many problems of optimal design lead to problems with unknown boundaries, where a choice of the shape of Ω is one of the unknown design variables. In a rigorous formulation of shape optimization for an elastic body we must deal with a design problem with an unknown boundary. Boundary "control" problems also arise when one attempts to find the location of surfaces separating materials with different mechanical properties when inhomogeneous or piecewise-homogeneous materials are used.

We shall follow [3.6] in deriving necessary extremum conditions in problems of optimizing the shapes of elastic bodies. Let us consider a deformed elastic body occupying a region Ω whose boundary consists of a surface Γ. Displacements are assigned on the part Γ_u of this surface, while tractions are given on the part Γ_σ, with $\Gamma = \Gamma_\sigma \cup \Gamma_u$ and $\Gamma_\sigma \cap \Gamma_u = \varnothing$. The corresponding boundary conditions are

$$u_i = U_i \quad \text{on} \quad \Gamma_u, \tag{3.53}$$

$$\sigma_{ij} n_j = T_i \quad \text{on} \quad \Gamma_\sigma, \tag{3.54}$$

where σ_{ij}, u_i, and n_j denote, respectively, the components of the stress tensor, the displacement vector, and the unit vector pointing in the direction of an outward normal to the surface of the body, while U_i and T_i are given functions of the coordinates x_i. We use a Cartesian reference frame $\{0, x_1, x_2, x_3\}$. The indices i and j in (3.53) and (3.54) assume the values 1, 2, 3. The summation convention is used, and summation is performed over repeated indices.

The body is deformed by forces $q(x)$, which are combinations of external loads applied to the body surface and body forces. The state of a stressed and deformed body is modeled by a system of equations of the theory of elasticity.

The basic equations

$$\sigma_{ij,j} + q_i = 0, \qquad \varepsilon_{ij} = 1/2(u_{i,j} + u_{j,i}), \qquad \sigma_{ij} = C_{ijkl}\varepsilon_{kl}, \tag{3.55}$$

are derived under the assumption that the laws of statics are obeyed and that the deformations are "small." The constants C_{ijk_ℓ} are components of the elastic moduli's tensor.

The stress-deformed state of the body (which could be a structural element) is determined by the values of the integral functionals

$$J_\nu = \int_\Omega f_\nu(u_k, \sigma_{ij})\mathrm{d}\Omega. \tag{3.56}$$

Here f_ν is a given function of the components of the displacement vector and the stress tensor; $\nu = 1, 2, \ldots, r$.

The optimization problem consists in finding an admissible surface Γ_v, $\Gamma_v \subset \Gamma$ such that the total volume (or weight of the body) attains a minimum:

$$V = \int_\Omega \mathrm{d}\Omega \to \min_{\Gamma_v}, \tag{3.57}$$

while satisfying prescribed bounds on some properties of the stressed state,

$$J_\nu \leq c_\nu, \qquad \nu = 1, 2, \ldots, l, \tag{3.58}$$

where c_ν are designated constants. For the sake of convenience we shall introduce auxiliary (slack) variables μ_i and replace inequalities (3.58) by the equality-type, constraints (3.10).

In this type of a problem we shall allow boundary variations as long as the connectivity properties of Ω are not affected. This means that we shall compare only such regions that can be continuously mapped into each other. We also assume that for each admissible body there exists a solution to a basic problem in the theory of elasticity. Finally, to formulate an optimization problem, we must be more specific in describing the manner in which the boundary conditions change with changes in the shape of the region. In the simplest case the vector fields $U_i(x_1, x_2, x_3) \equiv U_i(x_k)$ and $T_i(x_1, x_2, x_3) \equiv T_i(x_k)$ are given.

In that case, for each point of the admissible surface the right-hand sides of the boundary-value conditions (3.53) and (3.54) are known. Unless otherwise stipulated, we shall adopt exactly this manner of specifying boundary conditions as we vary the shape of the region.

To formulate an optimization problem, let us write the Lagrangian functional

$$J = \int_\Omega [1 + \psi_i(\sigma_{ij,j} + q_i) + \chi_{ij}(\sigma_{ij} - 1/2C_{ijkl}u_{k,l} - 1/2C_{ijkl}u_{l,k})]\mathrm{d}\Omega$$

$$+ \sum_{v=1}^{r} \lambda_v\left(\left(\int f_v \mathrm{d}\Omega + \mu_v^2 - c_v\right) = \int_\Omega f^L \mathrm{d}\Omega + \sum_{v=1}^{r} \lambda_v f_v, \qquad (3.59)$$

where

$$f^L = 1 + \psi_i(\sigma_{ij,j} + q_i) + \chi_{ij}(\sigma_{ij} - C_{ijkl}u_{k,l}) + \sum_{v=1}^{r} \lambda_v f_v.$$

The Lagrangian functional is a function of the state $\{\sigma_{ij}, u\}$ and also a function of the (constant) Lagrangian multipliers λ_v and of the adjoint functions χ_{ij} and ψ_k and their spatial derivatives. The functions that occur in the formula for f^L may be regarded as forming an 18–dimensional vector:

$$\{v_\alpha\} = \{u_1, u_2, u_3, \psi_1, \psi_2, \psi_3, \chi_{11}, \ldots, \chi_{33}, \sigma_{11}, \ldots, \sigma_{33}\}, \alpha = 1, 2, \ldots, 18.$$

The expression for the variation δJ of the functional defined by (3.59), which depends on the variation $\{\delta v_\alpha\}$ of the vector $\{v_\alpha\}$ and simultaneous variation of the boundary $\{\delta x_i\}$, can be expressed as follows:

$$\delta J = \int_\Omega \{(\partial f^L/\partial v_\alpha - \partial/\partial x_i(\partial f^L/\partial v_{\alpha,i}))(\delta v_\alpha - v_{\alpha,j}\delta x_j)\}\mathrm{d}\Omega$$

$$+ \int_\Gamma [(\partial f^L/\partial v_{\alpha,i})n_i](\delta v_\alpha - v_{\alpha,j}\delta x_j)\mathrm{d}\Gamma$$

$$+ \int_{\Gamma_v} f^L n_i \delta x_i \mathrm{d}\Gamma + 2\sum_{v=1}^{r} \lambda_v \mu_v \delta\mu_v. \qquad (3.60)$$

For all repeated indices α we sum from 1 to 18.

On Γ the variation of v_α becomes $(\delta v_\alpha)_\Gamma = \delta v_\alpha - v_{\alpha,i}\delta x_i$. We substitute the expression for f^L from (3.59):

$$\delta J_i = \int_\Omega \left(\chi_{ij} - \psi_{i,j} + \sum_{v=1}^{r} \lambda_v(\partial f_v/\partial\sigma_{ij})\right)(\delta\sigma_{ij} - \sigma_{ij,k}\delta x_k)\mathrm{d}\Omega$$

$$+ \int_\Omega \left[(C_{ijkl}\chi_{kl}), j + \sum_{v=1}^{r} \lambda_v(\partial f_v/\partial u_i)\right](\delta u_i - u_{i,k}\delta x_k)\mathrm{d}\Omega$$

$$+ \int_\Gamma \psi_i n_j(\delta\sigma_{ij} - \sigma_{ij,k}\delta x_k)\mathrm{d}\Gamma - \int_\Gamma C_{klij}n_j(\delta u_i - u_{i,k}\delta x_k)\mathrm{d}\Gamma$$

$$\int_{\Gamma^v}\left(1 + \sum_{v=1}^{r} \lambda_v f_v)n_k\delta x_k \mathrm{d}\Gamma + 2\sum_{v=1}^{r} \lambda_v\mu_v\delta\mu_v. \qquad (3.61)$$

We do not want variation δJ of J to be a function of the variations $\delta\sigma_{ij}$, δu_k and $\delta\mu_v$. To attain this objective, we define the adjoint variables ψ_i and χ_{ij} to be solutions of the boundary-value problem

$$\left.\begin{aligned} \chi_{ij} - \psi_{i,j} + \sum_{\nu=1}^{r} \lambda_\nu(\partial f_\nu/\partial \sigma_{ij}) = 0, \\ (C_{klij}\chi_{kl})_{,j} + \sum_{\nu=1}^{r} \lambda_\nu(\partial f_\nu/\partial u_i) = 0, \end{aligned}\right\} \tag{3.62}$$

$$(\psi_i)_{\Gamma_u} = 0, \qquad (C_{kli}n_j)_{\Gamma_\sigma} = 0, \tag{3.63}$$

while the Lagrangian multipliers λ_ν and the auxiliary (slack) quantities μ_i are to satisfy the condition

$$\lambda_\nu \mu_\nu = 0, \qquad \nu = 1, 2, \ldots, r. \tag{3.64}$$

Taking into account (3.62) to (3.64) and observing that on $\Gamma_\sigma - \Gamma_v \cap \Gamma_\sigma$, $\delta x_k = 0$ and $\delta\sigma_{ij}n_j = 0$, while on $\Gamma_u - \Gamma_v \cap \Gamma_u$, $\delta x_k = 0$ and $\delta u_i = 0$, we can represent (3.61) as a sum of two terms that depend on variations restricted to the parts of the boundary $\Gamma_v \cap \Gamma_\sigma$ and $\Gamma_v \cap \Gamma_u$, respectively.

$$\left.\begin{aligned} \delta J_\sigma &= \int_{\Gamma_v \cap \Gamma_\sigma} \left[\psi_i n_j(\sigma_{ij} - \delta\sigma_{ij} - \sigma_{ij,k}\delta x_k) + \left(1 + \sum_{\nu=1}^{r} \lambda_\nu f_\nu\right) n_k \delta x_k\right] d\Gamma \\ \delta J_u &= \int_{\Gamma_v \cap \Gamma_u} [C_{klij}n_j\chi_{kl}(\delta u_i - u_{i,p}\delta x_p) + \left(1 + \sum_{\nu=1}^{r} \lambda_\nu f_\nu\right) n_k \delta x_k \Bigg] d\Gamma \end{aligned}\right\} \tag{3.65}$$

Boundary variations may be written in the form $\delta x_i = a_i \delta t'$, where $a_i a_i = 1$. The vector $\{a\}$ defines a spatial direction, and t' is the absolute value of a component of the surface displacement pointing in a given direction. For admissible boundary variations we derive the following expression for the variation of $U_i : \delta U_i = U_{i,j}a_j\delta t'$. Therefore, the variations of the boundary values u_i are given by $\delta u_i = \delta U_i = U_{i,j}a_j\delta t'$. We substitute these values in (3.65):

$$\delta J_u = J_{\Gamma_v \cap \Gamma_u}\left[C_{klij}\chi_{kl}(U_i - u_i)_{,m}a_m + \left(1 + \sum_{\nu=1}^{r} \lambda_\nu f_\nu\right) n_k a_k\right] dt' d\Gamma.$$

It can be shown that variation of this functional does not depend on the direction of a displacement computed at a boundary point. To show that the gradient of this functional does not depend on the direction of such variation, we observe that

$$\begin{aligned} (U_i - u_i)_{,m}a_m &= (U_i - u_i)_{,m}(a_m - a_j n_j n_m) + (U_i - u_i)_{,m}a_j n_m n_j \\ &\equiv (U_i - u_i)_{,m}a_j n_j. \end{aligned}$$

(Here $n_m n_j$ is regarded as a dyadic.)

But, the product $(U_i - u_i)_{,m}(a_m - a_j n_m n_j)$ is equal to 0, because the first term $(U_i - u_i)_{,m}$ represents a vector parallel to the surface Γ, while the second term $(a_m - a_j n_m n_j)$ is a vector orthogonal to Γ.

Therefore, the first term in the integrand may simplified to the form $C_{klij}\chi_{kl}(U_i - u_i)_{,m}n_m n_j a_p n_p$. The dyadic $n_j n_m$ may be eliminated using the relation $\xi_j\xi_k + \eta_j\eta_k + n_j n_k = \delta_k^j$, which is valid if ξ_j and η_j are mutually orthogonal unit vectors that are tangential to coordinate curves spanning the boundary surface. A proof of this relation may be found in the preprint [3.6a]. Observing that that the derivatives of the quantity $(U_i - u_i)$ in the direction tangential to the boundary surface are all equal to 0, we derive

$$C_{klij}\chi_{kl}(U_i - u_i)_{,m}(\delta_{jm} - \xi_j\xi_m - \eta_j\eta_m)a_p n_p = C_{klij}\chi_{kl}(U_i - u_i)_{,j}a_p n_p.$$

As a result, the expression for δJ_u assumes the form

$$\delta J_u = \int_{\Gamma_v \cap \Gamma_u} \left[C_{klij}\chi_{kl}(U_i - u_i)_{,j} + 1 + \sum_{\nu=1}^{r} \lambda_\nu f_\nu \right] \delta t \, d\Gamma \tag{3.66}$$

where $\delta t = a_p n_p = \delta t'$ is a component of the boundary displacement in the normal direction.

Now, let us manipulate the formula (3.65) for δJ_σ. It is not hard to show that if we vary a segment of the boundary $\Gamma_v \cap \Gamma_\sigma$, then without any loss of generality we need consider only variations that are normal to the surface of the body $\delta x_i = n_i \delta t$. A specific feature of the case presently considered is the implicit dependence of δj on the variation of the direction normal to the surface δn_j through the explicit dependence on the variation of the boundary values of the stress components. Variations of the normal unit vector and of the stress components are related as follows: $\delta(\sigma_{ij}n_j) = n_j\delta\sigma_{ij} + \sigma_{ij}\delta n_j = T_{i,j}\delta t$. We express $n_j\delta\sigma_{ij}$ using this relation and substitute it in the formula (3.65) for δJ_σ. Thus, we obtain

$$\delta J_\sigma = \int_\Gamma \left[\psi_i(T_{i,k}n_k\delta t - \sigma_{ij}\delta n_j - \delta t \sigma_{ij,k}n_j n_k) + \delta t\left(1 + \sum_{\nu=1}^{r} \lambda_\nu f_\nu\right)\right] d\Gamma.$$

We shall use the following equation derived in the preprint [3.6a]:

$$\delta n_j = (\xi/\sqrt{E})\partial\delta t/\partial\xi - (\eta_j/\sqrt{G})\partial\delta t/\partial\eta,$$

where E, G, ξ, and η are the coefficients of the first quadratic form and curvilinear coordinates, respectively. The formula $d\Gamma = (EG)^{1/2} d\xi d\eta$ allows us to transform the integral of $\psi_i\sigma_{ij} \cdot \delta n_j$ along $\Gamma_v \cap \Gamma_\sigma$:

$$\int_{\Gamma_v \cap \Gamma_\sigma} \psi_i\sigma_{ij}[(\xi_j/\sqrt{E})\partial\delta t/\partial\xi - (\eta_j/\sqrt{G})\partial\delta t/\partial\eta](EG)^{1/2} d\xi d\eta$$

$$= \int_{\Gamma_v \cap \Gamma_\sigma} \{\partial/\partial\xi(\psi_i\sigma_{ij}\xi_j\sqrt{G}) + \partial/\partial\eta(\psi_i\sigma_{ij}\eta_j\sqrt{E})\}\delta t \, d\xi d\eta$$

$$+ \int_{\Gamma^l} [\psi_i\sigma_{ij}(\xi_j\sqrt{G}n_k^l\xi_k + \eta_j\sqrt{E}n_k^l\eta_k)\delta t] d\Gamma^l.$$

Here Γ^l denotes a curve that forms the boundary of the varied segment of the surface $\Gamma_v \cap \Gamma_\sigma$; $\{n_k^l\}$ is the unit vector normal to the curve Γ^l and situated in the plane tangential to the surface $\Gamma_v \cap \Gamma_\sigma$.

The integral along Γ^l is equal to 0. This is true, because on the part of Γ^l that separates $\Gamma_v \cap \Gamma_\sigma$ from $\Gamma_v \cap \Gamma_u$ the variable ψ_i is identically equal to 0, and on the remaining part of Γ^l the variation δt vanishes, since for any admissible variation the transformation of the surface is continuous. Now we can consider the second integral and apply to the right side a differential transformation and use the formula $\partial/\partial\xi(\xi_j\sqrt{G}) + \partial/\partial\eta(\eta_j\sqrt{E} = 2Hn_j(EG)^{1/2}$, where H represents the average curvature of the surface. Thus,

$$\int_{\Gamma_v\cap\Gamma_\sigma} [\psi_i\sigma_{ij}\{(\xi_j/\sqrt{E})\partial\delta t/\partial\xi - (\eta_j/\sqrt{G})\partial\delta t/\partial\eta\}(EG)^{1/2}]\mathrm{d}\xi\mathrm{d}\eta$$

$$= \int_{\Gamma_v\cap\Gamma_\sigma} [\{(\psi_i\sigma_{ij})_{,k}(\xi_k\xi_j + \eta_k\eta_j) + 2H\psi_i\sigma_{ij}n_j\}\delta t(EG)^{1/2}]\mathrm{d}\xi\mathrm{d}\eta$$

$$= \int_{\Gamma_v\cap\Gamma_\sigma} [\{(\psi_i\sigma_{ij})_{,k}(\eta_k\eta_j - \delta_{kj}) + 2H\psi_i\sigma_{ij}n_j\}\delta t(EG)^{1/2}]\mathrm{d}\xi\mathrm{d}\eta.$$

Finally, we derive a formula for the variation of the surface $\Gamma_v \cap \Gamma_\sigma$:

$$\delta J_\sigma = \int_{\Gamma_v\cap\Gamma_\sigma} \left\{1 + \sum_{v=1}^{r} \lambda_v f_v - 2H\varphi_i T_i - (\psi_i\sigma_{ij})_{,j} + (\psi_i T_i)_{,k}n_k\right\}\delta t\mathrm{d}\Gamma \tag{3.67}$$

Equations (3.66) and (3.67) define the sensitivity of the optimized functional with respect to changes in control variables (the shape of the boundary in our case). This permits us to derive iterative numerical procedures for optimization. From (3.66) and (3.67) we can immediately deduce the following necessary optimality conditions:

$$\left.\begin{aligned} 1 + \sum_{v=1}^{r} \lambda_v f_v + C_{ijkl}(U_i - u_i)_{,j} = 0 \quad &\text{on } \Gamma_v \cap \Gamma_u \\ 1 + \sum_{v=1}^{r} \lambda_v f_v - 2H\psi_i T_i - (\psi_i\sigma_{ij})_{,j} + (\psi_i T_i)_{,k}n_k = 0 \quad &\text{on } \Gamma_v \cap \Gamma_\sigma. \end{aligned}\right\} \tag{3.68}$$

These conditions can be used to determine the shape of unknown segments of the boundary: Γ_u and Γ_σ.

3.6 Optimality Conditions for Dynamic Design Problems

A general formulation of dynamic problems in optimization of structures was already discussed in Section 1.5. Therefore, it will suffice in the present section to derive basic relations for this class of problems and to offer a proof of some conditions for the existence of an extremum (cf. [1.6]).

Let $x \in \Omega$ be a vector of spatial variables; $t \in [0, t_k]$ time; Ω a given domain occupied by a structure; and t_k some fixed constant denoting the span of time during which the structure will function; $u(x, t)$, $h(x)$, and $q(x, t)$ will denote the vector state function, the design vector, and the load, respectively.

We shall restrict our discussion to linear initial boundary-value problems:

$$Au_{tt} + Cu = q, \tag{3.69}$$

$$(Nu)_\Gamma = g(x, t), \qquad (Mu)_{t=0} = \chi(x). \tag{3.70}$$

These relations completely determine the behavior of the structure for a given choice of admissible design variables $h(x)$. Here A and C represent linear differential operators (inertial and elastic terms, respectively) whose coefficients are functions of $h(x)$; N and M are linear operators that depend on the boundary and initial conditions; $g(x, t)$ and $\chi(x)$ are given functions; and Γ is the boundary of Ω.

We consider the problem of optimal design for a structure subjected to dynamic interactions. The problem consists in finding a function $h(x)$ that minimizes the functional

$$J = \int_0^{t_k} \int_\Omega f(x, t, u, h, q) \mathrm{d}\Omega \mathrm{d}t \to \min_h \tag{3.71}$$

and satisfies the bounds

$$J_i - c_i \leq 0 \qquad (i = 1, 2, \ldots, r_1, r_1 + 1, \ldots, r_1 + r_2 = r) \tag{3.72}$$

that are assigned to some integral and local properties

$$\left.\begin{aligned} J_i &= \int_0^{t_k} \int_\Omega f_i(x, t, u, h, q) \mathrm{d}\Omega \mathrm{d}t, \quad i = 1, 2, \ldots, r_1, \\ J_j &= \max_{x,t} (f_j(x, t, u, h, q), \qquad j = r + 1, \ldots, r_1 + r_2, \quad \text{where } r_1 + r_2 = r. \end{aligned}\right\} \tag{3.73}$$

c_i in (3.73) are a priori given constants. The inequality bounds in (3.72) and those given in Section 3.1 can now be replaced using auxiliary variables by the equality-type constraints (3.10), while local properties may be approximated (as discussed in Section 2.3) by integral functionals:

$$J_j = \int_0^{t_k} \int_\Omega [f(x, t, u, h, q]^p \mathrm{d}\Omega \mathrm{d}t. \tag{3.74}$$

In this approximation by functionals, the constants c_i in (3.72) are also replaced by the quantities t_k, c_j^p, and meas Ω.

In order to introduce a basic formula for the analysis of sensitivity, we have to find necessary optimality conditions and introduce an adjoint system, and then use techniques of optimal control for systems with distributed parameters, as we have done in preceding sections of this chapter. Let us vary the design functions and the state function in (3.69) and (3.70):

$$A\delta u_{tt} + C\delta u + A_{\delta h} u_{tt} + C_{\delta h} u = 0, \tag{3.75}$$

$$(N\delta u)_\Gamma = 0, \qquad (M\delta u)_{t=0} = 0, \tag{3.76}$$

where $A_{\delta h} = (\partial A/\partial h)\delta h$ and $C_{\delta h} = (\partial C/\partial h)\delta h$.

We introduce a vector of adjoint variables $v(x, t)$. We form a scalar product of the left-hand side of (3.75) with $v(x, t)$ and then integrate by parts:

$$\int_0^{t_k}\int_\Omega v[A\delta u_{tt} + C\delta u + A_{\delta h}u_{tt} + C_{\delta h}u]\mathrm{d}\Omega\mathrm{d}t$$
$$= \int_0^{t_k}\int_\Omega [\delta u(A^*v_{tt} + C^*v]\mathrm{d}\Omega\mathrm{d}t$$
$$+ \int_0^{t_k}\int_\Omega [A^*_{\delta h}uv + C^*_{\delta h}uv]\mathrm{d}\Omega\mathrm{d}t. \tag{3.77}$$

The symbol $*$ denotes the adjoint operation. To make certain that the nonintegral terms resulting from integration by parts vanish, we use (3.76) and impose on the function $v(x, t)$ the boundary conditions $(N^*v)_\Gamma = 0$ and terminal conditions $(M^*v)_{t=t_k} = 0$.

The variations of the functionals (3.73) and (3.74) can now be written:

$$\delta J_i = \int_0^{t_k}\int_\Omega [(\partial f_i/\partial u)\delta u + (\partial f_i/\partial h)\delta h]\mathrm{d}\Omega\mathrm{d}t,$$
$$\delta J_j = \int_0^{t_k}\int_\Omega (f_j)^{p-1}[(\partial f_j/\partial u)\delta u + (\partial f_j/\partial h)\delta h]\mathrm{d}\Omega\mathrm{d}t. \tag{3.78}$$

An analogous expression can be obtained for the functional (3.71).

Let us state optimality conditions for the extended Lagrangian functional by setting equal to 0 the factors multiplying the (arbitrary) variations $\delta\mu_i$ and δu, that is, the variations of the auxiliary variables μ_i and the state function u. We obtain a system of algebraic equations that may be used to determine the quantities μ_i,

$$\lambda_i\mu_i = 0, \qquad i = 1, 2, \ldots, r, \tag{3.79}$$

and an equation for v, which together with the assigned boundary and terminal conditions, constitutes a boundary-value problem for the adjoint variable:

$$A^*v_{tt} + C^*v + \partial f/\partial u + \sum_{i=1}^{r_1}\lambda_i(\partial f_i/\partial u) + \sum_{j=1}^{r_1+r_2}\lambda_j f_j^{p-1}(\partial f_j/\partial u) = 0, \tag{3.80}$$

$$(N^*v)_\Gamma = 0, \qquad M^*(v)_{t=t_k} = 0. \tag{3.81}$$

The symbols λ_i ($i = 1, 2, \ldots, r$) in eqs. (3.79) and (3.80) denote Lagrangian multipliers that are introduced to account for the modified constraints (3.10).

After eliminating from the expression for the extended variation of the quality functional the terms that depend on the variations $\delta\mu_i$ and δu, we arrive at a basic formula of sensitivity analysis in relation to the variation δJ of the quality functional J, which depends on the variation δh of the design variable:

$$\delta J = \int_\Omega \delta\Bigg\{\int_0^{t_k}\Bigg[\partial f/\partial h + \sum_{i=1}^{r_1+r_2}\lambda_i(\partial f_i/\partial h) + \sum_{j=r_1+1}^{r_1+r_2}\lambda_j p(f_j)^{p-1}(\partial f_j/\partial h)$$
$$+ (\partial A^*/\partial h)u_{tt}v + (\partial C^*/\partial h)uv\Bigg]\mathrm{d}t\Bigg\}\mathrm{d}\Omega. \tag{3.82}$$

The equation $\delta J = 0$ provides the necessary optimality condition, since δh is arbitrary:

$$\int_0^{t_k} \left[\partial f/\partial h + \sum_{i=1}^{r_1} \lambda_i(\partial f_i/\partial h) + \sum_{j=1}^{r_1+r_2} \mu_j p(f_j)^{p-1}(\partial f_j/\partial h) \right.$$
$$\left. + (\partial A^*/\partial h)u_{tt}v + (\partial C^*/\partial h)uv \right] \mathrm{d}t = 0. \tag{3.83}$$

Equations (3.82) and (3.83) differ from corresponding formulas in static structural optimization since we are also integrating with respect to time.

Earlier we considered the case when the vector function $u(x, t)$ describing the response of the designed structure was defined as a solution to an initial-boundary-value problem (3.69) and (3.70). This way of modeling stress-deformed thin-walled structures and systems of beams (such as plates, shells, trusses, and frames) is quite typical. In the more general cases of three-dimensional elastic bodies or structures when the thin-wall assumptions cannot be made, the state function may still satisfy the following system of equations describing the dynamics of an elastic medium with the boundary and initial conditions given below:

$$\sigma_{ij,j} + q_i = \rho u_{i,tt}, \qquad \sigma_{ij} = \tfrac{1}{2} Cijkl(u_{k,l} + u_{l,k}),$$
$$(u_i)_{\Gamma_u} = U_i(x, t), \qquad (\sigma_{ij} n_j)_{\Gamma_u} = T_i(x, t),$$
$$(u_i)_{t=0} = u_i^0(x), \qquad (\dot{u}_i)_{t=0} = v_i^0(x). \tag{3.84}$$

This system is defined in the domain: $\Omega \times [0, t_k]$. The symbols ρ and q denote, respectively, the material density and the body forces, while $\Gamma = \Gamma_u + \Gamma_\sigma$ denotes the boundary of the domain Ω occupied by the structure. The assigned functions $u_i^0(x)$ and $v_i^0(x)$ denote the initial distribution of displacements and velocities, respectively. The functions $U_i(x)$ $(x \in \Gamma_u)$ and $T_i(x)$ $(x \in \Gamma_\sigma)$, $q(x, t)$ $(x \in \Omega)$, defined on the time interval $[0, t_k]$ describe the process of deforming and loading for the structure. The indices i, j, k, and l in (3.84) assume values 1, 2, 3. The finite instant of time t_k is fixed.

In addition to the system of equations and the boundary and initial conditions (3.84), the relations of an optimal design project for a dynamic system must include some constraints of the form (3.58) in which the quantities J_v $(v = 1, 2, \ldots, r)$ are defined by the formula

$$J_v = \int_0^{t_k} \int_\Omega f_v(x, t, u_k, \sigma_{ij}, q_l) \mathrm{d}\Omega \mathrm{d}t, \tag{3.85}$$

and a formula for the optimized functional, which could be the volume of material or, equivalently, the weight of a structure. As the unknown control function (or design variable) we choose the shape of a part of the boundary Γ_v instead of the vector function h, which occurs in the design coefficients (as in (3.69)). The design problem now consists in minimizing the quality criterion

$$V = (1/t_k) \int_0^{t_k} \int_\Omega d\Omega \to \min_{\Gamma_v} \tag{3.86}$$

by choosing the best (in this sense) segment of the boundary Γ_v while obeying the relations (3.58), (3.84), and (3.85). Below we shall assume that all the inequality constraints (3.58) have been replaced by equality constraints of the form (3.10) by means of auxiliary (slack) variables μ_i^2.

To derive an analytic formula for sensitivity and necessary conditions of optimality, we construct the extended Lagrangian functional

$$J = \int_0^{t_k} \int_\Omega (f^L) d\Omega dt + \lambda_v(\mu_v^2 - c_v), \tag{3.87}$$

where

$$f^L = (1/t_k)[1 + \lambda_v f_v + \psi_i(\sigma_{ij,j} + q_i - \rho\dot{v}_i) + \chi_{ij}(\sigma_{ij} - C_{ijkl}u_{k,l}) + \zeta_i(\dot{u}_i - v_i)],$$

and λ_v are (constant) Lagrangian multipliers, while $\psi_i(x, t)$, $\chi_{ij}(x, t)$, and $\zeta_i(x, t)$ are adjoint variables. The summation convention is used in (3.87) with respect to all repeated indices.

The variation of J may be accomplished using the technique explained in the preceding section, which is designed for problems with unknown boundaries. Omitting any further exposition we shall present only the final result:

The boundary-value problem for the adjoint variable is

$$\left.\begin{aligned} &\psi_{i,j} = \chi_{ij} + \lambda_v(\partial f_v/\partial\sigma_{ij}), && \rho\dot{\psi}_i - \zeta_i = 0, \\ &(C_{ijkl}\chi_{kl})_{,j} + \lambda_v(\partial f_v/\partial u_i) - \zeta_i = 0, && \\ &(\psi_i)_{\Gamma_u} = 0, && (C_{ijkl}\chi_{kl}n_j)_{\Gamma_\sigma} = 0, \\ &(\psi_i)_{t=t_k} = 0, && (\zeta_i)_{t=t_k} = 0. \end{aligned}\right\} \tag{3.88}$$

We remarked earlier when deriving necessary optimality conditions in problems of optimal control of the coefficients that the boundary–value problem (3.88) for adjoint variables differs from the boundary–value problem (3.84) for the state function in one important respect. The values of the spatial distribution are assigned in Ω at the final instant $t = t_k$ instead of at the initial value of time. Also, only first–order derivatives occur in the boundary-value system (3.88) both with respect to the spatial and the time variables. However, this is not an essential feature, because if we eliminate the variable ζ from these equations and from the initial conditions in (3.88), we arrive at a system which is identical to (3.84).

The adjoint variables determined by solving problem (3.88) permit us to produce a formula relating δJ to the variation of the boundary:

$$\delta J = \delta J_u + \delta J_\sigma = \int_{\Gamma_v \cap \Gamma_u} \left\{ (1/t_k) \int_0^{t_k} [1 + \lambda_v f_v + C_{ijkl} \chi_{kl} (U_i - u_i)_{,j}] \mathrm{d}t \right\} \delta l \mathrm{d}\Gamma$$

$$+ \int_{\Gamma_v \cap \Gamma_\sigma} \left\{ (1/t_k) \int_0^{t_k} [1 + \lambda_v f_v - 2H \psi_i T_i - (\psi_i \sigma_{ij})_{,j} + (\psi_i T_i)_{,k} n_k] \mathrm{d}t \right\} \delta l \mathrm{d}\Gamma. \tag{3.89}$$

We use (3.89) and necessary extremum condition for J: $\delta J = 0$ to derive the necessary optimality condition:

$$\int_0^{t_k} [1 + \lambda_v f_v + C_{ijkl} \chi_{kl} (U_i - u_i)_{,j}] \mathrm{d}t = 0, \qquad x \in \Gamma_v \cap \Gamma_u,$$

$$\int_0^{t_k} [1 + \lambda_v f_v - 2H \psi_i T_i - (\psi_i \sigma_{ij})_{,j} + (\psi_i T_i)_{,k} n_k] \mathrm{d}t = 0, \qquad x \in \Gamma_v \cap \Gamma_\sigma. \tag{3.90}$$

3.7 Conditions for an Extremum in Singular Optimization Problems

In many problems of optimal design it is common to encounter various kinds of singularities. For example, coefficients of the highest-order derivatives in the basic differential equation may vanish.

In the case of interior singular points (or lines in two-dimensional structural elements), the number of unknowns that must be determined to solve the problem increases, because the location (or coordinates) of these points is also unknown. Indeed, the coordinates of such singular points cannot be determined from the equations of equilibrium or from the Euler–Lagrange equations for the “control” variables.

In some articles the appearance of the “free” parameters (coordinates of the singular points) is mistakenly labeled “multiple-extremum” optimization, in other works these parameters are eliminated by introducing some additional hypotheses. Such is the case in [3.17] which investigated maximization of rigidity for a beam in pure bending. Here this “nonuniqueness” is taken care of by demanding that the “new” derivative of the displacement function must be continuous. In [3.11] a similar condition on continuity is derived without making additional constraints or hypotheses, instead as a result of maximizing the quality functional by varying a free parameter.

This “nonuniqueness” is in reality a direct consequence of the failure to make full use of the necessary conditions for stationary behavior. Thus, if we wish to obtain a “closed” system of necessary conditions for stationary behavior, we must include the Erdmann–Weierstrass criterion for corner points on broken extremals. As will be shown later (Section 4.5), it is possible to determine the location of singular points from these conditions and derive as a consequence the continuity of the derivatives of the displacement function in beam or column design problems. Here, we shall derive necessary conditions for stationary behavior in problems with nonadditive functionals and non-smooth extremals.

We shall consider the following optimization problem. We wish to determine the state vector function $u(x) = \{u^1_{(x)}, \ldots, u^n(x)\}$ and the design vector function $h(x) = \{h^1(x), \ldots, h^m(x)\}$ on the interval $[x_0, x_1]$ from the condition of stationary behavior of the functional

$$J = F(J_1, J_2, \ldots, J_s), \tag{3.91}$$

while satisfying the constraint

$$G(J_1, J_2, \ldots, J_s) = G_0. \tag{3.92}$$

Here

$$J_k = \int_{x_0}^{x_1} f_k(x, h, u, u_x, u_{xx})\mathrm{d}x, \qquad k = 1, 2, \ldots, s. \tag{3.93}$$

As before u_x denotes $\mathrm{d}u/\mathrm{d}x$, etc.

To be specific we shall assume that the vector function $u(x)$ satisfies at the endpoints of the interval $[x_0, x_1]$ the $(4n)$ conditions $u^i(x_0) = \alpha^i_0$, $u^i_x(x_0) = \beta^i_0$, $u^i(x_1) = \alpha^i_1$, $u^i_x(x_1) = \beta^i_1$, $(i = 1, 2, \ldots, n)$, where α^i_0, β^i_0, α^i_1, β^i_1, m, n, s, and G_0 are given constants, while F, G, and f_k are known functions.

In problem (3.91) to (3.93) we seek an extremum in the set of continuous state functions having piecewise-continuous first derivatives. We assume that the jump in the first derivative of the function $u(x)$ occurs at some point $x = x_*$ $(x_0 \le x_* \le x_1)$ whose location is not initially known but needs to be determined from a condition for stationary behavior of the functional J.

Before we derive the necessary conditions for stationary behavior using only techniques of classical calculus of variations, we observe that these conditions can also be obtained from the fundamental equations of optimal control theory. (cf. [3.5], [3.11], and previous comments in this section).

In order to derive the necessary conditions for stationary behavior, we introduce the functional $F^\lambda = F + \lambda G$ (where λ is a Lagrangian multiplier), and represent each integral J_k $(k = 1, 2, \ldots, s)$ as a sum of two integrals $J_k = J^1_k + J^2_k$, which are defined respectively on the intervals $[x_0, x_*]$ and $[x_*, x_1]$. We write the expression for the first variation of each of these integrals. We assume here that only one singular point exists on the interval $[x_0, x_1]$. The expression for δJ_k is obtained by adding δJ^1_k and δJ^2_k, while assuming u and h are continuous at the endpoint $x = x_*$.

Further, the difference $F(J_1 + \delta J_1, \ldots, J_s + \delta J_s) - F(J_1, \ldots, J_s)$ is written as series in powers of δJ_k and only the first-order terms are retained.

After some algebra we derive

$$\delta J = \sum_{i=1}^{n} \left\{ \int_{x_0}^{x_*} \sum_{k=1}^{s} (\partial F^\lambda/\partial J_k)[\partial f^k/\partial u^i - \mathrm{d}/\mathrm{d}x(\partial f^k/\partial u_x^i) \right.$$

$$+ \mathrm{d}^2/\mathrm{d}x^2(\partial f^k/\partial u_{xx}^i)]\delta u^i \mathrm{d}x + \int_{x_*}^{x_1} \sum_{k=1}^{s} (\partial f^\lambda/\partial J_k)[\partial f^k/\partial u^i$$

$$- \mathrm{d}/\mathrm{d}x(\partial f^k/\partial u_x^i) + \mathrm{d}^2/\mathrm{d}x^2(\partial f^k/\partial u_{xx}^i)]\delta u^i \mathrm{d}x$$

$$- \delta u^i \sum_{k=1}^{s} (\partial F^k/\partial J_k)[\partial f^k/\partial u_x^i - \mathrm{d}/\mathrm{d}x(\partial f^k/\partial u_{xx}^i)]_-^+$$

$$\left. - \left[\delta u_x^i \sum_{k=1}^{s} (\partial F^\lambda/\partial J_k)(\partial f^k/\partial u_{xx}^i) \right]_-^+ \right\}$$

$$- \delta x_* \sum_{k=1}^{s} (\partial F^\lambda/\partial J_k) \left[f^k - \sum_{i=1}^{n} (u_x^i \partial f^k/\partial u_x^i) + u_{xx}^i(\partial f^k/\partial u_{xx}^i) \right.$$

$$\left. - u_x^i \mathrm{d}/\mathrm{d}x(\partial f^k/\partial u_{xx}^i) \right]_-^+ + \sum_{i=1}^{m} \left\{ \int_{x_0}^{x_*} \sum_{k=1}^{s} [(\partial F^\lambda/\partial J_k)\partial f^k/\partial h^i]\delta h^i \mathrm{d}x \right.$$

$$\left. + \int_{x_*}^{x_1} \sum_{k=1}^{s} [(\partial F^\lambda/\partial J_k)\partial f^k/\partial h^i]\delta h^i \mathrm{d}x \right\}.$$

The notation «+» and «−» denotes the values computed at $x = x_* + 0$, and at $x = x_* - 0$, respectively, while the square brackets $[\ldots]_-^+$ denote the difference between the limiting values, that is, $[\ldots]_-^+ = (\ldots)^+ - (\ldots)^-$. Equating δJ to 0 and recalling that on $[x_0, x_1]$, the variations $\delta u^i(x)$, $\delta h^i(x)$ and the variations δx_*, $(\delta u_x^i)^+$, $(\delta u_x^i)^-$, $\delta u^i = (\delta u^i)^+ = (\delta u^i)^-$ are arbitrary, we derive the Euler–Lagrange equations

$$\sum_{k=1}^{s} (\partial F^\lambda/\partial J_k)[\partial f^k/\partial u^i - \mathrm{d}/\mathrm{d}x(\partial f^k/\partial u_x^i) + \mathrm{d}^2/\mathrm{d}x^2(\partial f^k/\partial u_{xx}^i)] = 0, \; i = 1, 2, \ldots, n; \tag{3.94}$$

$$\sum_{k=1}^{s} (\partial F^\lambda/\partial J_k)\partial f^k/\partial h^j = 0, \qquad j = 1, 2, \ldots, m, \tag{3.95}$$

and also the Erdmann–Weierstrass conditions for discontinuities in the derivatives at the point $x = x_*$:

$$\left[\sum_{k=1}^{s} (\partial F^\lambda/\partial J_k)(\partial f^k/\partial u_{xx}^i) \right]^+ = \left[\sum_{k=1}^{s} (\partial F^\lambda/\partial J_k)(\partial f^k/\partial u_{xx}^i) \right]^- = 0, \; i = 1, 2, \ldots, n, \tag{3.96}$$

$$\sum_{k=1}^{s} (\partial F^\lambda/\partial J_k)[\partial f^k/\partial u_x^i - \mathrm{d}/\mathrm{d}x(\partial f^k/\partial u_{xx}^i)]_-^+ = 0, \tag{3.97}$$

$$\sum_{k=1}^{s} (\partial F^\lambda/\partial J_k)\left[f^k - \sum_{i=1}^{n} \{(u_x^i \partial f^k/\partial u_x^i) + u_{xx}^i(\partial f^k/\partial u_{xx}^i) - u_x^i \mathrm{d}/\mathrm{d}x(\partial f^k/\partial u_{xx}^i)\} \right]_-^+ = 0. \tag{3.98}$$

The derivatives $\partial F^{\lambda}/\partial J_k = F^{\lambda}_{J_k}(J_1, J_2, \ldots, J_s)$ occuring in the Euler equations (3.94) and (3.95) and in conditions (3.96) to (3.98) are computed for values of the integrals $J_1, J_2, \ldots, J_s$ corresponding to the extrema of the optimization problem. Conditions (3.96) to (3.98) can be routinely generalized when the discontinuities in the derivatives of the vector $u(x)$ occur at more than one point of the independent variable interval $[x_0, x_1]$.

Euler's equations (3.94) and (3.95) defined on the intervals $[x_0, x_*]$, and $[x_*, x_1]$ form a system of $(2(n + m))$ integrodifferential equations. Equations (3.94) contain derivatives of functions u^i of orders up to 4, while eq. (3.95) may be regarded as boundedness conditions for the design variables h^i. To determine the $(8n)$ constants of integration for this system and also the unknown quantities x_* and λ, we need $4n$ boundary conditions assigned to the endpoints of the interval $[x_0, x_1]$, as well as $(3n + 1)$ Erdmann–Weierstrass conditions (3.96) to (3.98) and n conditions of continuity for the vector function $u(x)$ at the point $x = x_*$, and the constraint (3.92), that is, a total of $(8n + 2)$ conditions.

In closing this chapter, we note that specific types of singularities that occur in eigenvalue optimization problems have been discussed in [3.1] to [3.5], [3.7], [3.9], [3.12], [3.13], [3.15], and [3.16]. The existence of an optimal solution has been studied in [3.10].

References for Chapter 3

3.1. Armand, J.L., Application of the optimal control theory to structural optimization; an analytical and numerical approach. In: *Proc. IUTAM Symp. Optimization in Structural Design*, Warsaw, 1973, Springer-Verlag, Berlin 1975, pp. 15–39.

3.2. Armand, J.L., Lur'e, K.A., and Cherkaev, A.V., On solution of problems in optimization of eigenvalues, which occur in designs of elastic structures, *Izv. Akad. Nauk U.S.S.R., M.T.T.*, 1978, No. 5, pp. 159–162.

3.3. Armand, J.L., Lur'e, K.A., and Cherkaev, A.V., On the solution of optimization problems with singularities. In: *New Directions in Optimal Structural Design*, E. Atrek et al. (eds.), Chichester, Wiley, 1984, pp. 211–229.

3.4. Banichuk, N.V., and Karihaloo, B.L., On the solution of optimization problems with non-smooth extremals. *Intern. J. Solids Structures*, 1977. *13*, No. 8, pp. 725–733.

3.5. Banichuk, N.V., and Karihaloo, B.L., On the solution of optimization problems with non-smooth extremals. In: *Proc. Sixth Canadian Congress Appl. Mech.*, Vancouver, Canada, 1977, pp. 959–960.

3.6. Banichuk, N.V., Bel'skii, V.G., and Kobelev, V.V., Optimization in problems of the theory of elasticity, with unknown boundaries, *Izv. Akad. Nauk U.S.S.R., M.T.T.*, 1984, No. 3, pp. 46–52.

3.6a. Banichuk, N.V., Bel'skii, V.G., and Kobelev, V.V. *Problems with Unknown Constraints and The Theory of Optimal Structural Design*, Preprint No. 246, Institute of Problems of Mechanics, USSR Acedemy of Sciences, Moscow, 1985, 63 pp.

3.7. Bratus, A.S., and Seyranian, A.P., Bimodal solutions in problems of eigenvalue optimization, *Prikl. Mat. Mekh.*, 1983, *47*, No. 4, pp. 546–554.

3.8. Euler, L., A method for finding curves possessing a maximal or minimal property, or solving an isoperimetric problem in the broadest possible sense, Moscow–Leningrad, Gostekhizdat, 1934, 569 pp.

3.9. Haug, E.J., Choi, K.K., and Komkov, V., *Design Sensitivity Analysis of Structural Systems*, Academic, New York, 1986, 381 pp.

3.10. Litvinov, V.G., The problem of optimal control of the natural frequency for a plate with variable thickness, *Zh. Vichisl. Matem. i Matem. Fiziki*, 1979, *19*, No. 4, pp. 866–877.

3.11. Masur, E.F., Written discussion. In: *Proc. IUTAM Symp. Optimization in Structural Design*, Warsaw, 1973, A. Sawczuk and Z. Mróz (eds.), Springer-Verlag, Berlin, 1975, pp. 99–103.

3.12. Masur, E.F., Singular problems of optimal design. In: *Optimization of Distributed Parameter Structures*, Iowa City, 1980, E.J. Haug and J. Cea (eds.), Sijthoff & Noordhoff, Alphen an den Rijn, 1981, pp. 200–218.

3.13. Masur, E.F., and Mróz, Z., Singular solutions in structural optimization problems. In: *Proc. IUTAM Symp. Variational Methods in the Mechanics of Solids*, S. Nemat-Nasser (ed.), New York, Pergammon, 1980, pp. 344–348.

3.14. Miele, A., Drag minimization as the extremization of products of powers of integrals. In: *Problems of Hydrodynamics and Mechanics of Solids*, Moscow, Nauka, pp. 311–316.

3.15. Olhoff, N., on singularities, local optima and formation of stiffeners in optimal designs of plates. In: *Proc. IUTAM Symp. Optimization in Structural Design*, Warsaw, 1973, A. Sawczuk and Z. Mróz (eds.), Springer-Verlag, Berlin, 1975, pp. 82–99.

3.16. Olhoff, N., and Niordson, F.I., Some problems concerning singularities in optimal beams and columns, *Zeitschrift angew. Math. Mechanik*, 1979, *59*, pp. T16–T26.

3.17. Prager, W., and Taylor, J.E., Problems of optimal structural design, *J. Appl. Mech., Trans. ASME*, 1968, *35*, No. 1, pp. 102–106.

4 Analytic and Numerical Techniques for Distributed Parameters Structural Optimization Problems

4.1 On Some Applications of Analytic Techniques

Optimal design is closely related to solutions of some difficult mathematical problems. For example, in many instances the finding of optimal designs may be reduced to solving of variational problems with unknown boundaries. Also, some essential difficulties may be caused by the nonlinear nature of many optimal design problems.

The complexity of problems in structural design optimization is the main reason for the present heavy emphasis on numerical approaches and the use of computational techniques. Adoption of techniques based on numerical methods of optimizing (such as linear or nonlinear programming) and computer modeling permits a derivation of optimal (or near-optimal) solutions for very complex structures.

However, it is not true that numerical approaches to optimization are always effective. For a large class of problem the numerical techniques fail to give a clue to the general characteristics, or to even indicate some typical properties of optimal structures. These approaches turn out to be completely ineffective in displaying the laws that are obeyed in describing the totality of optimal structures and in studying the dependence of optimal solutions on specific parameters. As an example, we can offer multi-parameter optimal design problems in which the study of properties of optimal structures and the sensitivity analysis with respect to variations of parameters involve very large numbers of computations. One should also be familiar with the essential difficulties in tabulating and data storage of intermediate solutions, when one is optimizing for a large number of parameters. Thus, one should not neglect the advantages provided by analytical methods. In order to invent some more effective techniques for structural optimization it appears obvious that one should combine numerical computation and analytical techniques.

One of the most powerful techniques is the perturbation, or the small parameter technique. It is most effectively applied to a wide variety of research projects (generally outside the optimal design theory). It makes it possible to obtain an approximate analytical representation of solutions to some very complex linear, or nonlinear boundary-value problems for systems represented by either ordinary or partial differential equations.

Perturbation techniques are widely used to derive asymptotic solutions, to analyze singularities, or derive analytic solutions to some "test problems." It should be noted that these techniques may serve as foundations for constructing numerical algorithms. In fact, almost all techniques of successive approximations are based on perturbation theory (cf. [4.16]).

The application of perturbation techniques to optimization theory and, in particular, to optimization of structural design, is of relatively recent origin, and at the present time these techniques are still not generally acknowledged and are not widely utilized. We shall list a few publications related to applications of perturbation techniques to optimal design theory.

Apparently, one of the first articles on the subject of applying perturbation techniques to optimization of a design was the paper of Wu [4.33]. The application of perturbation techniques to optimization problems modelled by systems of partial differential equations with unknown boundaries are offered in [4.2] and [4.3] to [4.5]. In [4.10] and [4.29], perturbation techniques are used to solve problems of hydro- and aeroelasticity. Other applications of these techniques to structural design can be found in the monograph [5]. These references and other research articles indicate that the first two terms of the series (i.e., the zeroth- and first-order approximation term) found by perturbation techniques generally offer a very good approximation to the exact solution.

Inclusion of corrections for a second-order term can be, in general, neglected because of its very small magnitude. Therefore, in solutions of practical problems it is generally possible to limit such expansions to a single term, or to a two-terms approximation. As a rule, this ensures a sufficient degree of accuracy.

Another important feature, which is revealed in applying the perturbation techniques to optimal design problems, consists in the following. Even though decomposition into power series is carried out under the assumption that a parameter is "small," the approximation so obtained remains to be good (i.e., approximates closely the exact solution) even when the parameter used in the power series turns out to have values that are not "small."

In the articles referenced above, the perturbation technique was applied to optimization of structural designs. We comment that application of this technique to very complex structures depends on an individual. Indeed, quite frequently a "real" structure may be subdivided, in a fairly arbitrary manner, into smaller parts, which interact weakly with each other. Then one could use perturbation techniques to reduce the original problem to a sequence of much simpler problems for substructures, and to synthetize the structure from such substructures. Determining the "weak" interactions between the substructures of a complex structure and breaking up the structure into such substructures constitutes a very important problem of decomposition. At the present time theoretical approaches to designing of complex structures based on the ideas of decomposition and utilizing a perturbation technique are in their initial stages of development. However we have no doubt about their effectiveness.

Also, the perturbation techniques are not restricted to problems where the governing equations contain a clearly displayed "small" parameter. There are many examples when it is possible to introduce artificially a small parameter into a problem. One sufficiently general technique for introduction of a small parameter is discussed in [4.16]. Effective schcmes featuring small parameters were specifically designed for computer processing of numerical solutions.

To explain some early steps in forming the general features of an optimal structure, the authors of [4.12] introduced a technique of artificially inserting a small parameter into the definition of the design variable.

Numerical techniques have found very general applications for constructing regular types of solutions. However, analytical techniques are particularly effective in studies of singular points, or singular regions of solutions.

When analytical methods are used to produce optimal designs, one could find, for example, asymptotic behavior of design variables (such as the thickness of thin-walled structures, or volumetric density of reinforcing materials) in some neighborhood of the region occupied by the structural reinforcement, or in the vicinity of a concentrated load (see [3.15] and [3.16]).

Another type of singularity that can be investigated analytically is related to the formation of internal hinges (see [3.5] and [3.11]).

Recently, some developments took place in analytical construction of quasi-optimal solutions. In particular, it is important to identify the class of optimal design projects, for which it is possible to construct effectively some admissible solutions, and to obtain two-sided estimates for the optimized functionals.

We also should mention the technique introduced in [4.28] for constructing quasi-optimal solutions for problems with homogeneous functionals.

4.2 Perturbation Techniques

We shall consider some fundamental aspects of the perturbation techniques as applied to optimal designs of structures. Let $u(x)$ be the state vector function and $h(x)$ the vector function of design variables. The functions $u(x)$ and $h(x)$ are defined in some region Ω. The integral functionals: $J_1(u, h, q)$, $J_2(u, h, q)$, $\ldots, J_r(u, h, q)$ arc functions of the vector functions u, h, and q. We choose either one of them, or a function of these functionals, to be the designated quality functional:

$$\left.\begin{aligned} J &= \mathbf{F}(J_1, J_2, \ldots, J_r) \\ J_i &= \int_\Omega f_i(x, u, u_x, h, q)\mathrm{d}\Omega. \end{aligned}\right\} \tag{4.1}$$

The demands made regarding the properties and response of the structure can be reduced to constraints assigned to the design and phase variables:

$$\mathbf{F}_j(J_1, J_2, \ldots, J_r) = 0, \qquad j = 1, 2, \ldots, k, \tag{4.2}$$

where $\mathbf{F}$, $\mathbf{F}_j$, and f_j are known functions. The state functions and the control variables are related to each other through the equations and boundary conditions of a boundary value problem:

$$L(h, \varepsilon)u = q, \qquad (N(h, \varepsilon)u)_\Gamma = 0. \tag{4.3}$$

$L(h, \varepsilon)$ and $N(h, \varepsilon)$ denote differential operators, containing derivatives with respect to the spatial coordinate variables, while q is a vector function representing external forces acting on the structure. To be specific, we shall assume that the structure is made from linear-elastic materials, and is subjected only to small deformations, and therefore that the operators L and N are linear.

The coefficients of the operators L and N depend on the control function h and on the small parameter ε. The optimization problem consists in finding a function $h(x)$ that assigns a minimum (maximum) to the functional (4.1) and satisfies constraints (4.2) and (4.3).

To formulate the optimization problem (4.1) to (4.3) as defined in the preceding chapter, we need to introduce the optimality conditions and the boundary value problem equations for the adjoint variable. Using the same symbolism (see (3.7), (3.8), and (3.22)), we display the necessary conditions for optimality:

$$\left.\begin{aligned} &M^*(u, h, q, \varepsilon)v + \varphi(u, h, q, \lambda) = 0, \\ &\varphi \equiv \sum_{l=1}^{r} \frac{\partial \mathbf{F}^\lambda}{\partial J_l} \frac{\partial f_l}{\partial h}, \qquad \mathbf{F}^\lambda = \mathbf{F} + \sum_{i=1}^{k} \lambda_i \mathbf{F}_i, \end{aligned}\right\} \tag{4.4}$$

while, if we take care of the differential relations (4.3), we conclude that the vector function of adjoint variables $\mathbf{v}$, must solve the boundary-valued problem

$$\begin{gathered} L^*(h, \varepsilon)v + \chi(u, h, q, \lambda) = 0, \qquad (N^*(h, \varepsilon)v)_\Gamma = 0, \\ \chi \equiv \sum_{l=1}^{r} \frac{\partial \mathbf{F}^\lambda}{\partial J_l}\left[\frac{\partial f_l}{\partial u} - \sum_{j=1}^{e} \frac{\partial}{\partial x_j} \frac{\partial f_l}{\partial u_{x_j}}\right], \end{gathered} \tag{4.5}$$

where $L^*(h, \varepsilon)$ is the adjoint of the operator $L(h, \varepsilon)$, and e is the dimension of the space spanned by coordinates x_j. The dimension of the vector v is the same as that of u. Equations (4.2) to (4.5) form a "closed" system of equations, so that it is possible to determine h, u, v, and λ.

We shall assume that the solution to the optimization problem exists for any $\varepsilon < 1$, that the differential equations and the boundary conditions depend analytically on the parameter ε, and that the solution of the optimization problem for $\varepsilon = 0$ is either known, or can be derived very simply by analytic or numerical techniques.

Then for small values of ε it is possible to look for a solution by decomposing it into power series in ε. Since the terms of the differential equations and the boundary conditions depend analytically on the parameter ε, and derivation of the solution is accomplished by analytical procedures, it is reasonable to expect the solution to be an analytical function of ε, and to assert (heutistically) that the perturbation technique is justified.

We expand the quantities h, u, v, λ, J in power series of the small parameter:

$$\left.\begin{aligned} h &= h^0(x) + \varepsilon h^1(x) + \varepsilon^2 h^2(x) + \cdots \\ u &= u^0(x) + \varepsilon u^1(x) + \varepsilon^2 u^2(x) + \cdots \\ v &= v^0(x) + \varepsilon v^1(x) + \varepsilon^2 u^2(x) + \cdots \\ \lambda &= \lambda^0 + \varepsilon\lambda^1 + \varepsilon^2\lambda^2 + \cdots \\ J &= J^0 + \varepsilon J^1 + \varepsilon^2 J^2 + \cdots \end{aligned}\right\} \tag{4.6}$$

where the quantities h^i, u^i, v^i, λ^i and J^i are independent of ε and where h^0, u^0, v^0, λ^0, and J^0 solve the optimization problem for $\varepsilon = 0$.

To obtain relations that could be used to solve for h^i, u^i, v^i, λ^i and J^i we substitute the power series (4.6) in eqs. (4.2) to (4.5).

The next step consists in rewriting (4.2) to (4.5) as infinite series in powers of the small parameter ε, and equating coefficients of equal powers of ε. Since these equations must be satisfied for all (small) arbitrary values of ε, and different power terms are linearly independent of each other, it follows that the coefficient of every power of ε must vanish independently. In this manner we derive relatively simple equations featuring the unknown quantities h^i, u^i, v^i, λ^i, and J^i. These equations should be solved first for the zeroth order, then the first, and subsequent orders of approximation. Thus, the optimization problem has been reduced to solving a system of much simpler problems.

By solving these problems we can determine the optimal solution within an arbitrary degree of accuracy. Thus, the zeroth-order approximation satisfies the following relations:

$$\begin{aligned} &L(h^0, 0)u^0 = q, \qquad (N(h^0, 0), u^0)_\Gamma = 0, \\ &L^*(h^0, 0)v^0 + \chi(u^0, h^0, q, \lambda^0) = 0, \qquad (N(h^0, 0)v^0)_\Gamma = 0, \\ &M^*(u^0, h^0, 0)v^0 + \varphi(u^0, h^0, q, \lambda^0) = 0, \\ &\mathbf{F}_i(J_1^0, J_2^0, \ldots, J_r^0) = 0, \qquad J_j^0 = J_j(u^0, h^0), \end{aligned} \tag{4.7}$$

with $i = 1, 2, \ldots, k$ and $j = 1, 2, \ldots, r$.

The first-order approximation quantities are computed by solving the boundary-value problem

$$\begin{gathered} L(h^0, 0)u^1 + L^h(u^0, h^0, 0)h^1 + L^\varepsilon(u^0, h^0) = 0, \\ (N(h^0, 0)u^1 + N^h(u^0, h^0, 0)h^1 + N^\varepsilon(u^0, h^0))_\Gamma = 0, \\ L^*(h^0, 0)v^1 + \chi^h(u^0, h^0, q, v^0, \lambda^0)h^1 + \chi^u(u^0, h^0, q, \lambda^0)u^1 \\ + \chi^\lambda(u^0, h^0, q)\lambda^1 + \chi^\varepsilon(h^0, v^0, q) = 0, \\ (N^*(h^0)v^1 + N^{*h}(v^0, h^0)h^1 + N^{*\varepsilon}(v^0, h^0))_\Gamma = 0, \end{gathered}$$

$$M^*(u^0, h^0, 0)v^1 + \varphi^h(u^0, h^0, q, v^0, \lambda^0)h^1 + \varphi^u(u^0, h^0, q, v^0, \lambda^0)u^1$$
$$+ \varphi^\lambda(u^0, h^0, q)\lambda^1 + \varphi^\varepsilon(u^0, h^0, q, v^0) = 0,$$
$$\sum_{j=1}^{r} \frac{\partial \mathbf{F}_j}{\partial J_j}(J_1^0, J_2^0, \dots, J_r^0)J_j^1 = 0. \tag{4.8}$$

The symbols

$$L^h(u^0, h^0, 0)h^1, \qquad N^h(u^0, h^0, 0),$$
$$\varphi^h(u^0, h^0, q, v^0, \lambda^0)h, \dots, \varphi^\lambda(u^0, h^0, q)\lambda^1, \qquad \chi^\lambda(u^0, h^0, q)\lambda^1$$

denote the expressions obtained from expanding eqs. (4.3) to (4.5) and identifying the terms that are linear in h^1, u^1, and λ^1, respectively.

It is also possible to adopt a different scheme in applying the perturbation techniques to problems of structural optimization. Within the framework of that scheme the perturbation technique is accomplished by the following steps. First the unknown functions h, u, J, and J_i, for problems (4.1) to (4.3) are expanded in power series of the small parameter ε. Optimality conditions and the equations for the adjoint variables are not introduced at this stage.

The series for h, u, J, and J_i, are substituted in (4.1) to (4.3). Coefficients of the same powers of ε are collected into separate expressions, which are then set equal to zero. Then we formulate separate optimization problems for the zeroth-, first-, second-, and some higher-order approximations. The problem is reduced to optimization of the functionals $J^0, J^1, J^2, \dots$, for corresponding choices of the functions $h^0, h^1, h^2, \dots$. At this stage we introduce in each of these problems the Lagrangian multipliers and adjoint variables, then derive some necessary conditions for optimality and commence numerical computations. In principle, this scheme is completely equivalent to the preceeding one, and the only differences consist in methodology.

4.3 Application of the Perturbation Technique to Problems of Optimizing Coefficients of Differential Equations

In this section we shall offer as examples of application of perturbation techniques two typical problems concerning the optimization of coefficients of differential equations. In the first example the behavior of a structure is described by an inhomogeneous boundary-value problem for an ordinary differential equation. In the second example it is described by a homogeneous eigenvalue problem.

We shall consider the problem of optimal design of a beam lying on an elastic foundation of the Winkler type. The beam is freely supported at both ends $x = \pm l$ on the x-axis and is subjected to a concentrated load $q = P\delta(x)$, applied at the midpoint of its length. Let c denote the coefficient of proportionality between the reaction of the foundation and the deflection of the beam $w(x)$. We consider a beam with rectangular cross-section having a constant width b and a variable height $h = h(x)$, $-l \leq x \leq +l$. The length of the beam $2l$, and its volume V are regarded as known, which implies the existence of a constraint on the distribution of cross-sectional area $S(x) = bh(x)$.

The optimization problem consists in maximizing the integral describing the rigidity by designing the best possible profile for the beam, that is by choosing the function $h(x)$. In the case considered, that is for a single concentrated load applied to the beam, the maximization of the rigidity integral is equivalent to the maximization of the magnitude of the force P that produces a given deflection w_0 at the point of application of the load.

We recall the basic formulas for the problem (cf. [4.5]):

$$
\begin{gathered}
J \equiv P = \frac{1}{w_0}\int_{-l}^{l} (Dw_{xx}^2 + cw^2)dx \to \max_h, \\
(Dw_{xx})_{xx} + cw = q, \qquad D = EI = 1/12\, Ebh^3, \\
(Dw_{xx})_{x=\pm l} = (w)_{x=\pm l} = 0, \qquad w(0) = w_0, \\
\int_{-l}^{l} S dx = V, \qquad S = bh.
\end{gathered} \tag{4.9}
$$

Here E stands for Young's modulus, and I is the moment of inertia of the cross-sectional area of the beam. We observe that the problem of minimization of the weight (or volume) of the beam for a given magnitude of the rigidity integral is dual to the problem given by (4.9).

Knowing the solution of one of these problems allows us to obtain the other one by a straightforward computation (that is by a change of scale). For the problem (4.9) and for the dual problem a necessary condition for optimality is of the form given in [4.9]:

$$h^2(w_{xx})^2 = \lambda^2. \tag{4.10}$$

Making use of (4.10) we restate our problem if the value of c is "small":

$$
\begin{aligned}
h &= h^0(x) + ch^1(x) + \cdots, \qquad & w &= w^0(x) + cw^1(x) + \cdots, \\
\lambda &= \lambda^0 + c\lambda^1 + \cdots, & P &= P^0 + cP^1 + \cdots.
\end{aligned} \tag{4.11}
$$

Substituting this expansion in (4.11) for the quantities h, w, λ, and P whenever they occur in the basic equations of our problem, we compare terms having the same power of c. To find the quantities corresponding to the zeroth-order approximation, that is to the absence of interaction with the foundation, we examine this system of equations (with $0 < x < l, c = 0$), whose solutions are

$$
\begin{aligned}
&h^0 = (3V/4bl)\sqrt{1 - x/l}, \qquad P^0 = (81\delta A_3 V^3)/(64b^3 l^6), \qquad \delta = w_0, \\
&w^0 = \delta(1 - x/l)(3 - 2\sqrt{1 - x/l}), \qquad \lambda^0 = -9/8(\delta V/bl^3).
\end{aligned} \tag{4.12}
$$

Now using the zeroth-order approximation we write the boundary-value problem for the first-order approximation:

$$A_3((h^0)^2[h^0 w^1_{xx} + 3h^1 w^0_{xx}])_{xx} + w^0 = 0, \qquad 0 < x < l,$$

$$w^1(0) = u^1_x(0) = w^1(l) = 0,$$

$$(w^1_{xx}(h^0)^3 + 3(h^0)^2 h^1 w^0_{xx})_{x=l} = 0,$$

$$h^1 w^0_{xx} + h^0 w^1_{xx} = \lambda^1, \qquad 0 \le x \le l, \qquad \int_0^l h^1\,\mathrm{d}x = 0,$$

$$P^1 = \frac{2}{\delta}\left(\frac{\lambda^0\lambda^1 V A_3}{b} + \int_0^l (w^0)^2\,\mathrm{d}x\right). \tag{4.13}$$

We determine the unknown functions w^1 and h^1 from the first and fourth row in (4.13). The constants of integration and the constant λ^1 we determined from the boundary conditions and from the isoperimetric condition. As a result, we derive the following correction to the distribution of height and the magnitude of the force P (in the first approximation):

$$h^1 = \frac{8b^2 l^6}{27V^2 A_3}\left(1 - \frac{x}{l}\right)^{1/2}\left[\left(1 - \frac{x}{l}\right)^2\left(1 - \frac{16}{35}\left(1 - \frac{x}{l}\right)^{1/2}\right) - \frac{9}{35}\right],$$

$$P^1 = \frac{8}{7} l\delta, \qquad \lambda^1 = 0. \tag{4.14}$$

Note the following specific consequence of (4.14): If we increase the value of c, the thickness of the optimally designed beam increases in the middle and decreases near the ends (i.e., the ends become sharper).

This approximation of first order derived by perturbation technique already is very close to the exact solution of this problem that was derived numerically in [4.9], using a successive optimization algorithm. We remark that in the same article, a solution to this optimization problem was derived for values of c that are not "small."

As an example of an application of perturbation techniques that arise in structural designs for problems described by eigenvalue, boundary value equations, we shall consider the problem of minimizing the weight of a wing when a bound is assigned to velocity, and a specific value is realised for the elastic effectiveness of the ailerons (cf. [4.28]). We note that the decrease in elastic effectiveness of the ailerons is one of the important properties in aeroelasticity, affecting undesirable manoeuvrability characteristics of aircraft.

The purpose of this investigation is to determine the effects of this bound on the optimal distribution of the material along the span of the wing.

We assume that the wing is "very long" and to model mathematically its deformation we may apply the beam theory to the structure of the wing. The aileron is assumed to be a rigid body. The aerodynamic loads acting on the wing are computed according to the theory of a lifting strip, taking into account wing's deflections. As our unknown control function we choose the distribution of wing's profile thickness $h(x)$ along its span. We shall introduce dimensionless variables to describe the equations of equilibrium between the wing and the aileron in the environment of the flowing gas stream, with appropriate boundary conditions:

$$\left.\begin{aligned}(h\theta_x)_x + \mu\theta &= \kappa\mu\int_0^1 x\theta(x)\mathrm{d}x,\\ \theta(0) = (h\theta_x)_{|x=1} &= 0.\end{aligned}\right\} \tag{4.15}$$

Here θ denotes the angle of twist for the wing with respect to its elastic axis, μ is a dimensionless pressure head which in this case plays the part of an eigenvalue, and κ is an essential parameter depending on the indicator of the aileron's efficiency and on the aerodynamic coefficients. The formula for functional dependence of κ on these coefficients is given in [4.28], but will not be reproduced here because it is so unwieldy. In the same article it is shown that the parameter κ is "small."

The optimization problem consists in finding the thickness function $h(x)$ which minimizes the weight of the wing:

$$J = \int_0^1 h\mathrm{d}x \to \min_h, \tag{4.16}$$

and such that for a given value of κ the smallest positive eigenvalue μ is not smaller than an a priori assigned value μ_0, with $\mu_0 > 0$.

A necessary condition for optimality is

$$\theta_x\varphi_x = \lambda, \tag{4.17}$$

where the adjoint function φ satisfies the following equation and boundary conditions:

$$\left.\begin{aligned}(h\varphi_x)_x + \mu\varphi &= \kappa\mu x\int_0^1 \varphi\mathrm{d}x\\ \varphi(0) = (h\varphi_x)_{|x=1} &= 0.\end{aligned}\right\} \tag{4.18}$$

In solving this optimization problem we use the normalization condition

$$\int_0^1 \theta^2\mathrm{d}x = \int_0^1 \varphi^2\mathrm{d}x = \frac{1}{3}. \tag{4.19}$$

Regarding the parameter κ as "small," we apply a perturbation technique and seek the solution of the optimization problem by expanding it in power series of the small parameter κ, while setting $\mu = \mu_0$:

$$\begin{aligned}h &= h^0(x) + \kappa h^1(x) + \kappa^2 h^2(x) + \cdots\\ \theta &= \theta^0(x) + \kappa\theta^1(x) + \kappa^2\theta^2(x) + \cdots\\ \varphi &= \varphi^0(x) + \kappa\varphi^1(x) + \kappa^2\varphi^2(x) + \cdots\\ \lambda &= \lambda^0 + \kappa\lambda^1 + \kappa^2\lambda^2 + \cdots,\\ J &= J_0 + \kappa J^1 + \kappa^2 J^2 + \cdots.\end{aligned}$$

Substituting these expansions into (4.15) to (4.19) and setting coefficients of each power of κ equal to 0, we derive as the zeroth-order approximation to the necessary optimality conditions the following relation for the optimal functions, which coincides with the result of optimizing the divergence of the lifting surface (cf. [4.28]):

$$h^0 = \frac{1}{2}\mu_0(1 - x^2), \qquad \theta^0 = \varphi^0 = x. \tag{4.20}$$

Within the first-order approximation, we obtain relations for optimizing the functions

$$\left.\begin{aligned}
&(h^0\theta_x^1)_x + \mu_0\theta^1 = \mu_0\int_0^1 \theta^0 x\,\mathrm{d}x - (h^0\theta_x^0)_x,\\
&\theta^1(0) = (h^1\theta_x^0 + h^0\theta_x^1)_{|x=1} = 0,\\
&(h^0\varphi_x^1)_x + \mu_0\varphi^1 = \mu_0 x\int_0^1 \varphi^0\,\mathrm{d}x - (h^1\varphi_x^0)_x,\\
&\varphi^1(0) = (h^1\varphi_x^0 + h^0\varphi_x^1)_{|x=1} = 0,\\
&\int_0^1 \theta^0\theta^1\,\mathrm{d}x = 0, \qquad \int_0^1 \varphi^0\varphi^1\,\mathrm{d}x = 0, \qquad \theta_x^1\varphi_x^0 + \varphi_x^1\theta_x^0 = \lambda^1.
\end{aligned}\right\} \tag{4.21}$$

Now, using (4.20) and the last equation in (4.21), we conclude that $\theta_x^1 + \varphi_x^1 = \lambda^1$. Since $\theta^1(0) = \varphi^1(0) = 0$, we obtain $\theta^1 + \varphi^1 = \lambda^1 x$.

We multiply all terms of this equation by θ^0 and integrate from 0 to 1. Using (4.20) and the boundary conditions for φ^1 we conclude that $\lambda^1 = 0$. By using the boundary conditions for θ^1 and φ^1 at the point $x = 1$, we obtain $h^1(1) = 0$. Now we use the simultaneous equations for θ^1 and φ^1. Here we take advantage of the zeroth-order approximation and the relation $\theta^1 = -\varphi^1$. Integrating the resulting equations and recalling that $h^1(1) = 0$, we obtain

$$h^1 = -(\mu_0/24)(7 - 4x - 3x^2). \tag{4.22}$$

Similarly, making use of (4.21) and (4.22) we derive an expression for θ^1, which we write in the form

$$\frac{1}{2}((1 - x^2)\theta_x^1)_x + \theta^1 = \chi, \qquad x \in (0, 1),$$

$$\theta^1(0) = 0, \qquad \chi = \frac{1}{6} - \frac{x}{4}.$$

The solution for θ^1 is expanded in series of Legendre polynomials with odd indices:

$$\theta^1 = \sum_{i=1}^{\infty} \frac{(-1)^{i+1}(4i + 3)}{6i(2i + 3)(i + 1)!}\,\frac{1.3\cdots(2i - 1)}{2^{i+1}}P_{2i+1}(x).$$

In a similar manner, we could construct solutions for the boundary-value problem for the quantities h^2, θ^2, φ^2, λ^2, and J^2 (cf. [4.28]). To cut this discussion short, we display the final formula for the dimensionless weight computed within the second order accuracy:

$$J = J_0 + \kappa J^1 + \kappa^2 J^2 = \frac{1}{3}\mu_0(1 - 1/2\kappa - 0.0022\kappa^2).$$

Comparing this analytic result obtained by perturbation technique with a directly computed numerical result for our optimization problem we witness an excellent agreement between the analytical and numerical result, and observe that the effects of the second-order terms are negligible. Thus, the first-order approximation to the solution

$$h = \frac{l}{2}\left[\mu(1 - x)\left[1 + x - \left(\frac{l}{12}\right)\kappa(7 + 3x)\right]\right]$$

describes with sufficient accuracy the solution of the optimization problem for effectiveness of the aileron.

4.4 Application of the Perturbation Technique to Problems of Boundary Optimization

Perturbation techniques turn out to be effective techniques for solving two-dimensional problems in structural optimization with unknown boundaries.

Application of this technique follows an assumption that it is possible to expand the state functions (such as displacements, deformations and stresses), as well as the design functions that describe the shape of an object, in the power series of a small parameter. Then the problem of shape optimization can be reduced to the solution of much simpler problems, and such solutions will reveal the optimal shape of the object within a desired degree of accuracy.

In problems with unknown boundaries such expansions of unknown quantities are accomplished in a manner similar to that used in solving problems with unknown coefficients of differential equations. However, specific properties of such problems are revealed when one attempts to expand the boundary conditions and the optimality conditions while the shape of the boundary is not known.

There is a basic difficulty. The power series expansion determines not only the state variables, which are to be determined at unknown boundary points, but also the very functions that describe the boundary. Because of this we use double expansions. Let me explain this statement by offering an example.

Suppose that the condition $u(x, y) = 0$ is assigned on the unknown boundary curve $y = y(x)$. Here $u(x, y)$ is the unknown state function. Let us represent both y and u in power series of the small parameter ε:

$$\left.\begin{aligned} y &= y^0(x) + \varepsilon y^1(x) + \varepsilon^2 y^2(x) + \cdots \\ u &= u^0(x, y) + \varepsilon u^1(x, y) + \varepsilon^2 u^2(x, y) + \cdots \end{aligned}\right\} \tag{4.23}$$

Note that the argument y in the expression for $u(x, y)$ is expandable in the series given by first row of (4.23), and therefore, we can obtain an expansion of u in infinite series:

$$u = u^0(x, y^0) + \varepsilon(u^1(x, y^0) + u_y^0(x, y^0)y^1 + \varepsilon^2(u^2(x, y^0) + u_y^0(x, y^0)y^2 + u_{yy}^0(x, y^0)(y^1)^2 + u_y^1(x, y^0)y^1 + \cdots . \tag{4.24}$$

Thus, using the series (2.24) and $u(x, y) = 0$, we derive the following expressions for the zeroth, first-, and second-order approximations:

$$\left.\begin{aligned} &u^0(x, y^0) = 0, \qquad u^1(x, y^0) = -u_y^0(x, y^0)y^1, \\ &u^2(x, y^0) = -u_y^0(x, y^0)y^2 - u_{yy}^0(x, y^0)\frac{(y^1)^2}{2} - u_y^1(x, y^0)y^1. \end{aligned}\right\} \tag{4.25}$$

Note that various types of singular behaviour may arise in applications of this technique to problems with unknown boundaries. A study of such behaviour reduces to solving of some very difficult mathematical problems which generally have not been yet solved. We shall limit our discussion by illustrating availability of perturbation techniques to the well-posed problem of optimizing the shape of a transverse cross-section of a shaft subjected to pure torsion.

We examine the torsion of a homogeneous, isotropic shaft with doubly connected cross-section. Let Γ_0 and Γ denote, respectively, the inner and outer boundary contours that form the boundary of the cross-sectional region Ω lying in the plane perpendicular to the axis of the shaft. Let φ denote the stress function and K the torsional rigidity. In what follows we shall utilize a curvilinear coordinate system: $\{s, t\}$ related to the contour Γ_0. The coordinate s of an arbitrary point $P \in \Omega$ is measured along Γ_0 starting from some fixed point $O_1 \in \Gamma_0$ to the point O_2 of intersection of Γ_0 with the normal to Γ_0 passing through the point P. The coordinate t is equal to the length of the interval O_2P.

Let us denote by $R = R(s)$ and $h = h(s)$ the radius of curvature of the reference curve Γ_0, and the defining equation of the curve Γ, respectively, while l_0 will denote the length of the curve Γ_0. We assume that the shaft is thin-walled and the contour Γ_0 does not have any segments with "large" curvature, that is $\max_s h(s) \leq H \ll l_0$ and $\min_s R(s) \approx l_0$. Consequently, $\varepsilon = H/l_0$ may be regarded as a small parameter (that is $\varepsilon \ll 1$).

The inner contour Γ_0 is regarded as given, while the outer contour Γ must be determined by solving an optimization problem. Thus the problem consists in finding a function $h = h(s)$, that satisfies the isoperimetric condition of keeping a constant value for the cross-sectional area of Ω (implying that the area of the perpendicular cross-section is given) and such that the torsional rigidity assumes a maximum admissible value.

Optimality conditions for this problem have been derived in [4.2] and [4.3]. These conditions imply a constant value of the normal derivative of the stress function φ on the unknown boundary Γ.

The equations for the stress function, the boundary conditions on Γ_0 and on Γ, the optimality conditions assigned on the unknown boundary Γ, Bredt's formula, the isoperimetric condition of constant value for the cross-sectional area of perpendicular section of the shaft assume the following forms in dimensionless coordinates:

$$\left.\begin{aligned}&(T\varphi_t)_t + \varepsilon^2(T^{-1}\varphi_s)_s = -2\varepsilon T, \qquad T = 1 + \varepsilon t/R,\\&\varphi(s, 0) = C, \qquad \varphi(s, h) = 0, \qquad [\nabla\varphi(s, h)]^2 = \lambda^2,\\&\int_0^1 \varphi_t(s, 0)\mathrm{d}s = -2S_0, \qquad \int_0^1 [h + (\varepsilon h^2/(2R))]\mathrm{d}s = S,\\&K = 2\left(CS_0 + \varepsilon\int_0^1\int_0^h (T\varphi)\mathrm{d}t\,\mathrm{d}s\right),\end{aligned}\right\} \tag{4.26}$$

where S denotes the area of region Ω, S_0 is the area bounded by the contour Γ_0, C is a constant which is to be determined from the theory of torsion with the help of Bredt's formula; and λ is an unknown Lagrangian multiplier that helps to incorporate the isoperimetric condition.

To solve this problem we apply the perturbation technique, and attempt to find the functions $h(s)$ and $\varphi(s, t)$ and the unknown quantities C, λ, and K by expanding them in power series of the small parameter ε:

$$\left.\begin{aligned}&h = h^0(s) + \varepsilon h^1(s) + \varepsilon^2 h^2(s) + \cdots, C = C^0 + \varepsilon C^1 + \varepsilon^2 C^2 + \cdots\\&\varphi = \varphi^0(s, t) + \varepsilon\varphi^1(s, t) + \varepsilon^2\varphi^2(s, t) + \cdots\\&\lambda = \lambda^0 + \varepsilon\lambda^1 + \varepsilon^2\lambda^2 + \cdots, K = K^0 + \varepsilon K^1 + \varepsilon^2 K^2 + \cdots.\end{aligned}\right\}$$

Substituting these series in (4.26) and equating terms having the same power of ε, we arrive at a boundary-value problem whose solution allows us to determine all unknown quantities. To determine the quantities of the zeroth approximation we need to solve the following boundary-value problem:

$$\left.\begin{aligned}&\varphi^0_{tt} = 0, \qquad \varphi^0(s, 0) = C^0, \qquad \varphi^0(s, h^0) = 0,\\&\varphi^0_t(s, h^0) = -\lambda^0,\\&\int_0^1 \varphi^0_t(s, 0)\mathrm{d}s = -2S_0, \qquad \int_0^1 h^0(s)\mathrm{d}s = S.\end{aligned}\right\} \tag{4.27}$$

Using the properties of the functions in the zeroth approximation the relations corresponding to the boundary-value problem in the first-order approximation can be written as

$$\left.\begin{aligned}&\varphi^1_{tt} = -2 - (1/R)\varphi^0_t, \qquad \varphi^1(s, 0) = C^1,\\&\varphi^1(s, h^0) = \lambda^0 h^1, \qquad \varphi^1_t(s, h^0) = -\lambda^1,\\&\int_0^1 \varphi^1_t(s, 0)\mathrm{d}s = 0, \qquad \int_0^1 h^1(s)\mathrm{d}s = -\int_0^1 (h^0)^2/(2R)\mathrm{d}s.\end{aligned}\right\} \tag{4.28}$$

The solution of the problem in the zeroth approximation (i.e., problem (4.27)) may be derived in the following manner. The function φ^0 can be determined from the equations and boundary conditions in the first row of (4.27). Now using the expressions for φ^0 and optimality conditions, as given in the second row of (4.27) we can determine h^0. Furthermore, by using the isoperimetric conditions (in the third row of (4.27)) we determine λ^0 and C^0. Thus, we obtain expressions for the unknown quantities in the zeroth-order approximation: $h^0 = S$, $\lambda^0 = 2S_0$, $\varphi^0 = 2SS_0\ (1 - t/S)$, $K^0 = 4SS_0^2$, $C^0 = 2SS_0$. As can be seen by examining these formulas, in the case when the curvature of the contour is small, a constant thickness turns out to be the optimal distribution of shaft's thickness in the zeroth-order approximation.

Now we can determine the quantities of the first-order approximation. To do that we integrate relevant equations (4.28) containing φ^1 and use the boundary conditions to find the constants of integration. The function h^1 and the constants λ^1 and C^1 are found from the isoperimetric conditions and from the optimality conditions (4.27). As a result we derive the following expressions for the unknown quantities in the first-order approximation:

$$\left.\begin{aligned}
&\varphi^1 = t^2(S_0/R - 1) + 2SS_0\left[\int_0^1 (1/R)\mathrm{d}s - 1/R\right] + S^2 - 2S_0S^2\int_0^1 (1/R)\mathrm{d}s,\\
&h^1 = -S^2/(2R), \qquad K^1 = 4S_0S^2\left[1 - S_0\int_0^1\left(\frac{1}{R}\right)\mathrm{d}s\right],\\
&C^1 = S^2\left[1 - 2S_0\int_0^1 (1/R)\mathrm{d}s\right], \qquad \lambda^1 = 2S\left[1 - S_0\int_0^1 (1/R)\mathrm{d}s\right].
\end{aligned}\right\} \tag{4.29}$$

In the expression for h^1 in (4.29) we considered the effects of the curvature of the inner contour Γ_0 on the optimal shape of the outer boundary Γ.

The formula for the optimal distribution of the shaft's thickness,

$$h = h^0 + \varepsilon h^1 = S(1 - \varepsilon S/(2R)),$$

indicates that as we increase the curvature, the thickness h at corresponding points of the contour Γ_0 decreases.

Unknown quantities in the second-order approximation are found in an analogous manner (cf. [5] and [4.2]).

4.5 Solution of Singular Optimization Problems

A typical phenomenon in many optimal design projects involving structural elements is the occurrence of singular points (or singular lines in two-dimensional structures). The coefficients of the highest order derivatives in the basic differential equations vanish at such points. It is well known that the behavior of the optimal solution (i.e., the shape of a structural element) in the neighborhood of such critical points or lines has a profound influence on the integral characteristics.

To clarify this remark let us recall some results related to research in the theory of continuous elastic plates with sharp corners on the boundary. (cf. [1.27]). Then if the thickness of the plate converges to 0 (in the vicinity of a sharp corner) with some predetermined rate of convergence, the definite property of the plate-bending operator is destroyed, and the spectrum of the natural frequencies of vibrations ceases to be discrete.

The study of the asymptotic behavior of optimal solutions is also important from the perspective of developing effective numerical techniques in optimal design projects.

4.5.1 Optimization of Bending Rigidity for a Plate. We consider the bending problem for an infinitely long (in the y-direction) rectangular plate that is rigidly built-in along the edges $|x| = l$ and subjected to a load of magnitude P uniformly distributed along the axis of symmetry $x = 0$. The distribution of thickness $h = h(x)$ is independent of the coordinate y. Therefore, all our arguments may be restricted to the specific plane $y = 0$. We assign the magnitude of the concentrated load P_* that causes a prescribed deflection u_0 at the point $x = 0$, $y = 0$ to be the measure of rigidity.

The optimization problem we wish to solve consists in finding the distribution of thickness $h(x, y)$ that satisfies the isoperimetric condition

$$\int_{-l}^{+l} h(x)dx = S \tag{4.30}$$

and assigns a maximum value to the integral

$$P_* = \max_h P(h) = \max_h \min_u (K_m/u_0) \int_{-l}^{+l} h^m(u_{xx})^2 dx \tag{4.31}$$

with the boundary conditions

$$u(0) = u_0, \qquad u(-l) = u(+l) = u_x(-l) = u_x(+l) = 0, \tag{4.32}$$

where u_0 and S are given numbers. The quantities m and K_m depend on the type of plate being considered. In particular, the cases $m = 1$, $K_1 = EH^3/4(1 - \nu^2)$, and $m = 3$, $K_3 = E/12(1 - \nu^2)$ correspond to the three-layered and the uniformly continuous plate, respectively (see Chapter 1, Section 1.4).

The necessary conditions for the stationary behavior of u and h in problem (4.30) to (4.32) now assume the form

$$(h^m u_{xx})_{xx} = 0, \qquad h^{m-1}(u_{xx})^2 = \lambda^2. \tag{4.33}$$

Here λ^2 is an unknown constant. The first of these equations must be satisfied on the interval $[-l, l]$ with the exception of the point $x = 0$, and the points x_{*i} at which the derivatives of the displacement function have jump discontinuities. However, the second condition in (4.33) must be satisfied at every point in the interval $[-l, l]$. At the points x_{*i} of discontinuity in the derivatives of the displacement function $u(x)$, the optimal solution must satisfy the Weierstrass–Erdmann conditions (3.96) to (3.98), which for the problem considered may be written as

$$(h^m u_{xx})^+ = (h^m u_{xx})^- = 0, \qquad [(h^m u_{xx})_x]_-^+ = 0, \tag{4.34}$$

$$[-h^m(u_{xx})^2 + 2u_x(h^m u_{xx})_x]_-^+ = 0. \tag{4.35}$$

We should note that if the singular point coincides with the point of application of the concentrated load at $x = 0$ at which the prescribed deflection u_0 is assigned, the first condition in (4.34) and the entire condition (4.35) may be ignored, since in that case corresponding terms vanish in the variational expression for the minimized functional. For this reason the case $x_{*i} = 0$ shall be considered separately.

Conditions (4.34) have a simple physical interpretation. At the singular points the bending moment $M = (h^m u_{xx})$ is equal to 0, while the shear force $Q = (h^m u_{xx})_x$ is continuous.

It is a consequence of (4.34) and (4.35) that at a singular point, the distribution of thickness must satisfy the conditions

$$h^+ = h^- = 0, \qquad (h^{(m+1)/2})_x^+ = (h^{(m+1)/2})_x^-. \tag{4.36}$$

Let us analyze (4.34) to (4.36). Because of the optimality condition (4.33), the first term in (4.35) may be written in the form $h^m(u_{xx})^2 = \lambda^2 h$, and consequently, at this point it is continuous (recall that $h(x)$ is continuous there). Therefore, condition (4.35) can be written as

$$[u_x(h^m u_{xx})_x]_-^+ = 0. \tag{4.37}$$

Making use of the second relation in (4.34), we conclude that condition (4.37) is satisfied when

$$(u_x)^+ = (u_x)^- \tag{4.38}$$

or else if

$$(h^m u_{xx})_x^+ = (h^m u_{xx})_x^- = 0. \tag{4.39}$$

That eq (4.37) represents also a necessary optimality condition has been previously proved in [3.12].

Let us now show that condition (4.39) is not realized in the unknown optimal solution of our problem. Indeed, if we use (4.33), (4.36), and (4.39) we obtain a differential equation and associated boundary conditions:

$$\left.\begin{aligned} &(h^{(m+1)/2})_{xx} = 0 \quad \text{if} \quad x \neq x_{*i}, \\ &h^+ = h^- = (h^{(m+1)/2})_x^+ = (h^{(m+1)/2})_x^- = 0. \end{aligned}\right\} \tag{4.40}$$

It is easy to see that if we ignore the trivial solution, no solutions exist for the function $h(x)$. Consequently, condition (4.38) must be satisfied at all singular points. Thus, for the optimal distribution of thickness the displacement function must have continuous derivatives. However, derivatives of higher order may be discontinuous at such singular points. For the problem investigated in [3.11] it was shown that at most two singular points can exist.

Let us now investigate the case of a sandwich plate with $m = 1$. In this case the equations of equilibrium, combined with the optimality conditions (4.33), are reduced to the form

$$h_{xx} = 0, \qquad (u_{xx})^2 = \lambda^2. \tag{4.41}$$

In view of the second equation in (4.41) and conditions (4.33), we claim that no twice continuously differentiable solution can exist for the displacement function $u(x)$. Therefore, we shall look for the solution $u(x)$ in the class of functions having a discontinuity in the second derivative. From the symmetry of the problem we conclude that the singular points of the optimal solution must be symmetric with respect to the origin (i.e., $x_{*1} = -x_{*2}$). Thus, all our formulas can be derived for the region $[-l, 0]$. Now, using the second equation in (4.41) and conditions (4.32), we can determine the displacement function to within an accuracy determined by our knowledge of the value of x_{*1}. But (4.38) implies that $x_{*1} = -l/2$. Thus, we arrive at the following formulas for the displacement function and the magnitude of the applied load (which is the maximized functional):

$$\left.\begin{aligned} &u = 2u_0\left(1 + \frac{x}{l}\right)^2, && -l \le x \le -l/2;\\ &u = u_0\left(1 - \frac{2x^2}{l^2}\right), && -l/2 \le x \le 0;\\ &\lambda = 4u_0 l^{-2};\\ &P_* = K_1 S\lambda^2 u_0^{-1} = 16K_1 S u_0 l^{-4}. \end{aligned}\right\} \tag{4.42}$$

Note that it is not necessary to know the distribution of thicknesses $h(x)$ to find the value of P_* The thickness function can be determined from the first equation in (4.41) and from (4.36), which was derived from the singular point relations:

$$\left.\begin{aligned} &h = -(S/l^2)(x + l/2), && -l \le x \le -l/2;\\ &h = (S/l^2)(x + l/2), && -l/2 \le x \le 0. \end{aligned}\right\} \tag{4.43}$$

If we consider the case $x_{*1} = -x_{*2} = 0$ by itself and perform all the necessary calculations, we arrive at a load of magnitude $P_* = 4K_1 S u_0 l^{-4}$, which is smaller than the value previously obtained. Therefore, the solution given by (4.42) and (4.43) is optimal. This result was first derived in [3.17], where it was also shown that the conditions for stationary behaviour used in deriving (4.42) and (4.43) are not only necessary, but also sufficient optimality conditions.

We shall now consider a uniform solid plate ($m = 3$). As in the previously considered example, we shall restrict our investigation to the interval

$-l \le x \le 0$. Using (4.33) we derive the following equation describing the deflection of the optimal plate:

$$((u_{xx})^{-2})_{xx} = 0, \qquad -l \le x \le x_{*1} \quad \text{or} \quad x_{*1} \le x \le 0. \tag{4.44}$$

We integrate (4.44) on each subinterval $[-l, x_{*1}]$ and $(x_{*1}, 0]$ and determine eight constants of integration from the boundary conditions and the conditions at the singular point:

$$\left.\begin{aligned} &u(-l) = u_x(-l) = u_x(0) = 0, \qquad u(0) = u_0, \\ &[(u_{xx})^{-2}]^- = [(u_{xx})^{-2}]^+ = 0, \qquad [(u_{xx})_x^{-2}]^- = -[(u_{xx})_x^{-2}]^+, \\ &[u]_-^+ = 0. \end{aligned}\right\} \tag{4.45}$$

The last four relations are implied by (4.33) and (4.34) and the continuity of the displacement function $u(x)$. The displacement function does depend on the coordinate of the singular point x_{*1}, which can be trivially determined from condition (4.38) and turns out to be $x_{*1} = l/2$. Finally, wc derive the following formulas for the deflection function:

$$u = \frac{3\sqrt{2}u_0}{l^{3/2}}\left(\sqrt{\frac{l}{2}}(x + l) + \frac{2}{3}\left(\left(-\frac{l}{2} - x\right)^{3/2} - \left(\frac{l}{2}\right)^{3/2}\right)\right), \quad -l \le x \le -l/2,$$

$$u = \frac{3\sqrt{2}u_0}{l^{3/2}}\left(\sqrt{\frac{l}{2}}x + \frac{2}{3}\left(\left(\frac{l}{2}\right)^{3/2} - \left(x + \frac{l}{2}\right)^{3/2}\right)\right) + u_0, \quad -l/2 \le x \le 0. \tag{4.46}$$

The optimal distribution of thickness and the corresponding magnitude of the applied load can be derived from (4.30), (4.31), (4.33), and (4.46). We have

$$\begin{aligned} h &= \frac{3S}{4l}\sqrt{\left(-1 - \frac{2x}{l}\right)}, \qquad -l \le x \le -l/2 \\ h &= \frac{3S}{4l}\sqrt{\left(1 + \frac{2x}{l}\right)}, \qquad -l/2 \le x \le 0 \\ P_* &= (81K_3u_0S^3)/(4l^6) \end{aligned} \tag{4.47}$$

Carrying out analogous computations for the case $x_{*1} = x_{*2} = 0$, we obtain a smaller value of the applied load:

$$P_* = 81K_3u_0S^3/(64l^6).$$

Remark. If we derive the dependence of P on the position of the singular point x_{*1} (with $-l \le x \le 0$) without using the optimality condition (4.38), we find that

$$P = (81K_3S^3u_0/64)((x_{*1} + l)^{3/2} + (-x_{*1})^{3/2})^{-4}.$$

It is easy to show that the maximum for $P(x_{*1})$ is attained when $x_{*1} = -l/2$, and is given by $P = 81K_3S^3u_0/4l^3$. We also remark that as $x_{*1} \to 0$ or $x_{*1} \to$

$-l/2$, the magnitude of P is monotone decreasing and approaches the value $P = 81K_3S^3u_0/64l^6$.

4.5.2 As another example illustrating the application of extremality conditions at a singular point, we shall examine the optimization of the shape of a column of length l, compressed in the direction of its axis by loads P applied to its ends. We assume that the column is built-in at $x = 0$, and freely supported (i.e., free to rotate) at the end $x = l$. Let $h(x)$ denote the thickness of the column and $u(x)$ the deflection of its axis from the initial straight-line configuration. The problem consists in finding an admissible shape of the column that permits it to support a maximal load without losing stability.

Using Rayleigh's principle we reduce this problem to that of finding a continuous distribution of thickness that satisfies the isoperimetric condition

$$\int_0^l h(x)\mathrm{d}x = S \tag{4.48}$$

and maximizes the critical load functional

$$P_* = \max_h \min_u \left\{K_m \int_0^l h^m(u_{xx})^2\mathrm{d}x \Big/ \int_0^l u_x^2\mathrm{d}x\right\} = \max_h P(h), \tag{4.49}$$

with boundary conditions

$$u(0) = u_x(0) = u(l) = (h^m u_{xx})_{x=l} = 0. \tag{4.50}$$

For this problem the full system of optimality conditions (3.94) to (3.98) may be rewritten:

$$h^{m-1}(u_{xx})^2 = \lambda^2, \qquad K_m(h^m u_{xx})_{xx} + Pu_{xx} = 0, \tag{4.51}$$

$$(h^m u_{xx})^+ = (h^m u_{xx})^- = 0, \qquad [K_m(h^m u_{xx})_x + Pu_x]_-^+ = 0, \tag{4.52}$$

$$[P(u_x)^2 - K_m h^{m-1}(u_{xx})^2 + 2u_x(K_m h^m u_{xx})_x]_-^+ = 0. \tag{4.53}$$

The value of the parameter K_m differs from the value of the corresponding quantity in the previous problem of transverse bending. Conditions (4.52) and (4.53) must be satisfied at the points $x = x_{*i}$ of discontinuity in the derivatives. Using arguments analogous to those given above in this section, it is easy to show that the optimal solution has at most one singular point and that conditions (4.53), coupled with conditions (4.51) and (4.52), imply eq. (4.38), which in turn permits us to find the location of the singular point.

For the sake of simplicity we shall derive a solution corresponding to the case $m = 1$. From the optimality condition $u_{xx} = \lambda^2$ and the boundary conditions, it follows that there does not exist a nontrivial (i.e., $u \not\equiv 0$), twice continuously differentiable solution function $u(x)$. Therefore, these derivatives must have a discontinuity. To find it we make use of conditions (4.38) and (4.52) and eq. (4.51). It is a consequence of (4.52) and the first equation in (4.51) that at the point of discontinuity in the derivatives of $u(x)$,

$$h^+ = h^- = 0, \qquad (h_x^+) = -(h_x^-). \tag{4.54}$$

To determine the displacement function for the optimal column and find the location of the singular point, we make use of the fact that $(u_{xx})^2 = \lambda^2$ (i.e., $u_{xx} = -\lambda$ if $x \leq x_*$ and $u_{xx} = +\lambda$ if $x_* \leq x \leq l$), the first of the three conditions of (4.50), and the continuity conditions at the singular point $u^+ = u^-$ and $(u_x)^+ = (u_x)^-$. After elementary computations we derive the displacement function for the optimal shape:

$$\left.\begin{aligned} &u = -\lambda x^2/2 && \text{if } 0 \leq x \leq x_* \\ &u = (\lambda x/2)(x - 2(2-\sqrt{2})l) + (\lambda l^2/2)(2 - 2\sqrt{2}) && \text{if } x_* \leq x \leq l \end{aligned}\right\} \tag{4.55}$$

and the location of the singular point $x_* = l(\sqrt{2}-1)/\sqrt{2}$. The magnitude of the critical load that causes a loss of stability, is computed from eqs. (4.49), (4.51), and (4.55), and is given by the formula

$$P = 6K_m S/[(3\sqrt{2}-4)l^3].$$

The distribution of thickness $h(x)$ may be found on each interval $0 < x < x_*$ and $x_* < x < l$ using (4.51) with $m = 1$. To determine the constants of integration for the equation $h_{xx} + PK_1^{-1} = 0$, we have the three conditions (4.54) applied to the singular point. The fourth condition $h(l) = 0$ is a consequence of the fact that $(u_{xx})^2 = \lambda^2$ and $(hu_{xx})_{x=l} = 0$. Finally, we obtain

$$\begin{aligned} &h = (3S/[(3\sqrt{2}-4)l^3])(l(1-1/\sqrt{2}) - x)(x + l(\sqrt{2}-1)) && \text{if } 0 < x < x_*, \\ &h = (3S/[(3\sqrt{2}-4)l^3])(x - l(1-1/\sqrt{2})), && \text{if } x_* < x < l. \end{aligned} \tag{4.56}$$

4.6 Quasi-Optimal Solutions to Problems with Constraints: Estimates of Deviation from the Optimum

In Sections 4.2 and 4.3, we presented methods for constructing optimal solutions based on the perturbation technique. Approximate, or quasi-optimal, solutions may also be constructed by "artificially" introducing a small parameter (cf. [4.12]) while taking into account the specific properties of the functionals being studied. In [4.29] Seyranian introduced a technique for constructing admissible solutions and obtaining two-sided estimates by exploiting the homogeneous property of the relevant functionals.

Let us consider the following extremization problem with constraints: Find

$$\begin{aligned} &\min J(h), && h \in K; \\ &J_i(h) \geq c_i > 0, && i = 1, 2, \ldots, n, \end{aligned} \tag{4.57}$$

where c_i are given constants.

We assume that the functionals J and J_i $(i = 1, 2, \ldots, n)$ are positive on K, and are homogeneous with respect to h, with positive degrees of homogeneity β and β_i, respectively. First we consider problem (4.57) with only a single ith constraint, with the index "i" chosen arbitrarily. The solution of this problem shall be denoted by h_i^*, while the solution of problem (4.57) that satisfies all the constraints by h^*.

Let us compute the values of $J_j(h_i^*)$, where $j = 1, 2, \ldots, n$. Clearly, some constraints of problem (4.57) may be violated. We choose a multiplier γ_i such that the function $\gamma_i h_i$ satisfies all assigned constraints. It is easy to see that we can choose γ_i in the form

$$\gamma_i = \max_{j=1,2,\ldots,n} [c_j/J_j(h_i^*)]^{1/\beta_j}. \tag{4.58}$$

Indeed, we have

$$J_j(\gamma_i h_i^*) = \gamma_i^{\beta_j} J_j(h_i^*) \geq \{(c_j/J_j(h_i^*))^{1/\beta_j}\}^{\beta_j} J_j(h_i^*) = c_j.$$

Therefore $\gamma_i h_i^*$ is an admissible element (of K), since all the constraints in (4.57) are satisfied, however, generally speaking, it is not optimal. Consequently, the estimate $J(h^*) \leq J(\gamma_i h_i^*)$ is true. On the other hand, $J(h_i^*) \leq J(h^*)$, and therefore, the element h_i^*, which solves our problem with only a single constraint imposed on it, cannot assign to the minimized functional a larger value then the element h^*, which satisfies all n constraints. Combining these two inequalities, we obtain $J(h_i^*) \leq J(h^*) \leq J(\gamma_i h_i^*)$. Because the index "$i$" was chosen arbitrarily, we have

$$\max_{i=1,2,\ldots,n} [J(h_i^*) \leq J(h^*) \leq \min_{i=1,2,\ldots,n} J(\gamma_i h_i^*).$$

The admissible solution $h_q = \gamma_p h_p^*$ corresponding to a minimum of the right-hand side of the above inequality shall be called a quasi-optimal solution. Here p denotes the index for which this minimum was realized. In accord with this definition,

$$J(h_q) = J(\gamma_p h_p^*) = \min_{i=1,2,\ldots,n} J(\gamma_i h_i^*).$$

From the last two relations we derive the estimate

$$1 \leq \frac{J(h_q)}{J(h^*)} \leq \frac{\min_{i=1,2,\ldots,n} J(\gamma_i h_i^*)}{\max_{i=1,2,\ldots,n} J(h_i^*)}. \tag{4.59}$$

Thus, if we know the solution h_i^* of an extremal problem with a single constraint for all values of $i = 1, 2, \ldots, n$, we can compute the constants γ_i according to (4.58), construct a quasi-optimal solution and estimate how close this solution is to the "real" solution of an extremal problem with multiple constraints assigned to the minimized functional.

It is of interest to find out how the upper estimate in (4.59) depends on the values of the parameters c_i introduced in the statement of problem (4.57). To do so, we introduce the following representation of the optimal solutions h_i^* expressed in terms of h_i^0:

$$h_i^* = (c_i^{1/\beta_i})h_i^0, \tag{4.60}$$

where h_i^* obviously solves the optimization problem with a single ith constraint for $c_i = 1$. Making use of (4.60) and the homogeneity of all the functionals, we represent the constants γ_i in the form

$$\gamma_i = c_i^{1/\beta_i} \max c_j^{1/\beta_j} [J_j(h_i^0)]^{-1/\beta_j} = c_i^{1/\beta_i} c_{k_i}^{1/\beta_{k_i}} (J_{k_i}(h_i^0))^{-1/\beta_{k_i}}. \tag{4.61}$$

In this expression the index k_i stands for the value of j at which the expression in brackets in (4.61) attains a maximum. Using (4.60) and (4.61), we shall reduce (4.58):

$$\begin{aligned} 1 \le \frac{J(h_q)}{J(h^*)} \le \frac{\min\limits_{i=1,2,\ldots,n} J(\gamma_i h_i^*)}{\max\limits_{i=1,2,\ldots,n} J(h_i^*)} &= \min_{\substack{i=1,2,\ldots,n\\ j=1,2,\ldots,n}} \frac{J(\gamma_i h_i^*)}{J(h_j^*)} \\ &= \min_{\substack{i=1,2,\ldots,n\\ j=1,2,\ldots,n}} \left[\frac{c_{k_i}{}^{\beta/\beta_{k_i}} (J_{k_i}(h_i^0))^{-\beta/\beta_{k_i}} J(h_i^0)}{c_j^{\beta/\beta_j} J(h_j^0)} \right] \\ &\le \min_{i=1,2,\ldots,n} \frac{J(h_i^0) J_{k_i}^{-\beta/\beta_{k_i}}(h_i^0)}{J(h_{k_i}^0)}. \end{aligned} \tag{4.62}$$

The last inequality is true, because a minimum taken over a number of quantities cannot decrease if this number is decreased. Note that the estimate does not contain the parameters c_i, but in fact does depend on these parameters (cf. (4.61)). Note that, as we change the values of c_i, the components of the vector $\mathbf{k} = \{k_1, k_2, \ldots, k_n\}$ do not assume all possible values. To clarify this statement, we offer the following argument. Suppose that there exists a vector $\mathbf{k}$ satisfying (4.61) with components $\{k_1, \ldots, k_n\}$. Let us consider a system of inequalities for the parameters c_i^{1/β_j}:

$$\left.\begin{aligned} &(c_{k_i}/J_{k_i}(h_i^0))^{1/\beta_{k_i}} \ge (c_i/J_j(h_j^0))^{1/\beta_j}, \\ &c_i^{1/\beta_i} > 0, \qquad i, j = 1, 2, \ldots, n, \qquad j \ne k_i. \end{aligned}\right\} \tag{4.63}$$

In the n-dimensional space spanned by c_i^{1/β_i} $(i = 1, 2, \ldots, n)$, inequalities (4.63) are satisfied in a space R_k. If this space is nonempty, for an arbitrary vector $\mathbf{c} = \{c_1^{1/\beta_1}, \ldots, c_n^{1/\beta_n}\}$, $\mathbf{c} \in R_k$, (4.62) gives an absolute estimate of the closeness of the quasi-optimal solution to the optimal solution with respect to the minimized functional. If we take a maximum over all possible values of $\mathbf{k}$ of the right-hand side of (4.62) such that the corresponding space R_k does not turn out to be the empty set, we obtain an absolute estimate of closeness of the quasi-optimal solution in the quadrant:

$$c_i^{1/\beta_i} > 0 \qquad (i = 1, 2, \ldots, n).$$

Note that $h_i^* = c_i^{1/\beta_i} h^0$ is an optimal solution of the original problem (4.57) if the parameters c_i satisfy the system of inequalities

$$c_i^{-1/\beta_i} c_j^{1/\beta_j} \le J_j^{1/\beta_j}(h_i^*), \qquad j = 1, 2, \ldots, n, \qquad j \ne i \tag{4.64}$$

In fact, a direct check assures us that for values c_i, that satisfy the above inequalities, (4.57) is true and $\gamma_i = 1$, and consequently, the quasi-optimal solution is optimal, that is, $h_q = h_i^* = h^*$. In that case the index k_i of a component of the vector h is equal to i.

Note, too, that to analyze the respective regions in which we attain optimal and quasi-optimal solutions, respectively, we can introduce the substitution

$$t_i = c_i^{1/\beta_i}/c_n^{1/\beta_n} \qquad (i = 1, 2, \ldots, n-1),$$

and instead of conducting our analysis in the n-dimensional space of parameters c_i^{1/β_i} $(i = 1, 2, \ldots, n)$, transfer it to the $(n-1)$-dimensional space of parameters t_i $(i = 1, 2, \ldots, n-1)$.

4.7 Some Basic Approaches and Guiding Principles in Constructing Numerical Techniques: Algorithms for Iterative Optimization

As we have already stated, questions arising in the theory of optimal design lead to very complex mathematical problems. The complexity of design problems is conditioned by essential nonlinearity, high dimensionality, multiple-criteria formulation, and by the presence of a large number of constraints, particularly in the case of multicriteria designation of a structure. Therefore, one cannot expect an effective solution to the design of very complex projects "as a whole" without partitioning the structure into substructures and into separate elements, without first examining the effects of various demands imposed on the optimal solution, and without distinguishing the essential design variables whose variations have a definite effect on the overall shape of the structure. But even the most powerful supercomputers may not assure us of reasonable answers in an acceptable computational time when designing modern complex structures. Moreover, even in cases when a straightforward algorithm drives the computation (e.g., a nonlinear programming algorithm) and an optimal structural design can be obtained, such an approach is generally not regarded as adequate and may not represent the engineers' idea of what an optimal design should look like. It suffices to say that in designing an optimal structure the best choice of certain values for a selected (fixed) set of certain parameters is not necessarily more important than to have reliable information on the influence of each parameter, the constraints and principal functionals, and the optimal solution. Such information permits us to "review" a large number of possible variants (or "modifications") of the design without undertaking a "global" recomputation. It also permits us to study problems of stability of the optimal solution with respect to perturbations of the initially assigned data, and to handle additional requirements that were not formally stated in the early stages of design. Of particular importance is knowledge of the sensitivity of the design with respect to essential parameters. This may be clarified by observing that in designing a structure (particularly a unique or unusual type of structure), we frequently do not possess exact and complete information concerning parameters derived from the predicted working conditions. Therefore, in deriving additional information and establishing more reliable values of such parameters we need to make corrections to the perceived optimal solution. To avoid solving the entire optimal design problem from its very beginnings, we need to establish the sensitivity of the solution with respect to variations in the values of the parameters.

Generally, an effective solution of the entire problem of designing an optimal structure can be obtained only after a decomposition of the original problem into a sequence of simpler design problems for substructures and obtaining data concerning the sensitivity of certain functionals (defined for the substructures) with respect to changes in the values of certain essential design parameters. This information is then utilized at the higher levels of structural design.

For this reason, the rapid development of the theory of optimal design and the high effectiveness of its techniques in application to actual design problems is closely related to developments of computational algorithms that are applied to particular classes of design problems and make use of their specific properties.

This research activity is being intensively developed at the present time, and its results are being applied to the creation of computer-assisted design (CAD) systems (cf. [5] and [24]). Various numerical schemes are under study and are finding recognition in solving concrete problems. They may be based on finite difference, variational finite difference, or purely variational techniques.

Different variants of the finite-element method (FEM), boundary finite-element method (BFEM), or nonlinear programming techniques (NLPT) are finding broad applications. However, there is insufficient research into the basic properties and even no proof of convergence for many existing numerical schemes. There is no broad development in comparisons of effectiveness for numerical algorithms now applied even in solving certain test problems and certainly in applied problems of actual optimal design. As a result, we shall not consider theoretical problems concerning the validity of numerical techniques, and describe only the most commonly applied numerical algorithms from the point of view of the applicability and usefulness of these numerical techniques. We shall pay particular attention to the iterative optimization algorithms (cf. [5] or [4.7]), which can be conveniently applied to several types of optimization of design problems discussed in the present volume.

Let us describe the simplest algorithm of iterative optimization and point out certain variants. To be specific, let us consider as our optimization problem the one defined in (3.1) to (3.5), and assume that we can use variational principles in determining the vector state function u and its adjoint vector function v. Our algorithm consists of obtaining successive approximations to the optimal solution and is founded on the idea of "small" perturbations of the design vector function h with a multidimensional solution of the "direct" problem of finding u and v with the values of the function h^k regarded as fixed. The diagram shown in Figure 4.1 displays two crucial steps in this procedure,

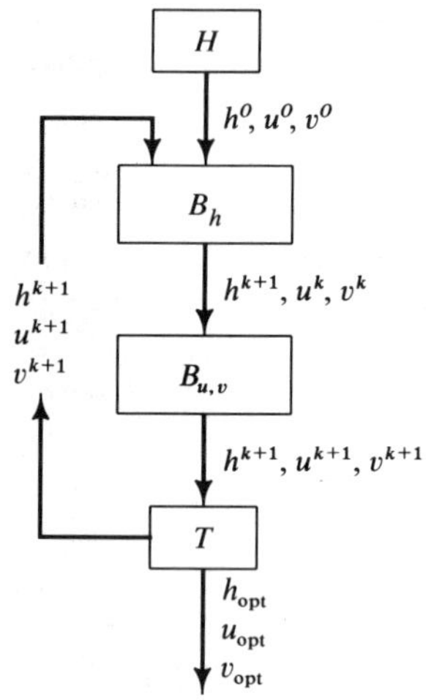

FIGURE 4.1. Iteration optimization algorithm.

symbolically shown as two blocks, a block $B_{u,v}$ that symbolizes the solution of the "direct" problem of deriving the functions u^k and v^k for fixed values of the functions h^k (here the superscript k denotes the kth approximation), and a block B_h that symbolizes derivation of a new approximation h^{k+1} for the design variable using the current values of the variables h^k, u^k, and v^k. The letters H and T denote the operations of introduction of the initial data, (i.e., specifying the initial approximation) and checking the accuracy of the derived approximations, respectively. The algorithm consists in specified steps. In each standard step with index $k + 1$, we perform the following operations. As a result of computations completed in the previous step, we regard the functions h^k, u^k, and v^k as known, and determine the value of the functional J_k. In the $(k + 1)$-th step we first determine the new values of the design functions $h^{k+1}(x)$ (working within B_h). Then using these values of $h^{k+1}(x)$, we proceed to solve the boundary-value problems (3.1), (3.2), (3.14), and (3.15) (now working within $B_{u,v}$) and we compute the values of the functions $u^{k+1}(x)$ and $v^{k+1}(x)$ and functional J^{k+1}. We estimate the error (or incompatibility) of our results in satisfying the necessary optimality conditions. If this error is sufficiently small, we terminate the program. If the error is too large, we proceed to the $(k + 2)$nd, that is next step and commence an analogous sequence of operations.

In B_h we may use different variational techniques of mathematical programming in deriving the h^{k+1} design variables. A very effective technique is gradient projection. To compute h^{k+1} by this method, we apply the formulas

$$\left.\begin{aligned} h^{k+1} &= h^k - \tau\Lambda^k, \\ \Lambda^k &= M^*(u^k, h^k)v^k + (\partial f^\lambda/\partial u)^k, \end{aligned}\right\} \tag{4.65}$$

where $\tau > 0$ is the size of the step in the direction of the gradient. The parameters λ_i and μ_i (see Section 3.1 for an explanation) are determined from (3.10) and (3.16).

Using (4.65) and the computed values of λ_i and μ_i, at each step we may frequently encounter certain difficulties. Let us consider the case of active constraints $\lambda_i \neq 0$ ($\mu_i = 0$) and describe an example in which we preserve the isoperimetric condition $J_i = c_i$ as we vary h.

We shall consider the case when exactly two integral constraints are assigned to the problem described by (3.1) to (3.5). The first variation of the functionals $J(h)$, $J_1(h)$, and $J_2(h)$ shall be written in the form

$$\delta J = (\chi, \delta h), \qquad \delta J_i = (\chi_i, \delta h), \qquad i = 1, 2, \tag{4.66}$$

where χ and χ_i are known functions and the parentheses denote the inner product of the enclosed functions. Let us set

$$\delta h = \{-\chi + \lambda_1 \chi_1 + \lambda_2 \chi_2\}\tau \tag{4.67}$$

and find the values of λ_1 and λ_2 from the relations

$$(\chi_i, \delta h) = 0, \qquad i = 1, 2, \tag{4.68}$$

which are a consequence of the constraints $J_1 = c_1$ and $J_2 = c_2$. We compute

$$\left.\begin{aligned} \lambda_1 &= \{(\chi, \chi_1)(\chi_2, \chi_2) - (\chi, \chi_2)(\chi_1, \chi_2)\}/\{(\chi_1, \chi_1)(\chi_2, \chi_2) - (\chi_1, \chi_2)^2\}, \\ \lambda_2 &= \{(\chi, \chi_2)(\chi_1, \chi_1) - (\chi, \chi_1)(\chi_1, \chi_2)\}/\{(\chi_1, \chi_1)(\chi_2, \chi_2) - (\chi_1, \chi_2)^2\}. \end{aligned}\right\} \tag{4.69}$$

If the optimization problem contains only one isoperimetric constraint $J_1 = c_1$, the expression for δh becomes

$$\delta h = \tau\{-\chi + [(\chi, \chi_1)/(\chi_1, \chi_1)]\chi_1\}.$$

It can be shown that for these variations δh, as defined by (4.67) and (4.69), it is true that

$$\delta J \leq 0, \qquad \delta J_i = 0, \qquad i = 1, 2. \tag{4.70}$$

Indeed, using (4.66) to (4.68) we derive

$$\delta J = (\chi, \delta h) = ((\chi - \lambda_1 \chi_1 - \lambda_2 \chi_2), \delta h) = -(1/\tau)(\delta h, \delta h) \leq 0.$$

The equalities in (4.70) are true if the multipliers λ_i are chosen by (4.69).

Consequently, if we choose functions δh in accordance with (4.67) and (4.69) and choose τ sufficiently small, the functional J will decrease in value, while the values of the functionals J_i will remain unaltered.

Note that this example may applied to the case of r constraints, but to determine the values of the r constants λ_i, it is necessary to solve a system of r linear algebraic equations obtained from the relations $(\chi_i, \delta h) = 0$, $i = 1, 2,$ at each iterative step of the algorithm.

To solve optimal design problems with inequality-type constraints, one may make effective use of the technique of ε-active constraints, as described in [4.19] to [4.22]. The basic idea of this technique consists in choosing some small number ε ($\varepsilon > 0$), and classifying in each iterative step all constraints into two types. One type consists of ε-active constraints that satisfy the inequalities

$$-\varepsilon \leq J_i - c_i \leq 0. \tag{4.71}$$

In each iteration, as we vary the design variables we regard the ε-active inequalities (4.71) as equality-type constraints. In particular, as we compute δh, we use (4.67) and (4.69).

The second type of constraints consists of inactive constraints, satisfying the condition

$$J_i - c_i < \varepsilon. \tag{4.72}$$

Constraints that are inactive during some iteration (i.e., constraints satisfying (4.72)) are ignored during that iteration in the process of improvement implemented by varying the design variables.

Remarks. If one applies an algorithm using successive approximations to minimize (or maximize) a local functional, then one should make use of the technique described in Section 2.3 and first reduce the problem to an equivalent problem with an integral functional.

The algorithm of successive optimization permits an adaptive improvement in the quality criterion. The process of solution may be discontinued after completing the computations in some arbitrarily chosen step indexed by an integer k.

Thus, we have a new distribution of the design variables h^k that is better from the point of view of the cost (or quality) functional than the initially assumed distribution h^0. Also, corresponding to the design h^k we have a variable u^k describing the state of the structure. Therefore, at each iterative step of the algorithm we obtain some useful information, which may be utilized to reveal certain tendencies that are present in the iterative formulation of the optimal shape and the internal composition of a structure.

For example, in optimal design problems that result in singular solutions, the use of sequential optimization algorithms permits us to avoid the appearance of a singularity before completing the algorithmic procedure by introducing additional constraints, thus effectively controlling the outcome of the procedure. In this manner use of iterative optimization algorithms may lead to a convenient variant of a search procedure, permitting active interaction with the computer software during the optimization process.

4.8 Applications of Finite-Difference Variational Methods

In the previous section, we introduced an algorithm of successive optimization. In the transition from one step to the next we effected certain small improvements by varying the design variables δh. If the kth step variation $\delta h = h^{k+1} - h^k$ is sufficiently small, then the computation of the new approximate functions u^{k+1}, v^{k+1} starts with a "good" initial guess u^k, v^k. It is this circumstance which makes the entire process of solving for some unknown quantities by iterating successive approximations a reasonable approach. Examples of such techniques that have been successfully used in solving certain "direct" problems include the finite-difference variational approaches. Let us assume that we have formulated certain variational principles for determining the changes $\delta u = u^{k+1} - u^k$ and $\delta v = v^{k+1} - v^k$. The effectiveness of numerical solution computer software if one of the "direct" minimization algorithms is used (e.g., sequential coordinate descent a variation of the gradient techniques, method of feasible directions, etc.) depends on the manner in which the continuous problem is approximated by a discrete analog. In constructing a computational scheme, two attributes are of paramount importance: accuracy and economy.

A very important aspect of actual computations using a numerical scheme is preservation of convexity (cf. [4.8]). That is, a substitute for a strictly convex functional should be a strictly convex approximating functional. The property of strict convexity is necessary for the uniqueness of a minimum and also speeds up the rate of convergence of the computational algorithm.

Let an integral functional be defined over a set of functions that are continuously differentiable in a region Ω:

$$J = \int_{\Omega}\int f(x, y, u, u_x, u_y)dxdy. \tag{4.73}$$

Here Ω denotes the domain for functions of the independent variables x and y, f is a given function, and the subscripts denote (as usual) partial derivatives. The function f is twice continuously differentiable with respect to u, u_x, and u_y, and satisfies the conditions

$$\begin{aligned} &f_{uu}(\xi_1)^2 + f_{u_x u_x}(\xi_2)^2 + f_{u_y u_y}(\xi_3)^2 + 2f_{uu_x}\xi_1\xi_2 \\ &\quad + 2f_{uu_y}\xi_1\xi_2 + 2f_{u_x u_y}\xi_2\xi_3 > 0 \end{aligned} \tag{4.74}$$

for an arbitrary vector $\xi = (\xi_1, \xi_2, \xi_3)$. Condition (4.74) ensures that the functional $J(u)$ is strictly convex.

We shall now construct finite difference schemes for approximating the functional given in (4.73). First we shall decompose Ω into elementary cells, and then replace the integral over each cell by approximate quadrature formulas and approximate the value of the function u and the values of its derivatives, by values of these functions at discrete points (vertices of the cells). The approximate nature of the three operations listed above determines the total error in the value of the approximating functional.

In what follows we shall assume that the decomposition of the region Ω into cells is compatible with the shape of the boundary of Ω (implying that this operation does not contribute to the total error).

To ensure that of solutions are unique and that the computational algorithm in a functional minimization problem converges, it is necessary that the approximating scheme be represented by a strictly convex function of the values of u at the vertices of each cell. As will be shown below, the construction of a strictly convex computational scheme may be attained by increasing the integer N ($N \geq 1$), which is a count of the number of operations performed on f in each cell when applying the quadrature formula.

Let κ denote the relative error of the approximating scheme in each cell, defined as the ratio of the total error in computing the integral over that cell to the area of that cell. The total error in computing the integral over the entire region Ω may be estimated by taking the maximum of this quantity over all cells. Let ε denote the maximal linear dimension of a cell. The maximum power of ε in the estimate of the relative error κ (as $\varepsilon \to 0$) does not exceed ε^2 for arbitrary $N \geq 1$ of operations of f, that is, $\kappa = O(\varepsilon^2)$. In fact, for such arbitrary N, if we know values of u only at the vertices of a cell, we cannot obtain a better accuracy than ε^2, and the value of the integral over that cell cannot be evaluated with accuracy of order higher than ε^4. Since the area of the cell is ε^2, the order of the relative error is $\kappa = O(\varepsilon^2)$.

For a given accuracy of approximation it makes sense to choose a scheme having the least number N of operations performed on the integrand function, that is, to choose the most economical scheme.

To meet these requirements, we shall attempt to construct a scheme that satisfies the following conditions:

1. The order of the relative error is $\kappa = O(\varepsilon^2)$.
2. The scheme preserves strict convexity.
3. We minimize the number N of operations within each cell that are necessary to compute f.

First let us investigate the case when the entire region Ω is subdivided into identical parallelograms. Without any loss of generality, we may consider only rectangular cells. This elementary modification is accomplished by an affine transformation with respect to the independent coordinates.

Let Δx and Δy denote the lengths of the sides of cells. We consider a single cell Q_{ij} in Ω, with vertices located at the points (x_i, y_j), (x_{i+1}, y_j), (x_i, y_{j+1}), and (x_{i+1}, y_{j+1}), where $x_i = i\Delta x$; $y_j = j\Delta y$; $i = 1, 2, \ldots$; and $j = 1, 2, \ldots$.

We shall denote by u^0, u^1, u^2, and u^3 the respective values of the discrete function u_{ij} at these vertex points. The integral J is represented as a sum of the integrals over the cells Q_{ij}, that is,

$$J = \sum_{i,j} J_{ij}.$$

To compute the integral J_{ij} over the cell Q_{ij}, we use the following finite difference approximation I_{ij}, necessitating N computations for the integrand function in that cell (with area $\Delta = \Delta x \Delta y$):

$$\left.\begin{aligned} &x_i^s = x_i + \lambda^s \Delta x, \quad y_j^s = y_j + \mu^s \Delta y, \\ &I_{ij} = \Delta \sum_{s=1}^{N} c_s f\Big(x_i^s, y_j^s, \sum_{m=0}^{3} \alpha_m^s u^m, \ (1/\Delta x) \sum_{m=0}^{3} \beta_m^s u^m, (1/\Delta y) \sum_{m=0}^{3} \gamma_m^s u^m\Big), \end{aligned}\right\} \tag{4.75}$$

where c_s, λ^s, μ^s, β_m^s, and γ_m^s, are the parameters of the scheme.

Let us show that in order to satisfy condition (1), it is necessary and sufficient that the parameters of the scheme satisfy the relations

$$\left.\begin{aligned}
&\sum_{s=1}^{N} c_s = 1, \sum_{s=1}^{N} c_s\lambda^s = \sum_{s=1}^{N} c_s\mu^s = \frac{1}{2}, \qquad \sum_{m=0}^{3} \alpha_m^s = 1,\\
&\sum_{m=0}^{3} \beta_m^s = \sum_{m=0}^{3} \gamma_m^s = 0,\\
&\beta_2^s = \beta_3^s = \gamma_1^s + \gamma_3^s = 0, \qquad \beta_1^s + \beta_3^s = \gamma_2^s + \gamma_3^s = 1,\\
&\sum_{s=1}^{N} c_s(\alpha_1^s + \alpha_3^s) = \sum_{s=1}^{N} c_s(\alpha_2^s + \alpha_3^s) = \sum_{s=1}^{N} c_s(1 + \beta_0^s)\\
&\qquad = \sum_{s=1}^{N} c_s(1 + \gamma_0^s) = \frac{1}{2}.
\end{aligned}\right\} \tag{4.76}$$

Making use of conditions (4.76) and setting $\beta_0^s = \beta_s$ and $\gamma_0^s = \gamma_s$, where β_s and γ_s are certain numbers (obeying these conditions), we arrive at an approximation for the integral over a cell:

$$\begin{aligned}
I_{ij} = \Delta x\Delta y \sum_{s=1}^{N} c_s f_s\{&x_i + \lambda^s\Delta x, y_j + \mu^s\Delta y, \alpha_0^s u^0 + \alpha_1^s u^1 + \alpha_2^s u^2 + \alpha_3^s u^3,\\
&(1/\Delta x)(\beta_s(u^0 - u^1) + (1 + \beta_s)(u^3 - u^2)),\\
&(1/\Delta y)(\gamma_s(u^0 - u^2) + (1 + \gamma_s)(u^3 - u^1))\}.
\end{aligned} \tag{4.77}$$

Let us consider this scheme with $N = 1$. In that case we find from (4.76) and (4.77) that

$$\begin{aligned}
I_{ij} = \Delta x\Delta y f(&x_i + \Delta x/2, y_j + \Delta y/2, \alpha(u^0 + u^3) + (1/2 - \alpha)(u^1 + u^2),\\
&(u^3 + u^1 - u^0 - u^2)/(2\Delta x), (u^3 + u^2 - u^1 - u^0)/(2\Delta y)),
\end{aligned} \tag{4.78}$$

where α is some number. Thus, we obtain a single-parameter family of finite difference quadrature schemes with $N = 1$, which approximate the integral with relative error $\kappa = O(\varepsilon^2)$. However, it is easy to show that these schemes do not satisfy condition (2); that is, they are not strictly convex. Let us clarify this statement by offering an example of a strictly convex functional

$$J = \iint_{\Omega} (u_x^2 + u_y^2 + u^2)dxdy, \tag{4.79}$$

which satisfies conditions (4.74). In the absence of boundary conditions and constraints, its absolute minimum is obviously equal to 0 and occurs for $u(x, y) \equiv 0$. Using the finite-difference approximation (4.77) for the functional (4.79) and setting $\alpha = 1/4$, we obtain the finite cell version

$$\begin{aligned}
I_{ij} = {}&(\Delta y/4\Delta x)(u^1 + u^3 - u^0 - u^2)^2 + (\Delta x/4\Delta y)(u^2 + u^3 - u^0 - u^1)^2\\
&+ (\Delta x\Delta y/16)(u^0 + u^1 + u^2 + u^3)^2.
\end{aligned} \tag{4.80}$$

It is clear that the form (4.80) is nonnegative; however, it is not positive-definite. It vanishes not only when $u^0 = u^1 = u^2 = u^3 = 0$, but also when we assign "chessboard" values to u^i; that is, $u^0 = u^3 = -u^1 = -u^2 = q$ ($q =$ const). Therefore, the minimum of (4.80) vanishes, but it is attained on the family of "chessboard" functions

$$u_{ij} = (-1)^{i+j} q. \tag{4.81}$$

We note that if Dirichlet-type boundary conditions are assigned, the situation changes radically. Here schemes with $N = 1$ may result in a strictly convex function. For example, in the above problem if we assign to the function $u(x, y)$ the condition $u = 0$ on the boundary of Ω, no solution of the problem satisfying (4.81) can also satisfy this boundary condition for $q \neq 0$. Moreover, for this boundary condition the quadratic form (4.80) vanishes only if $q = 0$ in (4.81); that is, only if $u_{ij} = u = 0$ for all values of i and j. Consequently, in that case the solutions are unique. However, even for Dirichlet-type boundary conditions it is more advantageous to use schemes satisfying condition (2) as it assures a better rate of convergence.

Let us now consider schemes of the form (4.77) with $N = 2$. For the sake of symmetry and simplicity, let us set $c_1 = c_2 = 1/2$. Then the parameters of the scheme must satisfy the relations

$$\begin{aligned}\lambda^1 + \lambda^2 = \mu^1 + \mu^2 = \alpha_0^1 + \alpha_1^1 + \alpha_2^1 + \alpha_3^1 = \alpha_0^2 + \alpha_1^2 + \alpha_2^2 + \alpha_3^2 = 1,\\ \beta_1 + \beta_2 = \gamma_1 + \gamma_2 = -1,\\ \alpha_1^1 + \alpha_3^1 + \alpha_1^2 + \alpha_3^2 = \alpha_2^1 + \alpha_3^1 + \alpha_2^2 + \alpha_3^2 = 1.\end{aligned} \tag{4.82}$$

We shall demonstrate that schemes of the form (4.77) and (4.82) with $N = 2$ satisfy condition (2) if at least one of the inequalities $\beta_1 \neq \beta_2$ or $\gamma_1 \neq \gamma_2$ is true. These inequalities imply that at least one of the partial derivatives u_x or u_y can be evaluated as a finite difference approximation by using values of the function f in a single cell. Thus, approximations of the form (4.77) or (4.82) with $N = 2$ solve our problem if either $\beta_1 \neq \beta_2$ or $\gamma_1 \neq \gamma_2$.

We offer as an example two of the simplest symmetric approximation schemes of the form (4.77) and (4.82) (where, as before, Δ denotes $\Delta x \Delta y$):

$$\begin{aligned}I_{ij} = \Delta[&f(x_i, y_j, u^0, (u^1 - u^0)/\Delta x, (u^2 - u^0)/\Delta y)\\ &+ f(x_i + \Delta x, y_j + \Delta y, u^3, (u^3 - u^2)/\Delta x, (u^3 - u^1)/\Delta y)],\end{aligned} \tag{4.83}$$

and

$$\begin{aligned}I_{ij} = \Delta[&f(x_i + \Delta x/3, y_j + \Delta y/3, (u^0 + u^1 + u^2)/3, (u^1 - u^0)/\Delta x, (u^2 - u^0)/\Delta y)\\ &+ f(x_i + 2\Delta x/3, y_j + 2\Delta y/3, (u^1 + u^2 + u^3)/3,\\ &(u^3 - u^2)/\Delta x, (u^3 - u^1)/\Delta y)].\end{aligned} \tag{4.84}$$

This time let us decompose Ω using a triangular mesh. We consider the specific case when each cell is a right-angled triangle and each pair of triangular cells forms a rectangular cell. Thus, the rectangular cell Q_{ij} contains the two triangular cells Q_{ij}^+, and Q_{ij}^-, with $I_{ij}^+ = \Sigma(c_s^+ f_+^s)$ and $I_{ij}^- = \Sigma(c_s^- f_-^s)$, $s = 1, 2, \ldots, N$. In computing f_+^s we only need the values u^0, u^1, u^2, and in computing f_-^s, we need u^1, u^2, u^3. It can be shown that in this case there exists a unique scheme for $N = 1$ satisfying conditions (1) and (2). This scheme may be derived from (4.84).

We see that the first term of the sum inside the brackets refers to the vertices of triangle Q_{ij}^+, while the second term refers to Q_{ij}^-.

Consequently, the approximation given by (4.84) may be regarded as a scheme with $N = 1$ corresponding to a decomposition into triangular cells.

4.9 Application of Finite-Element Techniques

The approaches to structural optimization discussed in Chapters 3 and 4 used techniques developed for optimization of systems with distributed parameters and were based on variational arguments. They allowed us to analyze sensitivity and obtain optimal solutions in the setting of a continuum.

These computations provide a reasonable agreement with direct computational methods used for estimating the stress-strain state (SSS) of a structure, such as the finite-element method (FEM) or the boundary integral equations method (BIEM). We offer a short exposition of FEM and BIEM with regard to potential applications to the design of shapes of elastic objects, emphasizing the shape of the region in which the equations of equilibrium are defined (cf. [4.6], [4.11], [4.14], [4.20], [4.25] to [4.27], [4.31], and [4.34]).

FEM is a very effective numerical technique used in algorithms of successive optimization. In these algorithms the construction of shape improvements is based on results of sensitivity analysis that are approximated by finite difference equations and then adopted to numerical computations of SSS.

The choice of the approximate representation of the sensitivity analysis relations must roughly agree with the outcome of computations for important functions in the direct SSS problem and with computed solutions of the adjoint boundary-value problem.

Note that in finding the optimal shapes of structural elements (generally these are problems with unknown boundaries) SSS computations can be accomplished in successive stages for a sequence of regions that originate with boundary variations. Initially, the geometry of these regions is not known and may turn out to be extremely complex. Therefore, it is necessary to apply certain very general numerical techniques that would be suitable for computations in a broad class of regions. FEM is a technique that satisfies this requirement, particularly, FEM using quadrangular finite-element nets. Because it is necessary to solve the SSS problem many times in the process of optimizing a structure, the formation of the rigidity matrix is an economical procedure from the point of view of machine time economy.

Moreover, the adopted approximation for the deflections (state of the system) should lead to sufficiently accurate predictions for different shape configurations and should be numerically stable. Accordingly, a combination of finite elements consisting of four triangles was used in the computations of SSS in [3.2] and [4.2].

An FEM algorithm combined with a successive approximation algorithm should be realized as a program that satisfies the following requirements at each iterative step:

1. The deformed object is decomposed into finite elements, that is, a finite-element net is created;
2. the rigidity matrix and then the vector of the right-hand (inhomogeneous) terms is derived for that finite-element net, and the rigidity matrix is decomposed into two triangular matrices;
3. displacements, stresses, and strains are computed, and the right-hand side vector is computed for the adjoint system;
4. the decomposition of the rigidity matrix into triangular matrices is used to determine the adjoint variables; and
5. a design improvement is effected by varying the boundary, and a new shape of the structure is derived.

Let us briefly review the process of formulating a rigidity matrix that corresponds to the combination of rectangular finite elements. For each of the two diagonals of a quadrangle we carry out the following construction. We assign linear shape functions in the two triangles obtained by dividing the quadrangle by a diagonal. The value of the shape function is equal to the arithmetic average of the values of the functions assigned to the different diagonals. This type of approximation combines the advantages of decomposing the region into quadrangular elements and the use of linear shape functions. As a result we should obtain some consistency in the representations of stresses and strains. In addition, the formulation of a rigidity matrix is convenient from a purely algorithmic point of view.

The computation of the elements of the rigidity matrix is accomplished by means of the following operations.

1. We sort the elements, assigning numbers to the vertices of the elements, and determine the position of the corresponding coefficients in forming the rigidity matrix.
2. We sort out the triangles, deciding which one belongs to a specific element and contains the required vertex.
3. We determine where this triangle belongs in our rigidity matrix.

In the third stage of this process, we must determine all the necessary relations of planar elasticity for the triangular finite-element net.

We introduce the following relations: Let x_i, y_i, x_j, y_j, x_k, and y_k be the coordinates of the vertices of one of the triangles. Let $a_i = x_j y_k - x_k y_j$, $b_i = y_j - y_k$, and $c_i = x_k - x_j$. The area Δ of this triangle is given by

$$\Delta = \frac{1}{2}\begin{vmatrix} 1 & x_i & y_i \\ 1 & x_j & y_j \\ 1 & x_k & y_k \end{vmatrix}.$$

The linear shape function N_i is equal to 1 at the point (x_i, y_i) and to 0 at the other vertices of the triangle, and is given elsewhere by the formula $N_i = (2\Delta)^{-1}(a_i + b_i x + c_i y)$. The displacement of an arbitrary point inside the triangle is given in terms of displacements at junction points by the formula

$$\begin{vmatrix} u_1 \\ u_2 \end{vmatrix} = \begin{bmatrix} N_i & 0 & N_j & 0 & N_k & 0 \\ 0 & N_i & 0 & N_j & 0 & N_k \end{bmatrix} (u_{2i-1}, u_{2i}, u_{2j-1}, u_{2j}, u_{2k-1}, u_{2k})^T.$$

We introduce a matrix $\mathbf{B} = \mathbf{AN}$, where $\mathbf{N}$ is the shape matrix given in the above formula. We obtain

$$\mathbf{B} = \frac{1}{2\Delta}\begin{bmatrix} b_i & 0 & b_j & 0 & b_k & 0 \\ 0 & c_i & 0 & c_j & 0 & c_k \\ c_i & b_i & c_j & b_j & c_k & b_k \end{bmatrix}.$$

The local rigidity matrix for a single element is given by

$$\mathbf{K} = \mathbf{B}^{\mathbf{T}}\mathbf{DB}\Delta, \quad \text{with } \mathbf{D} = \begin{bmatrix} \kappa + 2\mu & \kappa & 0 \\ \kappa & \kappa + 2\mu & 0 \\ 0 & 0 & \mu \end{bmatrix},$$

where the values of κ are $\kappa = (2\lambda\mu)/(\lambda + 2\mu)$ for plane stress, and $\kappa = \lambda$ for plane strain. Multiplying these matrices provides us with the values for embedding the rigidity of this triangular finite element into the overall rigidity matrix (cf. [4.15], [4.17], [4.24], [4.30], and [4.32]).

After determining the rigidity matrix and the vector of applied loads, we can solve the resulting system of algebraic equations, represented by a symmetric, positive-definite, multidiagonal matrix using the Cholesky technique, first decomposing this matrix into two triangular matrices.

Note certain characteristic properties of this finite-element representation. The scheme adopted above for approximating the displacements preserves the strict convexity of the strain energy functional, is computationally stable, allows us to choose a convenient algorithm, and does not necessitate numerical integration procedures.

This scheme for computing SSS for a two-dimensional elastic body subjected to stresses or strains is well suited to techniques of successive optimizations.

Let the boundary of a body consist of M line segments with coordinates of the vertices given by $\mathbf{X}_{i,\alpha}$ $(\alpha = 1, 2, \ldots, M)$. In each iteration the improvements in varying the coordinates of the vertices is determined by formulas of sensitivity theory. However, to simplify the computations implied by these formulas one should apply some algebra first. Since $(\delta\psi_i)_\Gamma = (\delta u_i)_\Gamma = 0$,

$$\delta J = \int_{\Gamma} (1 + \lambda f - (\psi_i \sigma_{ij})_{,j} + (T_i \psi_i)_{,k} n_k - 2H T_i \psi_i) t \, d\Gamma$$
$$= \delta \left\{ \int_{\Omega} d\Omega - \int_{\Omega} (\psi_{i,j} C_{ijkl} u_{k,l}) d\Omega + \int_{\Omega} (\psi_i q_i) d\Omega \right.$$
$$\left. + \lambda \int_{\Omega} f(C_{ijkl} u_{k,l}, u_m) d\Omega + \int_{\Gamma} (T_i \psi_i) d\Gamma \right\}. \qquad (4.85)$$

The computation of the gradient components for the Lagrangian functional using (4.85) is performed up to an approximation. Components of the vector $\delta \mathbf{x}$ of boundary variations may be written as a linear combination of the variations $\delta x_i = r_\alpha b_{i\alpha}$, where $b_{i\alpha}$ are components of the basic functions used for varying the boundary that are defined on the unknown surface of the elastic body.

The values of the components of the gradient given in terms of these basic functions $b_{i\alpha}$, which are to be computed in accordance with (4.85), are then evaluated as follows.

1. For a given initial domain, we determine the value of the unknown quantity enclosed inside the braces in (4.85), after the δ-symbol.
2. We now determine the value of the same quantity for the region obtained after moving the boundary of this initial region by $\varepsilon b_{i\alpha}$ ($\alpha = 1, 2, \ldots, M$, while ε is a "small" number).
3. The difference between the two quantities is divided by ε (for each value of α).

To carry out these computations we apply subroutines to compute the rigidity matrix, and the right-hand side vector of applied loads, and evaluate the stresses and values of the function f. If we use a local family of base functions for implementing variations of the boundary, the net of finite elements will vary only in a small neighborhood of the support of these functions. Therefore, it is necessary to recompute in the rigidity matrix only the set of entries that belong to that neighborhood, and the components of the gradient can be evaluated with sufficient rapidity.

4.10 Application of the Method of Boundary Integral Equations

When one applies variational finite-difference techniques and finite-element methods to solve problems of optimal design, the state functions describing the stress-strain state of the system and the adjoint variables are computed for every element, that is, for the entire structure. Some knowledge of the displacements and stresses at interior points of the body may be very useful, because this information can be used to perform a complete analysis of the structure. But, frequently in designing an optimal shape of a structure, such knowledge is not really necessary. Let us see why this is so. For a wide class of problems, when one formulates the sensitivity analysis and conditions for optimality, only the boundary values of the state functions and adjoint variables occur in these conditions. To compute SSS in such cases, it makes sense to apply the boundary integral equations method (BIEM) and the boundary finite-element method (BFEM). The application of BFEM based on a BIEM-type analysis permits us to reduce effectively the number of computations in FEM or in variational finite-difference schemes. The fundamental idea underlying the boundary integral equations method is a reduction of the boundary-value problems for a system of differential equations to an equivalent integral equation using only line (or surface) integrals defined over the boundary of the region (cf. [4.1] and [4.13]). The dimensionality of the original problem is then reduced by 1.

One such technique for deriving integral equations makes use of Green's third identity for elliptic-type partial differential equations. Green's identities applied to both the state function and the adjoint variables relate the values of a quantity inside a region to integrals of these functions and their derivatives over the region and along its boundary.

We can proceed to the limit by following a point with known values of the appropriate functions to the boundary of the region, thus obtaining a boundary integral equation that connects the values of the function with the values of its derivatives on the boundary. To solve a boundary integral equation, we utilize the boundary finite element method (BFEM), which entails the following operations. The boundary of the region is decomposed into distinct elements, while relevant functions defined on the boundary are expanded in terms of certain base functions. The coefficients of that expansion are regarded as unknown quantities. The influence matrix is computed by integrating the kernel of the boundary integral equation over all boundary elements. Next the system of algebraic equations is solved, and approximate values of the parameters are determined. Of course, there exist other methods for treating boundary-value equations (whether of singular- or regular-type) that may rely on techniques of distributing the singularities of local sources, potentials, and compensating loads.

Let us clarify the application of the boundary integral equations method (BIEM) by offering an example of maximizing the rigidity of a torsioned shaft. This problem has long been the subject of research (cf. [4.3] and [4.35]). The results given below were derived by V.V. Kobelev. We assume that the cross-section of the shaft is simply connected and that the area of the cross-section is fixed (meas $\Omega = \text{const}$). Part of the contour Γ is given, while the remaining part $\Gamma_v \subset \Gamma$ is to be found from the condition requiring the torsional rigidity to be maximized:

$$J = \int_{\Omega} ((w_{,1} - \vartheta x_2)^2 + (w_{,2} + \vartheta x_1)^2) d\Omega \to \max_{\Gamma_v}.$$

Here w denotes the "deplanar" function, that is, the deflection of the cross-section from the initial plane. It satisfies the boundary integral equation

$$\pi w(P) = \int_\Gamma w(Q)\partial/\partial n(\ln R_{PQ})\mathrm{d}\Gamma_Q - \int_\Gamma \ln R_{PQ}\partial w/\partial n \mathrm{d}\Gamma_Q. \tag{4.86}$$

The normal derivative of the deplanar function is uniquely determined by its location on the contour:

$$\partial w/\partial n \equiv w_n = \vartheta(x_2 n_1 - x_1 n_2). \tag{4.87}$$

The symbol R_{PQ} denotes the distance between two boundary points P and Q.

The variation of the minimized functional is given by

$$\delta J = \int_\Gamma (\lambda + \partial w/\partial s + (x_i n_i))\delta t \mathrm{d}\Gamma, \tag{4.88}$$

where δt denotes the variation of the boundary evaluated along the direction normal to Γ. As can be seen from (4.88), the variation of the rigidity depends on the value of the derivative of the deplanar function with respect to the parameter of the boundary curve. To compute this function, we utilize a discrete analog of eq. (4.86). The boundary curve is approximated by an M-polygon containing straight line intervals γ_i $(1 \le i \le M)$, while w and $\partial w/\partial n$ are replaced by piecewise-constant functions defined on the polygonal line segments γ_i. Their (piecewise-constant) values on γ_i are denoted by w_i and w_{ni}, respectively, with $1 \le i \le M$. The values of the normal derivative w_{ni} along the boundary are determined directly from conditions (4.87). The coefficients of the algebraic equations are expressed in terms of integrals of the kernel of this system of equations rewritten for the straight-line segments. Following these approximations, we restate eq. (4.86) in the form

$$\sum_{j=1}^{M} a_{ij} w_j = \sum_{j=1}^{M} b_{ij} w_{nj},$$

$$a_{ij} = h_j[(x_1^{(j)} - x_1^{(i)})n_{1j} + (x_2^{(j)} - x_2^{(i)})n_{2j}](R_{ij})^{-2}, \qquad i \ne j$$

$$b_{ij} = h_j(\ln R_{i,j-1} + 4\ln R_{ij} + \ln R_{i,j+1}), \qquad i \ne j$$

$$a_{ii} = -\sum_{i \ne k} a_{ik}, \qquad R_{ij} = \{(x_1^{(j)} - x_1^{(i)})^2 + (x_2^{(j)} - x_2^{(i)})^2\}^{1/2}.$$

Here R_{ij} represents the distance between the line segments γ_i and γ_j, h_i is the length of the line segment γ_i, while n_{1i} and n_{2i} are the x_1 and x_2 components, respectively, of the unit vector normal to γ_i. The matrix entries are computed by Gauss' formula:

$$\pi = \int_\Gamma \partial/\partial n(\ln R)\mathrm{d}s,$$

and it should be obvious how to compute the integrals b_{ii}. The system of

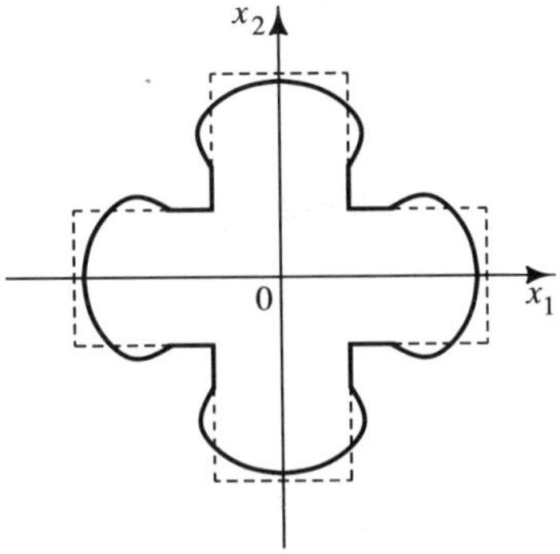

FIGURE 4.2. Torsioned shaft (BIEM algorithm).

algebraic equations is solved by using an iteration technique. Since the deplanar function is a solution of the Neumann problem, it can be determined only up to an arbitrary constant.

The coordinates of the vertices of the polygonal boundary corresponding to the endpoints of the line segments γ_i are written in the form $x_i = t_i \cos \alpha_i$ and $y_i = t_i \sin \alpha_i$, where t_i is the length of the radius vector and α_i are the polar angular coordinates of the junction points. The junction points (i.e., the vertices) are displaced along directions determined by these angles. The values of the parameters t_i may be improved using (4.88), and are assigned variations

$$\delta t_i = \varepsilon\left(-1 + \left(\tau_s \int_{\Gamma_v} ds \Big/ \int_{\Gamma_v} \tau_s ds\right)\right),$$

$$\tau_s = \partial w/\partial s + x_1 n_1 + x_2 n_2,$$

where ε is a "small" number. Varying the shape of Ω and computing the effects of torsion will continue until the norm of the error indicating noncompliance with the optimality criteria is sufficiently small. This algorithm was used in computing several optimal shapes of torsioned shafts. The results of one such computation are illustrated in Figure 4.2. The broken lines indicate the initially assumed shape, while the solid line describes the shape obtained from computations in accordance with this algorithm.

References for Chapter 4

4.1. Banerji, P.K., and Butterfield, R., *Boundary Element Methods in Engineering Science*, McGraw Hill, New York, 1981.

4.2. Banichuk, N.V., on a variational problem with an unknown boundary, and determination of the optimal shapes of elastic bodies. *Prikl. Matematika* Mekhan., 1975, *39*, No. 6, pp. 1082–1092.

4.3. Banichuk, N.V., Optimization of elastic bars in torsion, *Intern. J. Solids Struct.*, 1976, *42*, No. 4, pp. 275–286.

4.4. Banichuk, N.V., Application of the perturbation method to optimal design of structures. In: *New Directions in Optimal Design*; Atrek E. et al., (eds.) Chichester, Wiley & Sons, 1984, pp. 231–248.

4.5. Banichuk, N.V., *Metody perturbacyjne w optymalizacji konstrukcji* (Perturbation techniques in structural optimization), Warszawa, Ossolineum, 1983, pp. 13–35.

4.6. Banichuk, N.V., Bel'skii, V.G., and Kobelev, V.V., Investigation of the optimal shapes of elastic bodies. In: *Applied Problems of Elasticity and Plasticity*, Gorkii, Gorkii University Press, 1984, pp. 64–70.

4.7. Banichuk, N.V., Kartvelishvili, V.M., and Mironov, A.A., Methods of successive optimization for numerical solution of minimax problems in optimal designs of structures. In: *Numerical Techniques of Nonlinear Programming*, Lectures of the Second All-Union Seminar, Kharkov, 1976, pp. 54–59.

4.8. Banichuk, N.V., Kartvelishvili, V.M., and Chernous'ko, F.L., On finite difference quadratic approximations for convex integral functionals. *D.A.N. U.S.S.R.*, 1976, *231*, No. 2, pp. 269–272.

4.9. Banichuk, N.V., and Larichev, A.D., Optimization of beams taking into account the elasticity of supports. *Izv. Akad. Nauk Arm.S.S.R., Mekhanika*, 1980, *33*, No. 6, pp. 65–72.

4.10. Banichuk, N.V., and Mironov, A.A., Optimal design of plates, and dynamic problems of hydro-elasticity. In: *Proceedings of the Tenth All-Union Conference on the Theory of Shells and Plates.* Tbilisi, Matsniereba Publ., 1975, pp. 35–44.

4.11. Bel'skii, V.G., Computer assisted computation of shapes of two-dimensional bodies, *Izv. V.U.Z. Konstruk.* Arkhitec., 1983, No. 3, pp. 23–27.

4.12. Bratus, A.S., and Kartvelishvili, V.M., Approximate analytic solutions of problems in optimization of stability and frequency of vibrations for elastic thin-walled structures. *Izv. Akad. Nauk U.S.S.R., MIT*, 1981, No. 6, pp. 119–139.

4.13. Brebbia, K., and Walker, S., *Boundary Element Techniques in Engineering*, Newnes-Butterworth Publ., Boston, Mass. 1980.

4.14. Chun, I.W., and Haug, E.J., Two-dimensional shape optimal design, *Intern. J. Numer. Meth. Eng.*, 1978, *13*, pp. 311–336.

4.15. Ciarlet, F.G. *Applications of the Finite-Element Method to Elliptic Problems.* (in Russian), Mir, Moscow, 1980, 512 pp.

4.16. Dorodnitsin, A.A., Utilization of small parameter technique for numerical solutions of equations of mathematical physics, Lectures in the Summer school on Numerical Methods, Kiev, 1966, V.Ts. Akad. Nauk SSSR, Moscow, 1966, pp. 85–100.

4.17. Gallagher, R.H., *Finite Element Analysis: Fundamentals.* Prentice-Hall, Englewood Cliffs, New Jersey, 1973.

4.18. Haug, E.J., A gradient projection method for structural optimization. In: *Optimization of Distributed Parameter Structures*, Haug, E.J. and Cea, J. (eds.) Sijthoff and Noordhoff, Alphen an den Rijn, 1981, pp. 446–473.

4.19. Haug, E.J., Choi, K.K., and Komkov, V. *Structural Design Sensitivity Analysis*, Academic Press, New York, 1986.

4.20. Haug, E.J., Choi, K.K., Hou, J.W., and Yoo, Y.M., A variational method for shape optimal design of elastic structures. In: *New Directions in Optimal Design*; E. Atrek et al., (eds.) Chichester, Wiley, 1984, pp. 231–248.

4.21. Haug, E.J., Pan, K.C., and Streeter, T.D., Computational method for optimal structural design I: Piecewise uniform structures, *Intern. J. Numer. Met. Eng.*, 1972, *5*, pp. 171–184.

4.22. Haug, E.J., Pan, K.C., and Streeter, T.D., Computational method for optimal structural design II: Continuous problems, *Intern. J. Numer. Meth. Eng.*, 1975, *9*, pp. 649–667.

4.23. Haug, E.J., and Komkov, V., Sensitivity analysis in distributed parameter mechanical systems optimization, *J. Optimiz. Theory Appl.*, 1977, *23*, No. 3, pp. 445–464.

4.24. Oden, J.T., *Finite Elements in Nonlinear Continua*, MacGraw-Hill, New York, 1972, (Russian translation, Mir, Moscow, 1976, 464 pp.).

4.25. Pironneau, O., *Optimal Shape Design for Elliptic Systems*. Springer-Verlag, New York and Berlin, 1984, 168 pp.

4.26. Queau, I.P., and Trompette, P.H., Two-dimensional shape optimal design by finite element method, *Intern. J. Numer. Meth. Eng.* 1980, *15*, pp. 1603–1612.

4.27. Ricketts, R.E., and Zienkiewicz, O.C., Shape optimization of continuum structures. In: *New Directions in Optimal Design*; E. Atrek et al. (eds.), Chichester, Wiley, 1984.

4.28. Seyranian, A.P., An optimization problem concerning the effectiveness of an aileron, *Izv. Akad. Nauk Arm. SSR, Mekhan.*, 1980, *33*, No. 1, pp. 54–64.

4.29. Seyranian, A.P., Quasi-optimal solutions to the design problems with multiple constraints, *Prikl. Mekhanika*, 1977, No. 6, pp. 18–26.

4.30. Strang, G., and Fix, G.J., *An Analysis of the Finite Element Method*. Prentice Hall, Englewood Cliffs, New Jersey, 1973.

4.31. Tada, Y., and Seguchi, Y., Shape determination of structures based on the inverse variational principle: The finite element approach. In: *New Directions in Optimal Design*; E. Atrek et al. (eds.), Chichester, Wiley, 1984, pp. 197–209.

4.32. Washizu, K., *Variational Methods in Elasticity and Plasticity*, Pergammon, Oxford, 1975, 412 pp.

4.33. Wu, C.H., The strongest circular arch—A perturbation solution. *J. Appl. Mech., Trans. ASME*, 1968, *35*, No. 3, pp. 476–480.

4.34. Yoo, Y.M., Haug, E.J., and Choi, K.K., Shape optimal design for multiply loaded structural components. In: *Optimization Methods in Structural Design*, Zurich, Bibliogr. Institute, 1983, pp. 250–255.

4.35. Hou, J.W., Haug, E.J., and Benedict, R.L., Shape optimization of elastic bars in torsion, In: *Sensitivity of Functionals with Applications to Engineering Sciences*, Komkov, V. (ed.), Springer Verlag Lecture Notes in Mathematics Series, No. 1086, Berlin, 1984, 1984, pp. 31–55.

4.36. Mota-Soares, C.A., and Choi, K.K., Boundary elements in shape optimal design of structural components, In: *Computer Aided Optimal Design, Structural and Mechanical Systems*, Mota-Soares, C.A. (ed.), Springer Verlag, Berlin 1987, (NATO ASI Series F, Vol. 27), pp. 605–631.

5 Techniques for Optimization of Discrete Systems

5.1 Fundamental Concepts in Optimization Theory for Discrete Systems

Certain techniques widely used in the design of complex structures are founded on a finite-element description of the structure combined with the use of well-established techniques of mathematical programming and sensitivity analysis (cf. [2], [10], [13], [24], [28], [29], and [39]). Furthermore, there exist general techniques for formulating optimal design projects and reducing them to problems of algebra. The creation of finite-dimensional models for structures is discussed in [5.9], [5.10], [5.26], and [5.27].

Let a structure be completely described by the values of the design parameters h_i $(i = 1, 2, \ldots, n)$, which can be regarded as components of a vector $\mathbf{h} = \{h_1, h_2, \ldots, h_n\}$. We need to choose the design variables in such a manner that the minimum of the cost functional is attained,

$$J(\mathbf{h}) \to \min \tag{5.1}$$

when the constraints

$$\psi_j(\mathbf{h}) \leq 0, \qquad j = 1, 2, \ldots, k \tag{5.2}$$

are satisfied. Here $J(\mathbf{h})$ and $\psi_j(\mathbf{h})$ are known functions of the design variables. There are many articles dealing with problems of parametric optimization using the notation of (5.1) and (5.2) e.g., [24], [41], [5.4], [5.16], [5.25], and [5.28].

The problem (5.1) and (5.2) represents the extremization of a functional with certain inequality constraints. The presence of the inequality constraints raises the degree of difficulty in finding the optimal solution. Therefore, it is advisable to replace inequality-type constraints by equality constraints.

We can introduce auxiliary (slack) variables α_j^2 $(j = 1, 2, \ldots, k)$, so that the inequality constraints are replaced by the constraints

$$\psi_j + \alpha_j^2 = 0. \tag{5.3}$$

After introducing the slack variables, the optimization problem consists of finding the values of the $n + k$ quantities h_i, α_j (instead of the n quantities h_i) using the minimizing condition for the cost function (5.1), together with the constraints (5.3). We apply the classical Lagrangian multiplier technique directly to problem (5.1) and (5.3). Denoting by λ the vector of Lagrangian multipliers, we introduce the Lagrangian function

$$\Psi(\mathbf{h}, \boldsymbol{\alpha}, \boldsymbol{\lambda}) = J + \sum_{j=1}^{k} \lambda_j(\alpha_j^2 + \psi_j).$$

Directly differentiating Ψ with respect to h_i, λ_j, and α_j, we obtain a system of $(n + 2k)$ algebraic equations:

$$\partial\Psi/\partial h_i = \partial J/\partial h_i + \sum_{j=1}^{k} \lambda_j(\partial\psi_j/\partial h_i) = 0, \qquad i = 1, 2, \ldots, n, \tag{5.4}$$

$$\partial\Psi/\partial\lambda_j = \psi_j + \alpha_j^2 = 0, \qquad j = 1, 2, \ldots, k, \tag{5.5}$$

$$\partial\Psi/\alpha_j = 2\lambda_j\alpha_j = 0, \qquad j = 1, 2, \ldots, k, \tag{5.6}$$

which may be solved for the $(n + 2k)$ unknown quantities h_i, λ_j, and α_j.

This approach to optimal design (introduction of slack variables) increases the dimensionality of the problem. A somewhat more effective (in this sense) approach to this problem consists in applying the Kuhn–Tucker optimality condition. Before we discuss this condition, let us write out condition (5.20) in more detail and offer some definitions that are essential to the subsequent development.

First, let us separate the inequalities (5.2) into equality and "genuine" inequality conditions

$$\psi_j(\mathbf{h}) = 0, \qquad j = 1, 2, \ldots, r, \tag{5.7}$$

$$\psi_j(\mathbf{h}) \leq 0, \qquad j = r + 1, \ldots, k. \tag{5.8}$$

The set Ω of all "points" $\mathbf{h}$ satisfying conditions (5.7) and (5.8) shall be called the admissible set of designs. If the function $J(\mathbf{h})$ is continuous in the admissible set Ω and the set Ω closed and bounded, then it is known that the corresponding problem of nonlinear programming has a unique solution, which assigns an absolute minimum to the functional $J(\mathbf{h})$ on the admissible set Ω (cf. [24]).

The problem (5.1), (5.7), and (5.8) is called a convex programming problem if the set Ω is convex and $J(\mathbf{h})$ is a convex function. In a convex programming problem, a local minimum turns out to be the absolute minimum.

The point $h^0 \in \Omega$ is called a regular point of Ω if the function $J(h^0)$ is differentiable at $h = h^0$ and if all gradients $\nabla\psi_j(h^0)$ that vanish at the point $h = h^0$ (i.e., $\nabla\psi_j(h^0) = 0$) are linearly independent for $j = 1, 2, \ldots, k$. Using these definitions we can now formulate the Kuhn–Tucker optimality condition.

In the problem (5.1), (5.7), and (5.8) with differentiable functions $J(\mathbf{h})$, $\psi_i(z)$, for a regular point h^0 in Ω, which is a local minimum of the cost functional, there exist Lagrangian multipliers λ_i, $i = 1, 2, \ldots, k$, such that

$$\lambda_i \geq 0, \qquad i = r + 1, \ldots, k, \tag{5.9}$$

$$\lambda_i \psi_i(h^0) = 0, \qquad i = r + 1, \ldots, k, \tag{5.10}$$

$$\nabla J(h^0) + \sum_{i=1}^{k} \lambda_i \psi_i(h^0) = 0. \tag{5.11}$$

The Kuhn–Tucker conditions (5.9), (5.10), and (5.11) are extremely important in nonlinear programming. Note that the nonnegative multipliers λ_i that appear in these conditions correspond to inequality constraints, while equality constraints admit negative, 0, or positive multipliers.

In order to conduct sensitivity analysis and to apply in successive optimization computations, it is convenient to reformulate the optimal design problem and to separate the unknoqn parameters into two distinct classes.

One class consists of the design parameters h_i, $i = 1, 2, \ldots, n$, while the other contains the state parameters u_i, $i = 1, 2, \ldots, m$. The design problem now assumes the form

$$J(\mathbf{u}, \mathbf{h}) \to \min, \tag{5.12}$$

$$\psi_i(\mathbf{u}, \mathbf{h}) = 0, \qquad i = 1, 2, \ldots, r, \tag{5.13}$$

$$\psi_i(\mathbf{u}, \mathbf{h}) \leq 0, \qquad i = r + 1, \ldots, k. \tag{5.14}$$

Here u is an m-dimensional vector of state variables, h, an n-dimensional vector of design variables, and $J(\mathbf{u}, \mathbf{h})$ and $\psi_i(\mathbf{u}, \mathbf{h})$ (with $i = 1, 2, \ldots, k$) are given functions. All the functions are differentiable. We also demand that the (higher-dimensional) vectors $\{\partial\psi_i/\partial\mathbf{u}, \partial\psi_i/\partial\mathbf{h}\}$ should be linearly independent for all values of the index i for which $\psi_i(\mathbf{u}, \mathbf{h}) = 0$ and the matrix $\partial\psi_i/\partial u_j$, $i = 1, 2, \ldots, r$, $j = 1, 2, \ldots, m$, should have maximal rank. We shall consider the problem of optimal design from the viewpoint of (5.12) to (5.14) in Sections 5.5 to 5.7 below. Note that the optimal design problem described in (5.12) to (5.14) can be put into a typical linear programming form by introducing a new vector variable $\mathbf{z} = \{h_1, h_2, \ldots, h_n, u_1, \ldots, u_m\}$.

5.2 Techniques Based on Optimality Criteria

A whole class of techniques based on optimality criteria can be found in the literature on optimal design (see [5.18], [5.19], and [5.28]). Typically, in employing such techniques, one first derives an optimality condition, applying techniques that range from rigorous mathematical proofs to purely intuitive arguments. Next, a computational algorithm is derived, which may deliberately alter the actual project, in order to comply with the optimality criterion. We remark that it would make better sense to use the terminology *optimality conditions* rather than *optimality criteria.* However, this terminology is commonly used in the literature, and we may just as well continue its use.

Techniques using the optimality criteria belong to the class of indirect optimization techniques. Mathematically, these "optimization criteria," or conditions for optimality, are equivalent to the Kuhn–Tucker condition of nonlinear programming. These conditions lead to iterative procedures in optimization algorithms because of the nonlinearity of the problem, and in particular, the nonlinearity of the constraints with respect to the design variables. The computations performed at each iterative step may be divided into two parts. In the first part, we perform direct structural computations and determine the response of the structure under prescribed loads. In the second part, we modify the design variables using recurrent relations derived from the optimality criteria. If the internal forces do not depend on the thickness or any other parameter describing the design of its elements, or at least are almost independent of the design, as is the case with the great majority of well-designed structures, after only a few iterations in which changes are made in the size or the shape, one arrives at an optimal, or near-optimal design (see [5.19]).

The number of iterations necessary to arrive at such optimal, or near-optimal design, is virtually independent of the number of elements comprising the structure. This fact allows us to select techniques that are advantageous in dealing with very complex structures. However, if stresses in structural members are sensitive to size changes in the structural elements, as is the case with certain trusses, a larger number of iterations may be necessary to assure convergence of the algorithm (to near-optimal values).

In the process of solving a design problem, the variation of the variables obeys appropriate recurrence relations. These relations treat the gradients of constraints and Lagrangian multipliers as unknown quantities. These gradients can be estimated from variational principles, while explicit formulas can be derived for the Lagrangian multipliers. In problems with a large number of constraints and in nonlinear problems this is, in general, impossible.

The application of techniques founded on necessary conditions for optimality is particularly effective in problems having only a single active constraint.

Let us begin the discussion with problems having constraints assigned to the strength. Such constraints are very important in structural optimization, played a crucial role in early developments in optimization theory, and are tied to the concept of a fully stressed structure. In designing a fully stressed structure, one adopts the principle that each structural element must carry the maximum admissible stress at least for one admissible system of loads.

Perhaps it would be of interest if we explain the relation between a fully stressed structure and a structure having the smallest weight. In early research on this topic, authors would frequently fail to distinguish between structures having minimal weight and fully stressed structures. In later research articles it was finally agreed that fully stressed structural designs imply minimal weight only if the structure is statically determinate and subjected to a single concentrated load. Note that the insufficiency of the fully stressed condition to assure minimization of weight was first discussed by Gallagher in [5.9]. Nevertheless, because of many advantages of this approach and its simplicity, design techniques using the fully stressed structure condition are applied to statically indeterminate structures and to structures subjected to multiple loads.

If one applies this technique to statically indeterminate structures, one assumes implicitly that the internal forces in all of the elements are insensitive to changes in the sizes of these elements, that is, the relations between the magnitudes of stresses and values of geometric parameters are ignored. As a result, many numerical algorithms applied to such structures not only fail to converge to a correct solution, but also produce utterly nonsensical stress distributions.

Let us now describe the main subject of this section (i.e., techniques based on necessary optimality criteria) applied to a problem with only a single constraint. Let $j = d$ signify the index of the dominant constraint at the optimal point.

Then $\psi_d = 0$, and from (5.5) we deduce that $\alpha_d = 0$. Since the other constraints are not critical, it follows that $\alpha_j \neq 0$ if $j \neq d$. Consequently, the classical formulation of (5.4) to (5.6), which determine the design variables, Lagrangian mutlipliers, and slack variables, leads to

$$\psi_d = 0, \qquad \partial J/\partial h_i - \lambda_d \partial\psi_d/\partial h_i = 0, \qquad i = 1, 2, \ldots, n. \tag{5.15}$$

The system of conditions (5.15) can be written as a single vector equation, $\Delta J = \lambda_d \nabla\psi_d$, which implies that for an optimal design with a single active constraint the gradient of the quality functional is proportional to the gradient of the constraint.

This system of relations can be rewritten as

$$\left[\frac{\partial\psi_d}{\partial h_i}\bigg/\frac{\partial J}{\partial h_i}\right] = \frac{1}{\lambda_d}. \tag{5.16}$$

By (5.16), the ratio of the rate of change of the constraint function and the rate of change of the quality functional are the same for every design variable. Thus, all the design parameters in an optimization project may be thought of as having a common rate of change.

In the theory of optimal design of structures, one frequently encounters the condition or "criterion" of constant energy density distribution throughout the structure. Using this criterion it is possible to establish optimal solutions for certain types of structures or for single structural elements.

Using this condition in solving minimal weight problems with constraints assigned to the compliance or maximization of an integral rigidity functional with given total weight has been justified in a number of articles, where the condition of constant energy density has been conceived as a heuristic law.

Note that in such assertions, such a "law" is a direct consequence of (5.16) if one assumes that there exists a linear relation (i.e., linear with respect to the design variables) between the quality criterion functional and the constraint functions.

Let us now consider a problem of optimal design with several inequality constraints assigned to the design variables:

$$J = \sum_{i=1}^{n} \rho_i l_i h_i = \min_{h_i}, \tag{5.17}$$

$$\psi_j = \chi_j - \chi_j^0 \leq 0, \qquad j = 1, 2, \ldots, k, \tag{5.18}$$

where ρ_i, l_i, and h_i are, respectively, the density of the material and the length and cross-sectional area of the ith element; here χ_j are known fuctions of the design variables $h_1, \ldots, h_n$, while χ_j^0 are assigned constants.

In this case the Kuhn–Tucker conditions (5.9) to (5.11) become

$$\left.\begin{aligned} &\sum_{j=1}^{k} [\lambda_j/(\rho_i l_i)]\partial\chi_j/\partial h_i = -1, \qquad i = 1, \ldots, n, \\ &\lambda_j \geq 0, \qquad \lambda_j \psi_j = 0. \end{aligned}\right\} \tag{5.19}$$

The optimality condition (5.19) implies that for all elements of the structure, the ratio of the sum of the derivatives of the constraint on the design variables to the derivative of the quality criterion remains the same. Relations (5.18) and (5.19), which can be used to derive the $(n + k)$ values of the variables $h_1, \ldots, h_n, \lambda_1, \ldots, \lambda_k$, are nonlinear. Consequently, the computation of the values of these variables is usually accomplished by iterative techniques.

In the simplest problems of designing statically determinate structures, the design variables may be directly given in terms of the optimality criteria. In more difficult cases of optimal design of statically indeterminate structural systems, methods based on recurrence relations have been developed and are now in common use (cf. [5.19]). We shall present some of these techniques. Let us assume that in the study of an optimization problem we find optimization criteria written in the form $1 = \varphi_i$, $i = 1, 2, \ldots, n$, with φ_i denoting the same expression for all elements of the structure. Then we can employ a simple device to establish the recurrence relation. It consists in multiplying both sides of the equation $1 = \varphi_i$ by $(h_i)^r$ and extracting the rth root of this product.

As a result, we obtain the recurrence relation

$$(h_i)^{v+1} = (h_i)^v[(\varphi_i)^v]^{1/r}, \tag{5.20}$$

where v is the iteration counter. The term φ_i^v in (5.20) denotes the value of φ_i computed on the vth iteration, while h_i^v and h_i^{v+1} denote the values of the design variable h_i on the vth and $v + 1$st iteration, respectively. In the last few iterations of the iterative process generated by (5.20), when the quantity φ_i is almost equal to 1, the values of the design variables should be close to some constant value.

Rewriting eq. (5.20) in the form

$$h_i^{\nu+1} = h_i^{\nu}(1 + [\varphi_i^{\nu} - 1])^{1/r}, \tag{5.21}$$

we assume that the values of the design variables differ by only a small amount from their optimal values. Close to the optimum design the value of $[\varphi_i - 1]$ is small compared to unity, and consequently the expression on the right-hand side of (5.21) may be expanded in series of powers of this quantity. Retaining only terms of order not higher than 1, we obtain a linear recurrence expression, which we can solve, thus obtaining the value of the design variable:

$$h_i^{\nu+1} = h_i^{\nu}\{1 + 1/r[\varphi_i^{\nu} - 1]\}. \tag{5.22}$$

The quantity $[\varphi_i^{\nu} - 1]$ in the recurrence formula (5.22) represents the error of the compliance with the criterion $\varphi_i = 1$.

We could also introduce the following slightly different formalism for the recurrence relation (5.22):

$$h_i^{\nu+1} = h_i^{\nu} + \Delta h_i^{\nu}, \qquad \Delta h_i = ((1/r)h_i^{\nu}[\varphi_i^{\nu} - 1]).$$

In research articles on optimization of structures one also finds other froms of the recurrence formulas, such as those using the inverses of the values of the design variables $z_i = 1/h_i$. After introducing the variables z_i, eq. (5.22) becomes $z_i^{\nu+1} = z_i^{\nu}[1 - (1/r)(\varphi_i^{\nu} - 1)]$. If in this formula we eliminate z_i and again substitute h_i, we arrive at the recurrence relation $h_i^{\nu+1} = h_i^{\nu}[1 - (1/r)(\varphi_i^{\nu} - 1)]^{-1}$. This formula has been used in articles dealing with design theory.

5.3 Design of Structures with Single-Type Constraints Assigned to the Structural Elements

In many design problems dealing with complex structures, the optimal design is derived by considering only a single (critical) constraint assigned to each structural element. In other problems of optimal structural design only a single type of constraint is taken into account, but later we have to check different types of constraints, and as a result of such correction we may arrive at a design that is close to the optimal, the "quasi-optimal design." For this reason, it is important to study techniques of structural optimization with only a single type of constraint, for example, a constraint on the maximal stress or the displacement.

Let us consider the problem of optimizing the weight of a structure composed of rods with constraints on the magnitudes of stresses in each element, with the number of constraints matching the number of structural elements. In more complex structures that contain beams, plates, and membranelike elements, the number of constraints will exceed the number of structural elements. We shall display the constraints on the stress σ_i present in the ith element as follows:

$$\psi_i = \sigma_i - \sigma_i^0 \leq 1, \quad \text{where } \sigma_i = T_i/h_i, \qquad i = 1, 2, \ldots, k,$$

where σ_i^0 denotes the maximum admissible stress, k is the number of elements in the structure, and T_i is the tensile or compressive stress in the ith member. Differentiating ψ_i with respect to h_i and substituting the first relation from (5.19) we obtain

$$\rho_i l_i - \lambda_i(T_i/h_i^2) + (1/h_i) \sum_{j=1}^{k} \lambda_j(\partial T_j/\partial h_i) = 0. \tag{5.23}$$

In the general case, the derivatives $\partial T_i/\partial h_j$ may be found by performing a sensitivity analysis, which includes computations of the stress-deformed state of the structure. Statically determinate structures constitute an exceptional case. Here the forces in each member are independent of the cross-sectional area, and all derivatives mentioned above are identically equal to 0.

However, in studying statically indeterminate structures we could make the additional assumption that forces depend only "slightly" on the magnitudes of the cross-sectional areas and set $\partial T_i/\partial h_j = 0$. In that case the summand in (5.23) (where the number of elements in that sum is equal to the number of active constraints, and does not exceed the number of members) is equal to 0. Therefore,

$$\rho_i l_i - \lambda_i(T_i/h_i^2) = 0. \tag{5.24}$$

If the stresses in each member attain the maximum admissible value (i.e., if $\sigma_i = \sigma_i^0$), the Lagrangian multipliers λ_i obey the formula

$$\lambda_i = \rho_i l_i h_i/\sigma_i^0,$$

while the optimality conditions (5.24) become

$$1 = T_i/(h_i \sigma_i^0). \tag{5.25}$$

Condition (5.25) represents the "fully stressed" criterion for the design project.

Note that fully stressed designs for statically indeterminate structures substantially differ from optimal structural designs if the sensitivity of the forces values with respect to changes in cross-sectional areas is not small.

Let us now discuss a class of structures designed for minimum weight, where the only integral constraint represents the rigidity:

$$\psi = \sum_{i=1}^{k} (\Pi_i/h_i) - \chi^0 = 0. \tag{5.26}$$

Here h_i is the cross-sectional area of the ith member, and Π_i/h_i is the energy density of the elastic deformation. We shall limit our discussion to the statically determinate case. In that case Π_i is independent of the design variables h_i, and as a result the partial derivatives of the function ψ with respect to h_j can be evaluated immediately: $\partial\psi/\partial h_i = -\Pi_i/h_i^2$. The optimality condition (5.19) may be rewritten as $(\lambda/l_i\rho_i)(\Pi_i/h_i^2) = 1$.

If we denote by e_i the energy density in an elastic deformation of the ith member ($e_i = \Pi_i/l_i h_i^2$) and determine the Lagrangian multiplier λ from (5.26), the optimality condition assumes the form

$$(e_i/\chi^0\rho_i)\sum_{j=1}^{k}\rho_j l_j h_j = 1. \tag{5.27}$$

From the optimality condition (5.27), the ratio of the energy density of an elastic deformation to the mass density is constant and exactly the same for all elements of the structure.

In optimizing the weight of a statically determinate structure, let us now demand that the displacements of the ends of the individual elements satisfy certain constraints. Such bounds on rigidity may be written as inequalities:

$$\psi_j = \sum_{i=1}^{n}(S_{ij}/h_i) - \chi_j^0 \leq 0, \qquad j = 1, 2, \ldots, k, \tag{5.28}$$

where S_{ij} are the compliance coefficients and χ_j^0 are given constants. The coefficients S_{ij} are independent of $h_i - s$, and therefore, the computation of the derivatives of the constraint functions with respect to the design variables is given by the formulas $\partial\psi_j/\partial h_i = -S_{ij}/h_i^2$ for $j = 1, 2, \ldots, k$, $i = 1, 2, \ldots, n$. Substituting these expressions for the derivatives of the constraint functions in (5.19), we derive

$$\left.\begin{aligned}&\sum_{j=1}^{k}(\lambda_j S_{ij})/(\rho_i l_i h_i^2) = 1, \qquad i = 1, 2, \ldots, n,\\&\lambda_i \geq 0, \qquad \lambda_j\psi_j = 0.\end{aligned}\right\} \tag{5.29}$$

Criteria (5.26) and (5.29) are obtained under the assumption that the structure is statically determinate. However, in practical applications, it has been found that satisfactory results are obtained using (5.26) and (5.29) in many cases of statically indeterminate structures, where typically there was only a weak dependence of force values on the cross-sectional areas.

5.4 Iterative Techniques for Designing Thin-Walled Structures with Discretized Equal Strength

In the practice of designing complex structures, most engineers use purely heuristic approaches (and criteria) that rely on the accumulated experience of optimizing some prototypes. Designs using discretized equal strength criterion reflect satisfactorily intuitive engineering approaches. Design techniques reflecting the discretized equal strength approach to structural systems are discussed in [28], [5.22] and [5.23].

Let us consider the problem of designing a thin-walled compound system with criteria concerning the weight and strength, with several variants of applied loads. Let us assume that the structure is subjected to static loads and consider k separate forces applied to it. The design variables are chosen so that the weight of the structure (which is to be minimized) depends linearly on the design parameters $h_i, i = 1, 2, \ldots, n$. We denote by $\psi_{ij}(h_1, \ldots, h_j, \ldots, h_n)$ a quantity that characterizes the stress intensity in the jth element (with $1 \leq j \leq n$) under the action of the ith load ($1 \leq i \leq k$).

The strength criterion has the form

$$\psi_{ij} \leq \sigma_{j^*}, \qquad i = 1, 2, \ldots, k, \qquad j = 1, 2, \ldots, n. \tag{5.30}$$

The symbol σ_{j^*} in inequality (5.30) denotes the bound on the admissible values of the stress intensity for the material of the jth element. The index i assumes all possible values $i = 1, 2, \ldots, k$.

Because the constant on the right-hand side of (5.30) depends on j, we can consider structures whose members are made from different materials.

It shall be said that a structure has discretely uniform strength if it is true that for each individual member the design variables obey the condition

$$\max_i \max_{h_j} \psi_{ij}(h_1, \ldots, h_j, \ldots, h_n) = \sigma_{j^*}. \tag{5.31}$$

For statically determinate structures, it is well known (e.g., [28]) that a structure with discretely uniform strength also has the least weight in the class of structures with a given constraint assigned to the magnitude of admissible stresses, such as (5.30).

The determination of the values of the design parameters for a structure with discretely uniform strength now reduces (as can be seen from (5.31)) to solving the system of equations

$$\mathbf{g}(\mathbf{h}) = \{g_1(\mathbf{h}), \ldots, g_n(\mathbf{h})\} = 0, \tag{5.32}$$

where $g_j(\mathbf{h}) = \max_i \max_j \{(\sigma_{j^*})^{-1} \psi_{ij}(h_1, \ldots, h_j, \ldots, h_n)\} - 1$. In computing these values, we use a method of successive approximations. An exact solution of the system of equations (5.32) may be written as a sum of two terms,

$$\mathbf{h} = \mathbf{h}^s + \xi^s, \tag{5.33}$$

where $\mathbf{h}^s$ is the value of $\mathbf{h}$ after the sth iteration and ξ^s is the error in the sth approximation. The functions $g_j(\mathbf{h})$ computed for the sth value of the dependent variable given in (5.33) may be expanded in Taylor's series. We retain only the linear terms

$$\mathbf{g}(\mathbf{h}^s + \xi^s) = \mathbf{g}(\mathbf{h}^s) + \mathbf{g}'(\mathbf{h}^s)\xi^s. \tag{5.34}$$

Now using (5.34) we obtain

$$\xi^s = -[\mathbf{g}'(\mathbf{h}^s)]^{-1} \mathbf{g}(\mathbf{h}^s).$$

The iteration scheme used for finding the values of the design variables for a discrete–equal strength system employs the formula

$$\mathbf{h}^{(s+1)} = \mathbf{h}^s - [\mathbf{g}'(\mathbf{h}^{(s)})]^{-1} \mathbf{g}(\mathbf{h}^{(s)}). \tag{5.35}$$

In computing statically indeterminate systems one discovers the relation

$$|\partial g_j / \partial h_j| > |\partial g_l / \partial h_j| \quad \text{if } j \neq l, \tag{5.36}$$

implying that if one alters the value of the design variable for the jth member, this would cause a substantive change in the magnitude of the stress intensity for that member. Therefore, to speed up convergence using (5.35), we could introduce the simplification

$$|\partial g_l/\partial h_j| = 0 \quad \text{if } j \neq l \quad \text{for } \mathbf{h} = \mathbf{h}^s. \tag{5.37}$$

For statically determinate structures, condition (5.37) must be strictly obeyed. However, condition (5.36) is applied to a broad class of thin-walled structures (including statically indeterminate structures) and produces quite satisfactory accuracy.

Substituting (5.37) in (5.35), we obtain a modification of (5.35):

$$h_j^{(s+1)} = h_j^{(s)} - g_j(\mathbf{h})[\partial g_j(\mathbf{h})/\partial h_j]^{-1}, \tag{5.38}$$

with $[\partial g_j(\mathbf{h})/\partial h_j] < 0$. The second term on the right-hand side of (5.38) is computed for the current value: $\mathbf{h} = \mathbf{h}^{(s)}$.

5.5 Sensitivity Analysis for Structural Design

We shall examine the problem of optimal design formulated in Section 5.1 (see also [24]):

$$J(\mathbf{u}, \mathbf{h}) \to \min, \tag{5.39}$$

$$\psi_i(\mathbf{u}, \mathbf{h}) = 0, \qquad i = 1, 2, \ldots, m, \tag{5.40}$$

$$\psi_i(\mathbf{u}, \mathbf{h}) \leq 0, \qquad i = m + 1, \ldots, k. \tag{5.41}$$

Here $\mathbf{u} = \{u_1, \ldots, u_m\}$ is an m-dimensional vector of state variables, and $\mathbf{h} = \{h_1, \ldots, h_n\}$ is an n-dimensional vector of design variables.

We shall examine the effect of a variation of the design variables $\delta\mathbf{h}$ on the properties of the structure. Hence, we shall observe both the state and the design variables: $\mathbf{u}$ and $\mathbf{h}$, and their perturbed values $\mathbf{u} + \delta\mathbf{u}$, $\mathbf{h} + \delta\mathbf{h}$. We linearize the relations between the variations of the quality functional, and the constraints and the vectors $\delta\mathbf{u}$ and $\delta\mathbf{h}$:

$$\begin{gathered} \delta J = (\partial J/\partial \mathbf{u})\delta\mathbf{u} + (\partial J/\partial \mathbf{h})\delta\mathbf{h}, \\ \delta\psi_i = (\partial\psi_i/\partial\mathbf{u})\delta\mathbf{u} + (\partial\psi_i/\partial\mathbf{h})\delta\mathbf{h}, \\ i = 1, 2, \ldots, k. \end{gathered} \tag{5.42}$$

The linearized equations that correspond to (5.40) have the general form

$$A\delta\mathbf{u} + B\delta\mathbf{h} = 0, \tag{5.43}$$

$$A = \begin{bmatrix} \partial\psi_1/\partial u_1, \ldots, \partial\psi_1/\partial u_m \\ \text{-----------} \\ \text{-----------} \\ \text{-----------} \\ \partial\psi_m/\partial u_1, \ldots, \partial\psi_m/\partial u_m \end{bmatrix}, \qquad B = \begin{bmatrix} \psi_1/\partial h_1, \ldots, \partial\psi_1/\partial h_n \\ \text{-----------} \\ \text{-----------} \\ \text{-----------} \\ \partial\psi_m/\partial h_1, \ldots, \partial\psi_m/\partial h_n \end{bmatrix}.$$

The quantities $\partial J/\partial \mathbf{u}$, $\partial J/\partial \mathbf{h}$, $\partial \psi_i/\partial \mathbf{u}$, and $\partial \psi_i/\partial \mathbf{h}$, which occur in (5.42) and (5.43), as well as the matrices A and B, must be computed for the unperturbed values of the arguments. Matrix A is assumed to be nonsingular. Therefore, we can solve this system with respect to $\delta \mathbf{u}$ and represent the variation of the state function in the form

$$\delta \mathbf{u} = -A^{-1} B \delta \mathbf{h}, \tag{5.44}$$

where, as usual, the superscript -1 denotes the inverse of a matrix.

We can substitute (5.44) in (5.42) and express the variations δJ, $\delta \psi_i$ $(i = m + 1, \ldots, k)$, in terms of $\delta \mathbf{h}$. At least in principle, we succeed by using this approach. However, a different approach for eliminating the dependence of these functionals on $\delta \mathbf{u}$ has found very wide acceptence. Let us discuss this technique. First, let us observe that $\delta \mathbf{u}$ occurs in all fundamental relations through expressions $(\partial J/\partial \mathbf{u})\delta \mathbf{u}$ and $(\partial \psi_i/\partial \mathbf{u})\delta \mathbf{u}$. It seems natural to try to eliminate all such expressions. To accomplish this, let us introduce $k + 1$ vectors λ^i

$$A^{\mathrm{T}} \lambda^0 = (\partial J/\partial \mathbf{u})^{\mathrm{T}}, \ldots, A^{\mathrm{T}} \lambda^i = (\partial \psi_i/\partial \mathbf{u})^{\mathrm{T}}, \qquad i = 1, 2, \ldots, k, \tag{5.45}$$

where the superscript T denotes the transpose of a vector or matrix. The functionals on the right-hand side of eq. (5.45) are computed for the current values of the vectors (points) $\mathbf{u}$ and $\mathbf{h}$.

Now we transpose terms on both sides of eq. (5.45) and multiply them on the right by $\delta \mathbf{u}$, deriving

$$(\lambda^0)^{\mathrm{T}} A \delta \mathbf{u} = (\partial J/\partial \mathbf{u})^{\mathrm{T}} \delta \mathbf{u}, \ldots, (\lambda^i)^{\mathrm{T}} A \delta \mathbf{u} = (\partial \psi_i/\partial \mathbf{u})^{\mathrm{T}} \delta \mathbf{u}. \tag{5.46}$$

The expressions on the right-hand side of (5.46) are exactly the terms that need to be eliminated from (5.42). The term $A\delta \mathbf{u}$ may be replaced using (5.43): $A\delta \mathbf{u} = -B\delta \mathbf{h}$. Thus, (5.46) may be replaced by

$$-(\lambda^0)^{\mathrm{T}} B \delta \mathbf{h} = (\partial J/\partial \mathbf{u})^{\mathrm{T}} \delta \mathbf{u}, \qquad -(\lambda^i)^{\mathrm{T}} B \delta \mathbf{h} = (\partial \psi_i/\partial \mathbf{u})^{\mathrm{T}} \delta \mathbf{u}. \tag{5.47}$$

Moreover, if we substitute from (5.47) the expressions for $(\partial J/\partial \mathbf{u})^{\mathrm{T}} \delta \mathbf{u}$ and $(\partial \psi_i/\partial \mathbf{u})^{\mathrm{T}} \delta \mathbf{u}$ in (5.42), we obtain

$$\delta J = [\nabla_{\mathbf{h}} J]^{\mathrm{T}} \delta \mathbf{h}, \tag{5.48}$$

$$\delta \psi_i = [\nabla_{\mathbf{h}} \psi_i]^{\mathrm{T}} \delta \mathbf{h}. \tag{5.49}$$

The gradients on the right-hand sides of eq. (5.48) and (5.49) can be computed from the formulas

$$[\nabla_{\mathbf{h}} J]^{\mathrm{T}} = (\partial J/\partial \mathbf{h})^{\mathrm{T}} - B^{\mathrm{T}} \lambda^0, \qquad [\nabla_{\mathbf{h}} \psi_i]^{\mathrm{T}} = (\partial \psi_i/\partial \mathbf{h})^{\mathrm{T}} - B^{\mathrm{T}} \lambda^i. \tag{5.50}$$

The components of the vectors $[\nabla_{\mathbf{h}} \psi_i]$, $i = m + 1, \ldots, k$, and of the vector $[\nabla_{\mathbf{h}} J]$ are called the sensitivity coefficients of the constraints and the quality functional, respectively, with respect to the corresponding design variables. The sensitivity coefficients in (5.50) are very useful in a design process, since they contain information as to how a change in design affects the quality functional and the constraints. In particular, the computation of the values of these coefficients permits us to identify those design variables whose changes have the most significant influence on important properties of a structure.

The analysis carried out in [5.14] indicates that the approach described above in computing the sensitivity coefficients is particularly effective if there is a large number of initially assigned constraints and varieties of loading, and for large dimensions of vectors **u** and **h**.

5.6 Second-Order Sensitivity Analysis

Well-established techniques of sensitivity analysis are based on different schemes for computing the gradients of the quality criteria and constraints with respect to the design variables. Sensitivity formulas and expressions for the gradients of the basic characteristics make it possible to develop highly effective techniques of iterative optimization and substantially speed up the search for an optimal design. A further improvement of this process may be attained by the use of effective estimates of the second-order derivatives (i.e., derivatives of structural characteristics with respect to design variables).

Following [5.11] we shall offer an exposition of certain techniques for computing these derivatives.

Let the behavior of a structure be described by a system of algebraic equations

$$L\mathbf{u} = \mathbf{q}, \tag{5.51}$$

where **u** and **q** are, respectively, m-dimensional vectors of displacements and loads; **h** will denote an n-dimensional vector of design variables, while L is an $(m \times m)$ matrix. The coefficients of $L(\mathbf{h})$ are twice differentiable with respect to the components of the design vector **h**. To simplify our analysis, we limit our discussion to derivatives with respect to only two (fixed) components h_i and h_j of the vector **h**. Therefore, in all subsequent discussion we shall consider only a dependence on h_i and h_j. The constraints assigned to the response of the structure are

$$\psi(\mathbf{u}, h_i, h_j) \leq 0. \tag{5.52}$$

Differentiating the left-hand side of (5.52) with respect to h_i, we have

$$\mathrm{d}\psi/\mathrm{d}h_i = \partial\psi/\partial h_i + \mathbf{Z}^{\mathrm{T}}\mathrm{d}\mathbf{u}/\mathrm{d}h_i, \tag{5.53}$$

where the components of the vector $\mathbf{Z}^{\mathrm{T}}$ are given by the formula $Z_s = \partial\psi/\partial u_s$, $(s = 1, 2, \ldots, m)$. Differentiating both sides of eq. (5.51) with respect to h_i, we derive

$$L\mathrm{d}\mathbf{u}/\mathrm{d}h_i = \mathrm{d}\mathbf{q}/\mathrm{d}h_i - (\mathrm{d}L/\mathrm{d}h_i)\mathbf{u}. \tag{5.54}$$

Solving the system of equations (5.54) for $\mathrm{d}\mathbf{u}/\mathrm{d}h_i$ and substituting these values in (5.53), we arrive at an expression for the derivative of the constraint function with respect to a design variable:

$$\mathrm{d}\psi/\mathrm{d}h_i = \partial\psi/\partial h_i + \mathbf{Z}^{\mathrm{T}}L^{-1}[\mathrm{d}\mathbf{q}/\mathrm{d}h_i - (\mathrm{d}L/\mathrm{d}h_i)\mathbf{u}]. \tag{5.55}$$

As before, the superscript -1 denotes the inverse of a matrix. We introduce an m-dimensional vector of adjoint variables $\boldsymbol{\lambda}$, designating it as a solution to the system of algebraic equations

$$L\boldsymbol{\lambda} = \mathbf{Z}. \tag{5.56}$$

Using eqs. (5.55) and (5.56) and the symmetry of the matrix L, we obtain a formula for first-order sensitivity analysis:

$$\mathrm{d}\psi/\mathrm{d}h_i = \partial\psi/\partial h_i + \boldsymbol{\lambda}^{\mathrm{T}}[\mathrm{d}\mathbf{q}/\mathrm{d}h_i - (\mathrm{d}L/\mathrm{d}h_i)\mathbf{u}]. \tag{5.57}$$

In this approach to the computation of derivatives using the adjoint vector $\boldsymbol{\lambda}$ we must know the solution to the system of equations (5.56) for each constraint assigned to the response of the structure. However, a direct computation of these derivatives using (5.54) and (5.55) necessitates solving the system (5.54) for each design variable. Therefore, computations that introduce the adjoint variables are more advantageous when the number of design variables is significantly higher than the number of structural constraints.

To compute the second-order derivatives, we differentiate both sides of eq. (5.53) with respect to h_j, obtaining

$$\begin{aligned}\mathrm{d}^2\psi/\mathrm{d}h_i\mathrm{d}h_j = {}& \partial^2\psi/\partial h_i\partial h_j + (\partial\mathbf{Z}/\partial h_i)^{\mathrm{T}}(\mathrm{d}\mathbf{u}/\mathrm{d}h_j) + (\mathrm{d}\mathbf{u}/\mathrm{d}h_j)^{\mathrm{T}}\mathbf{K}(\mathrm{d}\mathbf{u}/\mathrm{d}h_i)\\ &+ (\partial\mathbf{Z}/\partial h_j)^{\mathrm{T}}(\mathrm{d}\mathbf{u}/\mathrm{d}h_i) + \mathbf{Z}^{\mathrm{T}}(\mathrm{d}^2\mathbf{u}/\mathrm{d}h_i\mathrm{d}h_j),\end{aligned} \tag{5.58}$$

where the $(m \times m)$ matrix $\mathbf{K}$ has coefficients $K_{st} = \partial^2\psi/\partial u_s\partial u_t$. Differentiating (5.54) with respect to h_j, we obtain a formula for the quantity $\mathrm{d}^2\mathbf{u}/\mathrm{d}h_i\mathrm{d}h_j$ in (5.58):

$$\begin{aligned}L\mathrm{d}^2\mathbf{u}/\mathrm{d}h_i\mathrm{d}h_j = {}& \mathrm{d}^2\mathbf{q}/\mathrm{d}h_i\mathrm{d}h_j - (\mathrm{d}^2L/\mathrm{d}h_i\mathrm{d}h_j)\mathbf{u} - (\mathrm{d}L/\mathrm{d}h_i)(\mathrm{d}\mathbf{u}/\mathrm{d}h_j)\\ &- (\mathrm{d}L/\mathrm{d}h_j)(\mathrm{d}\mathbf{u}/\mathrm{d}h_i).\end{aligned} \tag{5.59}$$

Thus, computation of each second-order derivative of the constraint functions with respect to the design variables reduces to solving the systems of equations (5.59) and (5.54). If we wish to compute all second-order derivatives $\mathrm{d}^2\psi/\mathrm{d}h_i\mathrm{d}h_j$, with $i, j = 1, 2, \ldots, n$, it is necessary to solve the systems (5.59) and (5.54), $n(n + 1)/2$ and n times, respectively. Consequently, we have to solve a total of $[n + n(n + 1)/2]$ algebraic systems of equations. A more economic approach is offered in [5.13], where two adjoint vectors were introduced. Then if we wish to compute all second-order derivatives it is necessary to introduce $2n$ adjoint vectors and solve $2n$ systems of equatuions.

An even better rationale is offered in [5.11] for a computational algorithm to evaluate the second derivatives. The author of that article reduces formula (5.58) for the second derivative $\mathrm{d}^2\psi/\mathrm{d}h_i\mathrm{d}h_j$ applying (5.56) and (5.59), and transforms it into the form

$$\begin{aligned}\mathrm{d}^2\psi/\mathrm{d}h_i\mathrm{d}h_j = {}& \partial^2\psi/\partial h_i\partial h_j + (\mathrm{d}\mathbf{u}/\mathrm{d}h_j)^{\mathrm{T}}\mathbf{K}(\mathrm{d}\mathbf{u}/\mathrm{d}h_i) + (\partial\mathbf{Z}/\partial h_i)^{\mathrm{T}}(\mathrm{d}\mathbf{u}/\mathrm{d}h_j)\\ &+ (\partial\mathbf{Z}/\partial h_j)^{\mathrm{T}}(\mathrm{d}\mathbf{u}/\mathrm{d}h_i) + \boldsymbol{\lambda}^{\mathrm{T}}[\mathrm{d}^2\mathbf{q}/\mathrm{d}h_i\mathrm{d}h_j\\ &- (\mathrm{d}^2L/\mathrm{d}h_i\mathrm{d}h_j)\mathbf{u} - (\mathrm{d}L/\mathrm{d}h_i)(\mathrm{d}\mathbf{u}/\mathrm{d}h_j)\\ &- (\mathrm{d}L/\mathrm{d}h_j)(\mathrm{d}\mathbf{u}/\mathrm{d}h_i)].\end{aligned} \tag{5.60}$$

It must be assumed that the systems of equations (5.54) and (5.56) have been solved if one wishes to apply (5.60). If n is the dimension of the vector of design variables, to compute all second-order derivatives of the function ψ it is necessary to compute the vector $\boldsymbol{\lambda}$ from the system of equations (5.56) and to solve the system of equations (5.54) n times to find the values of the first-order derivatives $d\mathbf{u}/dh_s$, $s = 1, \ldots, n$. In this fashion second-order sensitivity analysis using this technique involves solving a total of $(n + 1)$ original systems of equations. This nearly halves the number of computations required if one follows the approach described in [5.13].

In the case when there are k constraints ($k \geq 1$), such an approach, based on relations (5.54), (5.56), and (5.60), reduces to the problem of solving $(n + k)$ systems of equations. If one uses the approach of [5.13], it is necessary to solve $n(k + 1)$ systems of equations.

5.7 Design Techniques Using Decomposition into Substructures

Techniques using decomposition of a structure into simpler substuctures are commonly used in computing complex structural systems. Several techniques of optimal design utilize concepts of structural decomposition. Such a decomposition does not only partition a structure, but also classifies all joints into boundary and internal joints and assigns the rigidity matrices accordingly. By definition, a joint is said to be a boundary joint if it belongs to more than one substructure. We shall denote by the subscripts B and I all quantities assigned, respectively, to boundary and internal structural joints. For example, displacements at boundary or internal joints, respectively, will be denoted by $\mathbf{u}_B$ and $\mathbf{u}_I$.

Let us assume that the structure that is being designed has been arbitrarily subdivided into l substructures and that the components of the vector $\mathbf{u}$ of displacements at the joints have been decomposed into the vectors $\mathbf{u}_B$ and $\mathbf{u}_I$.

The optimization problem now consists in computing the vector of design variables $\mathbf{h}$ and corresponding vectors $\mathbf{u}_B$ and $\mathbf{u}_I$ minimizing the quality functional

$$J = J(\mathbf{u}_B, \mathbf{u}_I, \mathbf{h}) \tag{5.61}$$

and satisfying the equations of equilibrium

$$\begin{bmatrix} L_{BB} & L_{BI} \\ L_{IB} & L_{II} \end{bmatrix} \begin{bmatrix} \mathbf{u}_B \\ \mathbf{u}_I \end{bmatrix} = \begin{bmatrix} \mathbf{q}_B \\ \mathbf{q}_I \end{bmatrix} \tag{5.62}$$

and the constraints

$$\psi_i(\mathbf{u}_B, \mathbf{u}_I, \mathbf{h}) \leq 0, \qquad i = 1, 2, \ldots, k. \tag{5.63}$$

The elements of the submatrices L_{BB}, L_{BI}, L_{IB}, and L_{II} and of the vectors $\mathbf{q}_B$ and $\mathbf{q}_I$ are regarded as functions of the design variables.

Now let us carry out a sensitivity analysis for the problem defined by (5.61) and (5.62) by formally differentiating the quality functional and the constraints with respect to design variables (cf. [5.3]).

By varying the quality functional and the constraint functionals, we obtain

$$\left.\begin{aligned} \delta J &= \partial J/\partial \mathbf{h}\delta\mathbf{h} + \partial J/\partial \mathbf{u}_B \delta \mathbf{u}_B + \partial J/\partial \mathbf{u}_I \delta \mathbf{u}_I, \\ \delta\psi_i &= \partial\psi_i/\partial\mathbf{h}\delta\mathbf{h} + \partial\psi_i/\partial\mathbf{u}_B\delta\mathbf{u}_B + \partial\psi_i/\partial\mathbf{u}_I\delta\mathbf{u}_I. \end{aligned}\right\} \tag{5.64}$$

The derivatives appearing in eq. (5.64) are computed for the current (i.e., unperturbed) values of the variables $\mathbf{h}$, $\mathbf{u}_B$, and $\mathbf{u}_I$. The variation $\delta\mathbf{h}$ is related to the variations $\delta\mathbf{u}_B$ and $\delta\mathbf{u}_I$, by the equations

$$L_{BB}\delta\mathbf{u}_B + L_{BI}\delta\mathbf{u}_I = \chi_1\delta\mathbf{h}, \tag{5.65}$$

$$L_{IB}\delta\mathbf{u}_B + L_{II}\delta\mathbf{u}_I = \chi_2\delta\mathbf{h}, \tag{5.66}$$

where $\chi_1 = \partial\mathbf{q}_B/\partial\mathbf{h} - \partial/\partial\mathbf{h}(L_{BB}\mathbf{u}_B) - \partial/\partial\mathbf{h}(L_{BI}\mathbf{u}_I)$, and $\chi_2 = \partial\mathbf{q}_I/\partial\mathbf{h} - \partial/\partial\mathbf{h}(L_{IB}\mathbf{u}_B) - \partial/\partial\mathbf{h}(L_{II}\mathbf{u}_I)$. Relations (5.65) and (5.66), which express the linear dependence of $\delta\mathbf{u}_B$ and $\delta\mathbf{u}_I$ on $\delta\mathbf{h}$, are derived by expanding (5.62) in series of powers of $\delta\mathbf{u}_b$, $\delta\mathbf{u}_I$, and $\delta\mathbf{h}$ and retaining only the linear terms, that is, terms of the first order of magnitude.

Solving eq. (5.66) for $\delta\mathbf{u}_I$ and substituting this expression in (5.65), we obtain

$$\begin{aligned} L_B\delta\mathbf{u}_B &= \chi\delta\mathbf{h}, & L_B &= L_{BB} + L_{BI}\mathbf{Q}, \\ \chi &= \chi_1 + \mathbf{Q}^{\mathrm{T}}\chi_2, & \mathbf{Q} &= -L_{II}^{-1}L_{IB}. \end{aligned} \tag{5.67}$$

We also substitute the expression for $\delta\mathbf{u}_I$ derived above in (5.64),

$$\begin{aligned} \delta J &= (\partial J/\partial\mathbf{h} + \partial J/\partial\mathbf{u}_I L_{II}^{-1}\chi_2)\delta\mathbf{h} + (\partial J/\partial\mathbf{u}_I\mathbf{Q} + \partial J/\partial\mathbf{u}_B)\delta\mathbf{u}_B, \\ \delta\psi_i &= (\partial\psi_i/\partial\mathbf{h} + \partial\psi_i/\partial\mathbf{u}_I L_{II}^{-1}\chi_2)\delta\mathbf{h} + (\partial\psi_i/\partial\mathbf{u}_I\mathbf{Q} + \partial\psi_i\partial\mathbf{u}_B)\delta\mathbf{u}_B. \end{aligned} \tag{5.68}$$

We also define vectors of the conjugate variables $\boldsymbol{\lambda}^0$, $\boldsymbol{\lambda}^i$, $\boldsymbol{\mu}^0$, and $\boldsymbol{\mu}^i$ by the relations

$$\begin{aligned} &L_{II}\boldsymbol{\lambda}^0 = (\partial J/\partial\mathbf{u}_I)^{\mathrm{T}}, & &L_{II}\boldsymbol{\lambda}^i = (\partial\psi_i/\partial\mathbf{u}_I)^{\mathrm{T}} \\ &L_B\boldsymbol{\mu}^0 = (\partial J/\partial\mathbf{u}_B)^{\mathrm{T}} + \mathbf{Q}^{\mathrm{T}}(\partial J/\partial\mathbf{u}_I)^{\mathrm{T}}, & &L_B\boldsymbol{\mu}^i = (\partial\psi_i/\partial\mathbf{u}_B)^{\mathrm{T}} + \mathbf{Q}^{\mathrm{T}}(\partial\psi_i/\partial\mathbf{u}_I)^{\mathrm{T}}. \end{aligned} \tag{5.69}$$

Using the definition of adjoint variables in solving the system of algebraic equations (5.69), and making use of the symmetry of the submatrix L_{II} and boundary rigidity matrix L_B (i.e., $L_{II} = L_{II}^{\mathrm{T}}$, $L_B = L_B^{\mathrm{T}}$), we can eliminate from (5.68) the components of $\delta\mathbf{u}_B$, thus deriving a fundamental sensitivity relation that expresses δJ and $\delta\psi_i$ directly in terms of $\delta\mathbf{h}$:

$$\begin{aligned} \delta J &= [\nabla_h J]^{\mathrm{T}}\delta\mathbf{h}, \\ \delta\psi_i &= [\nabla_h\psi_i]^{\mathrm{T}}\delta\mathbf{h}, \\ [\nabla_{\mathbf{h}}J]^{\mathrm{T}} &= (\partial J/\partial\mathbf{h})^{\mathrm{T}} + (\chi_2)^{\mathrm{T}}\boldsymbol{\lambda}^0 + \boldsymbol{\chi}^{\mathrm{T}}\boldsymbol{\mu}^0, \\ [\nabla_{\mathbf{h}}\psi_i]^{\mathrm{T}} &= (\partial\psi_i/\partial\mathbf{h})^{\mathrm{T}} + (\chi_2)^{\mathrm{T}}\boldsymbol{\lambda}^i + \boldsymbol{\chi}^{\mathrm{T}}\boldsymbol{\mu}^i. \end{aligned} \tag{5.70}$$

The sensitivity analysis formulas (5.70) can be effectively utilized in creating algorithms for sequential optimization.

A basic advantage of this methodology of utilizing a decomposition of a structure into substructures resides in the following feature. As the rigidity matrix L_B for boundary elements and the vector of the right-hand side of (5.67) are being computed, it is possible to simultaneously perform corresponding computations for each individual substructure and then sum up the results. Indeed, writing the equations of equilibrium for each substructure with the current index $j = 1, 2, \ldots, l$,

$$\begin{bmatrix} L_{BB}^{(j)} & L_{BI}^{(j)} \\ L_{IB}^{(j)} & L_{II}^{(j)} \end{bmatrix} \begin{bmatrix} \mathbf{u}_B^{(j)} \\ \mathbf{u}_I^{(j)} \end{bmatrix} \begin{bmatrix} \mathbf{q}_B^{(j)} \\ \mathbf{q}_I^{(j)} \end{bmatrix} \tag{5.71}$$

and solving for the quantity $\mathbf{u}_I^{(j)}$ from the first row in (5.71), we obtain

$$L_B^{(j)}\mathbf{u}_B^{(j)} = F_B^{(j)}. \tag{5.72}$$

Here we have introduced the substitutions

$$L_B^{(j)} = L_{BB}^{(j)} + L_{BI}^{(j)}\mathbf{Q}^{(j)}, \qquad \mathbf{Q}^{(j)} = -[L_{II}^{(j)}]^{-1}L_{IB}^{(j)},$$

$$F_B^{(j)} = \mathbf{q}_B^{(j)} + [\mathbf{Q}^{(j)}]^{\mathrm{T}}\mathbf{q}_I^{(j)}.$$

In solving (5.72) we solved for $\mathbf{u}_I^{(j)}$ from the second row in (5.71) and substituted the result in the first row.

Now each substructure can be regarded as a "superelement" and the relation between the displacements and forces acting on it is given by (5.72). Moreover, using well-known techniques (cf. [24] and [5.3]) and applying Boolean matrices $\beta^{(j)}$, we can compute the boundary rigidity matrix and the vector of effective boundary forces:

$$L_B = \sum_{j=1}^{l} [\beta^{(j)}]^{\mathrm{T}} L_B^{(j)} \beta^{(j)}, \qquad F_B = \mathbf{q}_B + \sum_{j=1}^{l} [\beta^{(j)}]^{\mathrm{T}} [\mathbf{Q}^{(j)}]^{\mathrm{T}} \mathbf{q}_I^{(j)}.$$

Here l denotes the number of substructures.

References for Chapter 5

5.1. Arora, J.S., and Haug, E.J., *Applied Optimal Design*, Wiley-Interscience, New York, 1979 (Russian translation: Mir, Moscow, 1983).

5.2. Arora, J.S., An algorithm for optimal structural design without line search. In: *New Directions in Optimal Design*; E. Atrek et al. (eds.), Chichester, Wiley, 1984, pp. 429–441.

5.3. Arora, J.S., and Govil, A.K., An efficient method for optimal structural design by substructuring, Intern. J. Computers and Structures, 1977, 7. No. 4, pp. 507–515.

5.4. Brandt, A.M., *Criteria and Methods of Structural Optimization*, PWN Publ. Warszawa, 1984, 538 pp.

5.5. Dajartnam, P., and Patnaik, S., Realization of uniformly stressed structures, *Raket. Tekhnika i Kosmonavtika*, 1969, 7, No. 4, pp. 238.

5.6. Eschenauer, H.A., Vector optimization in structural design and its application to antenna structures. In: *Optimization Methods in Structural Design*. H. Eschenauer and N. Olhoff (eds.), Mannheim, Bibliogr. Institute, 1983, pp. 146–155.

5.7. Fleury, C., Large scale structural optimization by finite elements. In: *New Directions in Optimal Design*, E. Atrek et al. (eds.), Chichester, Wiley & Sons, 1984, pp. 429–441.

5.8. Fleury, C., Remanathan, R.K., Salama, M., and Schmit, L.A. Access computer program for large structural systems. In: *New Directions in Optimal Design*; Editor E. Atrek et al, Chichester, Wiley & Sons, 1984, pp. 541–561.

5.9. Gallagher, R.H., Fully stressed design. In: *Optimum Structural Design Theory and Applications*, R.H. Gallagher and O.C. Zienkiewicz (eds.), Wiley and Sons, New York, 1973, pp. 19–32.

5.10. Gallagher, R.H., Computorized structural analysis and design—The next twenty years, *Intern. J. Comput. Struct.*, 1977, 7, No. 4, pp. 495–501.

5.11. Haftka, R.T., Second order sensitivity in structural analysis, *AIAA J.*, 1982, *20*, No. 12, pp. 1765–1766.

5.12. Haftka, R.T., and Prazard, B., Programs for analysis and resizing of complex structures, *Intern. J. Comput. Struct.*, 1979, *10*, pp. 323–330.

5.13. Haug, E.J., Second order design sensitivity analysis of structural systems, *AIAA J.*, 1981, *19*, No. 8, pp. 1087–1088.

5.14. Haug, E.J., and Arora, J.S., Design sensitivity analysis of elastic mechanical systems, *Comput. Methods in Appl. Mech. Eng.*, 1978, *15*, pp. 35–62.

5.15. Haug, E.J., and Choi, K.K., Structural design sensitivity analysis with generalized global stiffness and mass matrices, *AIAA J.*, 1984, *22*, No. 9, pp. 1299–1303.

5.16. Hoerner, S., Design of large steerable antennas, *Astron. J.* 1967, *72*, No. 1, pp. 35–47.

5.17. Hoerner, S., The derivative tensor of the stiffness matrix as a tool for optimizations. In: Optimization Methods in Structral Design. H. Eschenauer and N. Olhoff (eds.), Mannheim, Bibl. Institute, 1983, pp. 84–90.

5.18. Khot, N.S., Algorithms based on optimality criteria to design minimum weight structures. *Eng. Optim.* 1981, *5*, pp. 73–90.

5.19. Khot, N.S., and Berke, L., Structural optimization using optimality criteria methods. In: *New directions in Optimal Design*; E. Atrek et al. (eds.), Chichester, Wiley, 1984, p. 47–74.

5.20. Kiselev, V.G., An algorithm for level line search and its applications to optimization problems. In: *Applied Problems of Strength and Plasticity*, All-Union Univ. Collection, Gorkii Univ. Press, Gorkii, 1980, pp. 81–87.

5.21. Lac, P., and Morlier, P., Optimization of structures techniques, In: *Optimization Methods in Structural Design*. H. Eschenauer and N. Olhoff (eds.), Mannheim, Bibliogr. Institute, 1983, pp. 91–96.

5.22. Malkov, V.P., Achinova, G.D., Beh, L.P., Bol'shakova, A.T., Kiselev, V.G., Morozov, V.D., Povelikin, V.P., and Tarasov, V.L., Some results concerning Optimization of Elastic Systems. In: *Some Problems of Structural Mechanics; Optimization of Supporting Structures*. Proc. the IV-th All-Union Conference on structural statics and dynamics. Kiev, 1978, pp. 1–79.

5.23. Malkov, V.P., and Morozov, V.D., Combinatorial approach to multiparameter optimization of supporting structures. In: *Applied Problems of Strength and Plasticity*, Gorkii University Press, Gorkii, 1977, Vol. 7, pp. 85–90.

5.24. Malkov, V.P., and Tarasov, V.L., Uniformly stressed thin-walled structure and a minimum weight structure, *Izv. Akad. Nauk U.S.S.R., M.T.T.*, 1974, No. 5.
5.25. Ragsdale, K.M., The utility of nonlinear programming methods for engineering design. In: *New Directions in Optimal Design.* E. Atrek et al. (eds.), Chichester, Wiley, 1984, pp. 385–412.
5.26. Rozin, L.A., *Systems of Beams as finite Element Systems*, L'vov State University, L'vov, 1976, 232 pp.
5.27. Schmit, L.A., and Farshi, B., Some concepts of approximation for the synthesis of structures, *Raket. Tekhnika i Kosmonavtika* (Preprint), 1974, *4*, pp. 145–155.
5.28. Schmit, L.A., Structural optimization—Some key ideas and insights. In: *New Directions in Optimal Design*, E. Atrek et al. (eds.), Chichester, Wiley, 1984, pp. 1–45.
5.29. Thierauf, G., and Lawo, M., Optimal limit design-comparative studies for large scale. In: *Optimization Methods in Structural Design.* H. Eschenauer and N. Olhoff (eds.), Mannheim, F.R.G. Bibliogr. Institute, 1983, pp. 431–437.
5.30. Vanderplaats, G.N., Structural optimization—Past, present and future, *AIAA J.*, 1981, *20*, No. 7, pp. 992–1000.
5.31. Vinogradov, A.I., The optimal design problem, and its specific properties for a system of beams. *Stroit. Mekhanika i Raschet Sooruzhenii*, 1974, No. 4, pp. 55–60.

2
Strength, Rigidity, Stability, and Weight Criteria for Optimal Design

6
Application of Strength and Weight Criteria

6.1 Strength Criteria

Strength criteria constitute the fundamental constraints that must be taken into account in any optimal design. Such constraints are local in nature and are stated in terms of the components of the stress tensor. At the present time various theories of strength have been investigated and some of them have been extensively applied in optimal design research. The choice of any specific strength criteria, or the use of such approximate criteria, depends on the structural materials, the type of loads, the use of a specific computational scheme, and on other factors.

We shall introduce certain strength criteria that are commonly used in research on optimal design and consider their different formulations. In a majority of applicable criteria one considers a limiting surface in the stress space. Such a surface separates a region of the admissible stressed state of a material (it is an interior region containing the origin of the coordinate system) from the (exterior) region in which the material will either fail or will acquire inadmissible behavior. For an isotropic body all directions are equivalent, and the equation of such surface is invariant with respect to the orientation of the coordinate system and should depend only on the invariants I_1^σ, I_2^σ, I_3^σ of the stress tensor, where

$$I_1^\sigma = \sigma_{ii}, \qquad I_2^\sigma = -1/2[(\sigma_{ii})^2 - \sigma_{ij}\sigma_{ij}]$$
$$I_3^\sigma = 1/3(\sigma_{ij}\sigma_{jk}\sigma_{kl}) + 1/6(\sigma_{ii})^3 - 1/2(\sigma_{ii}\sigma_{ij}\sigma_{ij}). \tag{6.1}$$

As usual, the repeated indices in (6.1) imply summation. The strength criteria can now be written in the form

$$g(I_1^\sigma, I_2^\sigma, I_3^\sigma) \leq k^2, \tag{6.2}$$

where g is a given function and k^2 is a constant characterizing the properties of the material. The quantity appearing on the left-hand side of inequality (6.2) may be expressed as a symmetric function of the principal stresses $\sigma_1, \sigma_2, \sigma_3$. Thus, condition (6.2) may be rewritten in the equivalent form

$$g(\sigma_1, \sigma_2, \sigma_3) \le k^2. \tag{6.3}$$

Depending on the properties of the materials and the type of stressed state eq. (6.3) may be written in a more specific form. To describe the failure of a brittle material whose stressed state is close to uniaxial, one may use the Galileo–Rankin criterion of a maximal normal stress component. According to this criterion, in order to avoid failure the largest normal stress component must not exceed some given (constant) value,

$$\max\{|\sigma_1|, |\sigma_2|, |\sigma_3|\} \le k. \tag{6.4}$$

For materials with a pronounced slip mechanism of failure, one could use Coulomb's criterion

$$\max(\tau_n - |\sigma_n| \tan \chi) \le k, \tag{6.5}$$

where τ_n and σ_n are, respectively the tangential and normal stress component in the plane with the unit normal vector $\mathbf{n}$; k is the coefficient of cohesion, and χ is the angle of internal friction. A maximum in (6.5) is sought among all possible directions of the unit normal vector. Here, assuming that $\chi = 0$, we arrive at the Tresca–St. Venant criterion of maximum shear stress (see [6.37] and [6.71]):

$$\max\{|\sigma_1 - \sigma_2|, |\sigma_2 - \sigma_3|, |\sigma_3 - \sigma_1|\} \le 2k. \tag{6.6}$$

Here $2k = \sigma_s$ is the limiting value of the tensile stress. Inequality (6.6) differs from (6.5) because now we choose a maximum from three quantities. The limiting surface corresponding to (6.6) drawn in the space spanned by principal stresses is a regular hexagonal prism with axis $\sigma_1 = \sigma_2 = \sigma_3$. If we bisect this prism with a plane perpendicular to the hydrostatic stress axis $\sigma_1 = \sigma_2 = \sigma_3$, we obtain a regular hexagon. Note that according to this criterion, a uniform compression or tension (i.e., $\sigma_1 = \sigma_2 = \sigma_3$) will not cause a violation of the failure conditions.

To describe the failure of different materials having distinctly different responses to compression and tension, we may use Moore's criterion

$$\max[\tau_n - f(\sigma_n)] \le 0. \tag{6.7}$$

The function f in (6.7) is determined experimentally.

The maximum in (6.7) and (6.5) is taken over all possible orientations of the plane. In the plane σ_n, τ_n the relation $\tau_n - f(\sigma_n) = 0$ represents some curve (shown as a solid line in Figure 6.1). It is the boundary of the region of safe states. The stressed state of the material in that region is determined by Moore's circles. The limit curve is tangent to the circle having the largest radius. As can be seen from Figure 6.1, the limit curve has a negative derivative (τ_n decreases with increasing σ_n). The criterion of maximum shear stress (6.6) is a special case of Moore's criterion. In (6.6) the limiting curve degenerates into a straight line parallel to the σ_n axis.

All the criteria considered above become substantially simplified in the case of beams and depend only on two components of the stress tensor, σ_x and τ_{xy}

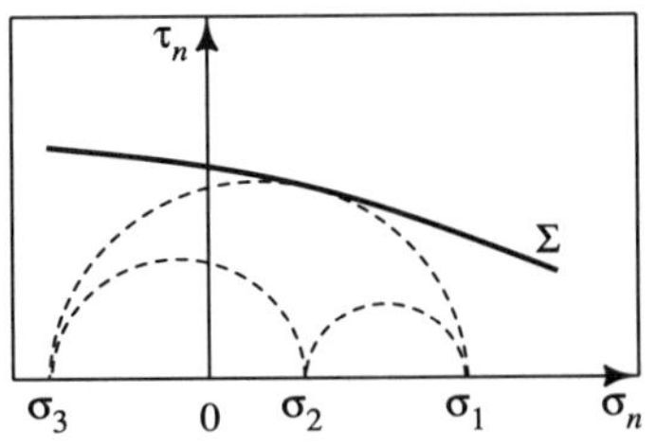

FIGURE 6.1.

(if we assume that the beam in its unstressed state lies along the x-axis, and the bending loads act parallel to the y-axis). Let us indicate one possible formula describing a failure criterion for a quite general class of beams (cf. [6.36]), which reduces to well-known criteria for certain specific values of α and β:

$$[(1 + \alpha)/2]\sqrt{(\sigma_x)^2 + \beta(\tau_{xy})^2} + [(1 - \alpha)/2]|\sigma_x| \leq k. \tag{6.8}$$

For $\alpha = 0$, $\beta = 4$, inequality (6.8) reduces to the Galileo–Rankin failure criterion. For $\alpha = \mu$ and $\beta = 4$, this becomes St. Venant's failure criterion; for $\alpha = 1, \beta = 4$, the Coulomb's criterion; for $\alpha = \Delta, \beta = 4$, Moore's criterion, and for $\alpha = 1$, $\beta = 3$, the von Mises–Hencky criterion. Here Δ denotes the admissible stress values in tension or compression. Formula (6.8) for failure criteria, which contains the criteria listed above, is convenient for incorporation into the optimal design problems.

Along with criteria whose noncompliance is supposed to assure an imminent structural failure, the theory of optimal design makes use of other types of conditions. Violation of these conditions may cause the appearance of inelastic deformations, which are inadmissible from the standpoint of the reliable behavior of the structure. We shall consider certain criteria used in the theory of plasticity and in the theory of bounds on the load capacity. To describe the transition of a material to the plastic state and to estimate the load-bearing capacity, we shall use, besides criteria (6.5) and (6.6), the Huber–von Mises criterion, which may be interpreted as a limitation on the maximum intensity of shear stresses. According to this criterion, the value of shear stresses must not exceed the limiting value

$$\begin{aligned} g = (1/\sqrt{6})[(\sigma_x - \sigma_y)^2 + (\sigma_y - \sigma_z)^2 + (\sigma_z - \sigma_x)^2 \\ + 6(\tau_{xy}^2 + \tau_{yz}^2 + \tau_{xz}^2)]^{1/2} \leq \tau_s, \\ \tau_s = \sigma_s/\sqrt{3}. \end{aligned} \tag{6.9}$$

Here τ_s and σ_s denote the plastic flow initiation limit in cases of pure shear or pure tension, respectively. In the space spanned by the principal stresses, condition (6.9) is represented by a circular cylinder having σ_1 as its axis, with $\sigma_2 = \sigma_3$. The surface of this cylinder touches an inscribed hexagonal prism, which represents condition (6.6). Condition (6.9) may be concisely written in terms of the components of the deviator of the stress tensor $s_{ij} = \sigma_{ij} - (1/3)\delta_{ij}\sigma_{ll}$:

$$g = [\tfrac{1}{2}s_{ij}s_{ij}]^{1/2} \leq \tau_s. \tag{6.10}$$

For the case of a two-dimensional stressed state with $\sigma_z = 0$, the criterion of maximum value for the shear stresses (6.9) and (6.10) may be restated thus:

$$(1/\sqrt{3})[(I_1^\sigma)^2 + 3I_2^\sigma]^{1/2} = (1/\sqrt{3})(\sigma_x^2 + \sigma_y^2 - \sigma_x\sigma_y + 3\tau_{xy}^2) \leq \tau_s, \tag{6.11}$$

where

$$I_1^\sigma = \sigma_x + \sigma_y, \qquad I_2^\sigma = \tau_{xy}^2 - \sigma_x\sigma_y.$$

Note that the maximum shear stress intensity as given by the function g and the greatest shear stress $\tau_{\max}$ do not differ from each other by a substantial amount. The following two-sided estimate is well known (see [6.33] and [6.34]):

$$1 \leq (g/\tau_{\max}) \leq (2/\sqrt{3}).$$

Therefore, assuming that the error does not exceed 7%, we may assume

$$g \approx 1.08\tau_{\max}.$$

For isotropic bodies we may apply also the von Mises–Schleicher–Nadai criterion

$$[1/2s_{ij}s_{ij}]^{1/2} - f(\sigma) \leq 0, \tag{6.12}$$

which relates the intensity of shear stresses to the hydrostatic pressure $\sigma = (1/3)\sigma_{ij}\delta_{ij}$. Here δ_{ij} denotes the Kronecker delta symbol. The function f in (6.12) is determined experimentally. If $f(\sigma) = \tau_s$, we obtain the Huber–von Mises criterion given in (6.10). The dependence of (6.12) on specific types of functions $f(\sigma)$ defines various shapes of the limiting surfaces (cf. [6.24] and [6.71]).

For anisotropic bodies there exist a considerable variety of corresponding failure criteria for isotropic bodies. The motivation of such criteria is a demand for invariance. The failure criteria must not be altered as we adopt a different system of coordinates. Thus the function determining a failure must depend only on the invariants

$$g(I_1^{\sigma c}, I_2^{\sigma c}, I_3^{\sigma c}, \ldots) \leq k^2. \tag{6.13}$$

Simultaneous invariants of the stress tensors and of the strength tensor coefficients (e.g., elastic cocfficients) are defined by the formulas

$$I_1^{\sigma c} = c_{ij}\sigma_{ij}, \qquad I_2^{\sigma c} = c_{ijkl}\sigma_{ij}\sigma_{kl}, \qquad I_3^{\sigma c} = c_{ijklmn}\sigma_{ij}\sigma_{kl}\sigma_{mn}, \ldots.$$

The strength tensor coefficients $c_{ij}, c_{ijkl}, c_{ijklmn}, \ldots$ satisfy the symmetry conditions $c_{ij} = c_{ji}$, $c_{ijkl} = c_{jikl}$, $c_{ijkl} = c_{ijlk}$, $c_{ijkl} = c_{klij}, \ldots$, etc. For plastic materials, instead of strength (or failure) criteria we can adopt flow criteria. One of the simplest conditions, using only the properties of the first two invariants $I_1^{\sigma c}$

and $I_2^{\sigma c}$ is given by the formula

$$I_2^{\sigma c} - f(I_1^{\sigma c}) \leq 0, \tag{6.14}$$

where f is some given function of the first invariant. The criterion (6.14) is a generalization of the von Mises–Schleicher–Nadai criterion. A classical condition for initiation of plasticity, which is given by a quadratic function of the stress components was originally proposed by von Mises. In the most general case, it contains 21 constants c_{ijkl}. If we assume that the material is orthotropic, that the limits determining plastic flow are identical for tension and compression, and that the hydrostatic pressure does not influence the onset of plastic flow, then this quadratic condition contains only the six constants A_{11}, A_{22}, A_{33}, A_{12}, A_{23}, and A_{13} whose values are determined by the properties of the material:

$$A_{11}(\sigma_{22} - \sigma_{33})^2 + A_{22}(\sigma_{33} - \sigma_{11})^2 + A_{33}(\sigma_{11} - \sigma_{22})^2 + A_{12}(\sigma_{12})^2 + A_{23}(\sigma_{23})^2 + A_{13}(\sigma_{13})^2 - 1 \leq 0. \tag{6.15}$$

6.2 Designing for the Minimum Weight, and the Problem of Reducing Stress Concentrations to an Admissible Level

Problems of weight minimization for structures with constraints assigned to the strength and related problems of lowering the highest stress concentrations while satisfying isoperimetric conditions on the total quantity of material used in the design are fundamental to the theory of optimal design. These problems arise both in designs of traditional structures and in computations of certain unique projects. Research in this area originated a long time ago (it is credited to Galileo), but was intensively advanced only in the past decade and produced a number of important results. Now we have available solutions for one-dimensional structural elements and certain results for two-dimensional elements. We have developed numerical and analytical optimization techniques, investigated various (heuristic) recommendations regarding optimal design, and justified several estimates for the "best" lowering of the volume of the material and for reducing stress concentrations. However, it should be stated that, despite the simplicity of the postulation and the natural way in which they arise, many problems in this area reduce to very complex mathematical problems and so far have not been solved. Mathematical problems in this category relate to several new and nonclassical variational problems.

Let us formulate several such problems and commence a discussion of optimal design. We consider the equilibrium for an elastic body subjected to static surface forces T and body forces q. This elastic body occupies a region Ω (as is shown in Figure 6.2). The body forces act on the interior points, while the surface loads are applied to a segment Γ_σ of the boundary. The displace-

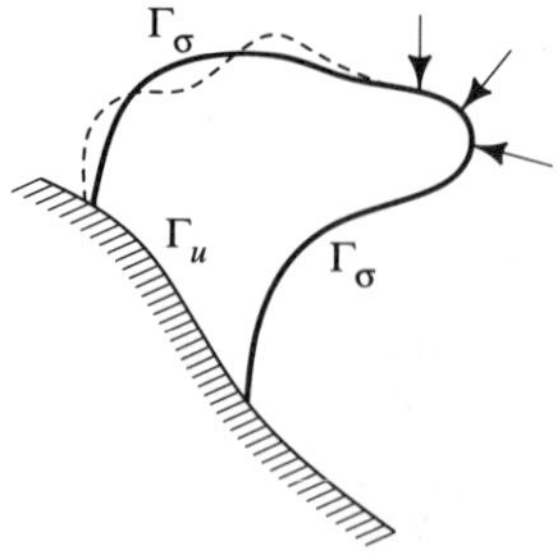

FIGURE 6.2. Static equilibrium for a deformed body.

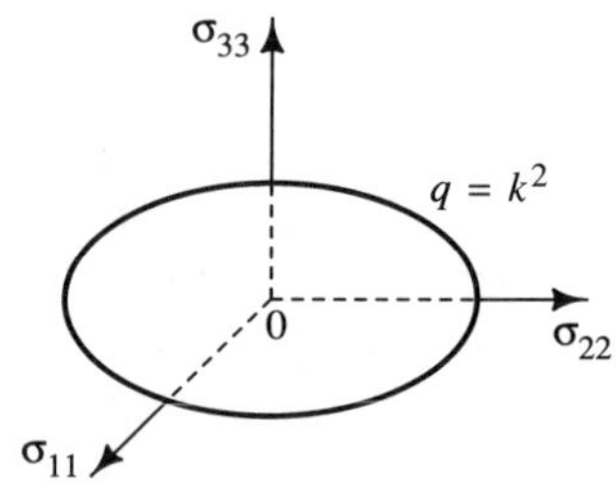

FIGURE 6.3. Strength constraint surface.

ments are prescribed on the segment Γ_u of the boundary. The stress-strain state of the body is described by equations of the linear theory of elasticity:

$$\left.\begin{aligned} \sigma_{ij,j} + q_i = 0, \qquad \sigma_{ij} = A_{ijkl}\varepsilon_{kl}, \\ \varepsilon_{kl} = 1/2(u_{k,l} + u_{l,k}). \end{aligned}\right\} \tag{6.16}$$

Here A_{ijkl} denote the elastic constants, q_i are components of the vector of body forces, u_k are the components of the displacement vector, and σ_{ij} and ε_{kl} are the components of the stress and strain tensors, respectively. An index following a comma denotes differentiation with respect to the indicated coordinate. The boundary conditions assigned to eqs. (6.16) are of the form

$$(u_i)_{\Gamma_u} = U_i, \qquad (\sigma_{ij}\mathbf{n}_j)_{\Gamma_\sigma} = T_i. \tag{6.17}$$

The constraint on strength is written as

$$g(\sigma_{ij}) \leq k^2, \tag{6.18}$$

where g is a given homogeneous function of the components of the stress tensor and k is a given constant. It is assumed that g is a convex homogeneous function of the stress tensor components (see Figure 6.3). Any of the functions discussed in the previous section may be chosen as the function $g(\sigma_{ij})$.

The optimization problem now consists in finding a segment Γ_v of the body surface that minimizes the volume (or weight) of the body:

$$V = \int_\Omega d\Omega \to \min_{\Gamma_v} \tag{6.19}$$

while the strength constraint (6.18) is satisfied.

In all known theories of strength, condition (6.18) has a local (i.e., non-integral) character. The function $g(\sigma_{ij})$ is a function of the spatial coordinates x_i and attains its maximum at some initially unknown internal point of the region Ω. An estimate of the maximum value of the function $g(\sigma_{ij})$ is closely associated with the computation of the values of σ_{ij}, which are to be determined from the solution of the boundary-value problem (6.16) and (6.17) of the static equilibrium for an elastic body. Here lies the source of well-known difficulties in computing the constraint (6.18).

In a number of cases in computing solutions of optimization problems with constraints on strength, it is possible to employ a computational device that consists in replacing the constraint (6.18) by the equality constraint

$$\int_\Omega \{(g - k^2) + |g - k^2|\} d\Omega = 0, \tag{6.20}$$

introduced in [4.23] to [4.26]. If inequality (6.18) is satisfied everywhere in Ω, at every point inside the body the expression inside the braces in (6.20) is identically equal to 0, and condition (6.20) is satisfied. Conversely, if at some point in Ω condition (6.18) is violated (i.e., if $g \geq k^2$), the integral (6.20) has a positive value. Since the expression inside the braces in (6.20) is nonnegative, we come to the conclusion that (6.18) and (6.20) are equivalent.

Another technique, which can be effectively used in computing the strength constraints, also consists in replacing the local constraint (6.18) by a global constraint. In this technique we utilize the closeness of the number $\max_x g$ to the p-norm of g, that is, the norm in the space of functions whose pth power is integrable, with the parameter p chosen sufficiently large (see also Section 2.3). Then condition (6.18) is replaced by the constraint

$$\left[(1/\text{meas }\Omega) \int_\Omega g^p d\Omega\right]^{1/p} \leq k^2. \tag{6.21}$$

Various questions arising in applications of the methodology of (6.21) to approximate strength criteria are discussed in [5], [2.3], and [4.3]. An effective technique consists in introducing slack (auxiliary) functions $\mu(x)$ to convert the strength constraint (6.18) into the equality constraint

$$g - k^2 + \mu^2 = 0. \tag{6.22}$$

The constraint (6.22) for solving an optimization problem may be implemented and the necessary conditions for optimality derived by the use of standard techniques.

Problems closely related to those of optimal design concern the lowering of critical values of stress concentrations. Let us explain the process of formulating such problems. Let an elastic body occupy some region Ω that, generally, may be multiply connected. It is subjected to the action of certain loads, which are proportional to a parameter t. To learn about the influence of the shape of the body on the stress concentrations, which are expressed by some quantitative measure of these concentrations (i.e., by a cost functional for the problem J), we choose as this functional the maximum value (with the maximum taken over the domain Ω occupied by the deformed body) of the function $g(\sigma_{ij})$, that is

$$J = \max_{x \in \Omega} g. \tag{6.23}$$

To be more specific, let us assume that the function g has degree of homogeneity 2. We then use the proportionality of the load and derive a formula expressing the critical value t_* for which condition (6.18) is violated:

$$t_* = k/\sqrt{J}. \tag{6.24}$$

In (6.24) the quantity J is computed for the value $t = 1$.

Consequently, it is possible to broaden the class of loads that are admissible according to the accepted criterion of strength by minimizing J. The minimization problem now consists in finding a segment of the boundary of Ω that produces the minimum of the maximum of the quantity g in the region $\Omega + \Gamma$, that is

$$J_* = \min_{\Gamma_v} J = \min_{\Gamma_v} \max_{x \in \Omega + \Gamma} g. \tag{6.25}$$

A study of techniques for formulating this problem has been conducted in [6.4], [6.5], [6.39], [6.40], [6.95], [6.96], and [6.99] with applications to bending and compression of elastic plates with holes, and also in [6.97] and [6.98] for determining the optimal shapes of strips inside a deformed material. In all of these research articles it was tacitly assumed that the domain Ω occupied by the material may be extended indefinitely (to an infinite domain).

Problems similar to those discussed in these studies arise in the search for the optimal shapes of elastic bodies of finite size and in computing optimal reinforcement in regions with sharp changes in their geometry and with high stress concentrations. However, in researching the problem of lowering the critical stress concentrations (problems with local cost functionals), one encounters serious mathematical problems caused by the finite size assumption. Therefore, research into such problems is carried out by using direct numerical techniques (cf. [6.6], [6.21], [6.25], [6.45], [6.68], [6.83], [6.95] and [6.101]).

6.3 Optimality and Uniform Strength

An important role has been played in optimal design theory by the concept of uniform strength. This concept is closely related to the idea of the simultaneous breakdown of strength and bearing capacity by all parts of a structure. The idea that no reserves of strength are left and that all elements of a structure fail at the same time has been associated in design practice with the minimal weight condition, and intuitively was assumed to represent an optimality criterion. The principle of uniform strength still remains important in modern design research. A problem of optimal design is greatly simplified if one uses the principle of uniform strength. For this reason it has been used in some "inverse" problems of the theory of elasticity.

Frequently, problems of optimality and uniform strength are regarded as equivalent. However, they are not always even close to being equivalent. Optimality conditions turn out to be far more general than the condition of uniform strength. To prove optimality for a uniform strength design, it is necessary to assign certain additional conditions implying that the constraints assigned to the strength and stability of the structure are inessential, and also that the structure possesses certain dynamic characteristics.

There exist different types of definitions of uniform strength. We shall reproduce some of them. A continuous body is said to have uniform strength if along any of its boundary surfaces the strength function $g(\sigma_{ij})$ is constant and equal to its critical (limit) value, while at any of the interior points the function g does not exceed its critical value, which is attained on the boundary:

$$(g(\sigma_{ij}))_\Omega \leq (g(\sigma_{ij}))_\Gamma = k^2. \tag{6.26}$$

In many cases the definition of uniform strength according to (6.26) is much too restrictive. We have in mind both cases when solutions do not exist, that is, when stresses have been "equalized" on the boundary, and cases when singularities arise in optimization projects. To enlarge the domain of applicability of the uniform strength concept, we introduce the concept of a structure with partially unform strength. For a structure with partially uniform strength, the function $g(\sigma_{ij})$ is constant only on certain segments Γ_l of the boundary Γ:

$$(g(\sigma_{ij}))_{\Omega+\Gamma} \leq (g(\sigma_{ij}))_{\Gamma_l}. \tag{6.27}$$

The problem of designing structures having partially uniform strength is quite topical for bodies of finite size and for structures with stress concentrations.

The greatest number of projects with partially uniform strength can be found in thin-walled structures, such as beams, curved rods, trusses, plates, and shells (cf. [6.1], [6.23], [6.26], [6.27], [6.31], [6.42], [6.65], [6.76], [6.77], [6.88], [6.90], and [6.92]. The fundamental problem consists of finding the distribution of thicknesses $h(s, t)$ (where s and t are coordinates of the median surface of a shell) and the shape of the median surface. The coordinate ζ, which measures the distance in the direction normal to the median surface, varies in the interval $-h/2 \leq \zeta \leq h/2$. For a structure subjected only to bending, the stresses normal to the median surface are in fact nonuniform. A thin-walled structure subjected to bending is said to have uniform strength if there exists a point (or an entire interval) on an arbitrary normal to the median surface on which the critical state of the material is attained, while at the other points the stressed state is characterized by smaller values of g, that is,

$$\max_{\zeta} g = k^2. \tag{6.28}$$

The maximum with respect to ζ is computed for $\zeta \in [-h/2, +h/2]$.

Let us examine cases of optimal design in which the uniform strength criterion may be used as the optimality condition. Let the integrand in the cost functional J,

$$J = \int_\Omega f(x, \mathbf{h}) d\Omega, \tag{6.29}$$

and the constraint function $g(\sigma_{ij}(x, h)) = g(x, \mathbf{h}, J(\mathbf{h}))$ be independent of the (arbitrary) design variables. Here $\mathbf{h} = \{h_1, \ldots, h_n\}$ and $J = \{J_1, \ldots, J_r\}$, while the integrals J_i are given by the formulas

$$J_i = \int_\Omega f_i(x, \mathbf{h}) d\Omega, \qquad i = 1, 2, \ldots, r. \tag{6.30}$$

In addition, we assume that f_i is a function of at least one of the h_i $(i = 1, 2, \ldots, n)$ and that it is monotone. To be more specific, we can replace the monotonicity condition by the requirement that as a function of each component h_κ, $1 \leq \kappa \leq n$, f is monotone, that is,

$$\partial f/\partial h_\kappa \neq 0. \tag{6.31}$$

Then the optimum of the functional J with constraint $g \leq k^2$ can be attained only if the structure obeys the uniform strength criterion

$$g(x, h_1, \ldots, h_n, J_1(h_1, \ldots, h_n), \ldots, J_r(h_1, \ldots, h_n)) \leq k^2. \tag{6.32}$$

This assertion has been proved in [7]. To prove it, we need to introduce a slack variable $\chi^2(x)$ (see Section 2.2) and rewrite the inequality constraint on strength $g(x, \mathbf{h}, J(\mathbf{h})) \leq k^2$ as an equality constraint:

$$g(x, \mathbf{h}, J(\mathbf{h})) - k^2 + \chi^2(x) = 0. \tag{6.33}$$

Now, let us assume to the contrary that at some point x inside the optimal structure $g < k^2$, and consequently that $\chi^2 \geq 0$. We list the necessary conditions for optimality at that point. To do so, we multiply the expression on the left-hand side of eq. (6.33) by λ (a Lagrangian multiplier) and integrate it over the region Ω. Adding this expression to the functional (6.29), we obtain a Lagrangian functional. The system of Euler–Lagrange equations represents a stationary behavior condition for the Lagrangian. The number of Euler–Lagrange equations is equal to the number of components of the vector of design variables. To these equations we need to add the necessary equation for extremum with respect to the slack variable χ. We present this condition and the Euler–Lagrange equations with respect to one of the design variables h_κ only:

$$\partial f/\partial h_\kappa + \lambda \partial g/\partial h_\kappa + \sum_{j=1}^{r} (\partial f_j/\partial h_\kappa) e_j = 0, \tag{6.34}$$

$$\lambda\chi = 0, \qquad e_j = \int_\Omega \lambda(\partial g/\partial J_j) d\Omega. \tag{6.35}$$

Since it has been assumed that at the point considered here $\chi > 0$, we can use (6.34) and (6.35) to obtain $\lambda = 0$ and $\partial f/\partial h_\kappa = 0$. However, this contradicts the assumption that f is a monotone function of h_κ and proves our assertion (6.31) concerning the optimality of a structure with the uniform strength property, as described by (6.29) to (6.31). Note that the assumptions made above are true, for example in the design of statically determinate structures.

As an example, we can consider the classic problem of designing an initially straight rod, subjected to the forces consisting of its own weight and a load P applied to the free end of the rod (at $x = l$). The other end at $x = 0$ is built-in, while the weight of the rod and the load P act in the direction of the x-axis. The unknown design variable, which is to be determined, is the function $S = S(x)$, describing the distribution of cross-sectional area along the length of the rod. The cost functional and constraint on strength assume the form

$$J = \int_0^l S(x)\mathrm{d}x \to \min_S, \tag{6.36}$$

$$\sigma_x \equiv P/S + (\gamma/S)\int_0^l S(x)\mathrm{d}x \leq k. \tag{6.37}$$

It is clear that the monotonicity assumption of the integrand with respect to the design variable is satisfied, and that this functional and the strength requirement are independent of the corresponding S.

The solution of the optimal problem (6.36) and (6.37) with the use of this assertion reduces to the much simpler problem of finding the shape of a uniformly stressed rod, that is, determining the function $S(x)$, from the condition $\sigma_x = k$. Integrating the equation of equilibrium for an infinitesimal element of the rod, we derive

$$\mathrm{d}S/S = -\gamma \mathrm{d}x/\sigma_x.$$

Now, using the condition $\sigma_x = k$ and the boundary condition $S(l) = P/k$, we arrive at the well-known solution

$$S = (P/k)\exp\{(\gamma/k)(l - x)\}. \tag{6.38}$$

The uniform strength solution (6.38) is discussed in several strength of materials textbooks (e.g., [6.10] and [6.93]).

6.4 Optimal Design of Beams with Constraints on Strength

Research into the optimization of structures dates back to the classical treatise of Galileo [6.22], which deals with the optimization of the shapes of beams. One result this work was the solution of several problems for beams subjected to pure bending. But even in a large number of contemporary works the beam model is used in optimal design research. The equations of beam bending are some of the simplest in strength of materials theory, and provide a convenient setting for the formulation of new problems and for comparing different algorithms and techniques. In the subsequent sections, we shall examine certain typical problems of optimal design within the framework of (the Euler–Bernoulli) beam theory.

6.4.1 If we choose either the volume of a beam or the maximum value of the stress as our cost functional, it is possible to pose the problem of minimizing the function representing the maximal stress intensity for a given volume of the beam, or minimizing the beam's volume for a given maximum value of the stress intensity function. These problems turn out to be dual, and consequently, we can follow [6.3] and examine only one of them, that is, the problem of minimizing the volume of a beam, subject to constraints on its strength.

We assume that the beam lies along the x-axis and that it has a rectangular cross-section with height $h = h(x)$ and width $b = b(x)$. The beam's length is l.

Here we consider the statically determinate case (simply supported beams subjected to the action of transverse loads), with bending moment function $M = M(x)$ acting on the cross-section of the beam independent of the elastic properties or the shape of the cross-section. We regard the function $M(x)$ as known for $0 \leq x \leq l$. The functions $b = b(x)$ and $h = h(x)$ determining the shape of the beam are the unknown quantities and are to be found from the solutions of the variational problem

$$\left.\begin{aligned} V &= \int_0^l bh\mathrm{d}x \to \min, & g &= (\sigma_x)^2 + \beta(\tau_{xy})^2 \leq k^2, & & \\ \sigma_x &= My/I, & \tau_{xy} &= (1/b)(Md/I)_x, & \beta &= 3, 4, \\ I &= bh^3/12, & d &= (b/2)(h^2/4 - y^2). & & \end{aligned}\right\} \tag{6.39}$$

Here I denotes the moment of inertia of the cross-sectional area with respect to the axis perpendicular to the plane of bending and passing through the neutral axis of the beam; d is the static moment, and k is a given constant. With the exception of the stress components, the subscript x denotes the operation of differentiation of the indicated quantities with respect to the variable x. The coordinate y measures the distance from the center of the cross-section and varies in the interval $-h/2 \leq y \leq +h/2$. We use the expressions for σ_x, τ_{xy}, I and d displayed in (6.39) to obtain a direct functional dependence of g on the variables b, h, and M:

$$g = (144/b^2h^6)\{M^2y^2 + \beta/4[(h^2/4 - y^2)M_x + (M/h)(3y^2 - h^2/4)h_x]^2\}.$$

The function g depends not only on the variables b, h, and M, but also on y, which varies with the height h of the cross-section ($|y| \leq h/2$). This variable may be excluded from future considerations by observing the explicit functional dependence of g on y, and recalling that the values of g are bounded only from above. We introduce an auxiliary function $\Psi = \max_y g$, with $-h/2 \leq y \leq +h/2$, and transform the strength constraint into the equivalent form $\Psi \leq k^2$. Let us derive a formula for Ψ. In computing Ψ we make use of the fact that g is a polynomial containing three terms quadratic with respect to y^2, with the coefficient of y^4 nonnegative. Therefore, the maximum of g with respect to y is attained either at $y = 0$ or else at $y = \pm h/2$. Computing these maxima, we arrive at the formula

$$\Psi = \max\{(36M^2/b^2h^4)(1 - (\beta/4)h_x^2), (9\beta/4b^2h^2)[M_x - (M/h)h_x]^2\}. \quad (6.40)$$

The operation max in (6.40) results in the selection of the larger of the two quantities inside the braces.

6.4.2 Let us find the shape of the beam in the case when $h = h(x)$ is given, and the optimal distribution of the width of the cross-section $b = b(x)$ has to be determined. We substitute eq. (6.40) for Ψ in the strength criterion $\Psi \le k^2$, and examine only the case when the maximum of (6.40) is attained by choosing the first of the two quantities inside the braces. As a result, we obtain constraints assigned to the function $b = b(x)$ that are generated by the strength criteria:

$$b \ge (6M/kh^2)(1 + \beta h_x^2/4)^{1/2}, \qquad b \ge (3\sqrt{\beta}/2k)|(M/h)_x|. \quad (6.41)$$

Now we deal with the problem of finding a function $b = b(x)$ that satisfies inequalities (6.41) and minimizes the integral (6.39). It should be clear that the required minimal solution will be of the form

$$b = \max[(6|M|/kh^2)(1 + \beta h_x^2/4)^{1/2}, (3\sqrt{\beta}/2k)|(M/h)_x|]. \quad (6.42)$$

In particular, for beams with constant height (i.e., $h = \text{const}$) we have $b = \max[(6M/kh^2)$ and $(3\sqrt{\beta})|M_x|/2kh]$. For the particular case when $\beta = 4$, $k = 1$, $h(x) \equiv 1$, $M = x(x - 1)$, and $l = 1$, the function $b(x)$ is illustrated in Figure 6.4.

6.4.3 Let us now discuss the problem of finding an optimal distribution of the height $h = h(x)$ for the case of constant width, that is, $b = \text{const}$. We consider a cantilevered beam built-in on its left end at $x = 0$ and loaded by a single concentrated force P at its right endpoint $x = l$. In this case $M = -P(l - x)$.

Now, we examine the inequalities assigned to the function $h(x)$ generated by the strength constraints. If the maximum in (6.40) is realized by the first term inside the braces, we can derive from the condition $\Psi \le k^2$ the two inequalities

$$h \ge \chi(x) \equiv [6P(l - x)/kb^2]^{1/2}, \quad (6.43)$$

$$|h| \le \{(4/\beta)[k^2b^2h^4/(36P^2(l - x)^2) - 1]\}^{1/2}. \quad (6.44)$$

We observe that the expression inside the brackets in (6.44) is positive if (6.43) is satisfied. Let $H(x)$ $(0 \le x \le l)$ denote a solution of the problem

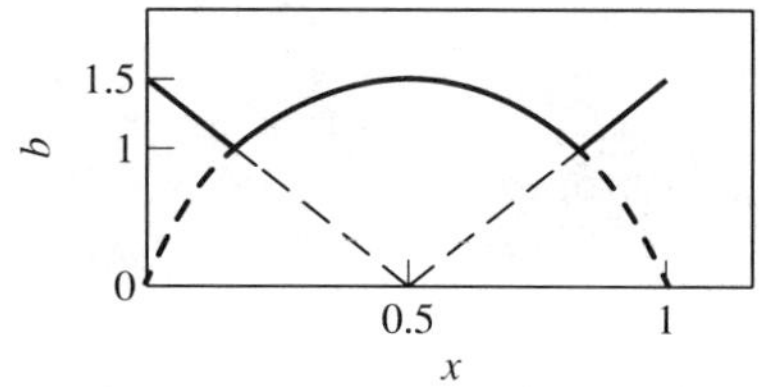

FIGURE 6.4. Optimal width function.

$$H_x = -(4/\beta)[k^2b^2H^4/(36P^2(l-x)^2) - 1]\}^{1/2}, \quad H(0) = [6Pl/kb^2]^{1/2}. \quad (6.45)$$

Using the theory of differential inequalities, it can be shown that $H(x) > \chi(x)$ and that $H(x) \to 0$ as $x \to l$, and that an admissible solution $h(x)$ (i.e., one that satisfies $\Psi \leq k^2$) also satisfies the inequality

$$h \geq (3P\sqrt{\beta})/(2kb). \quad (6.46)$$

These properties of the function $h(x)$ directly imply that

$$h = \max[(3P\sqrt{\beta})/(2kb), H(x)] \quad (6.47)$$

satisfies the strength constraint and also realizes a minimum for the integral (6.39), and therefore is the required optimal solution. This optimal solution may be rewritten in the form $h = H(x)$ if $0 \leq x \leq x_*$, and $h = (3P\sqrt{\beta})/(2kb)$ if $x_* \leq x \leq l$. The value of x_* is found from the requirement that $h(x)$ be continuous. Note that on the interval $0 \leq x \leq l$, the function $H(x)$ attains its maximum at $x = 0$. Therefore, if the inequality $H(0) \leq (3P\sqrt{\beta})/(2kb)$ is true, which happens if $P \geq 8kbl/3\beta$, the optimal beam has a constant height given by $h = 3P\sqrt{\beta}/(2kb)$.

Note, too, that our derived optimal solution is also a uniform strength solution, that is, $\max_y(\sigma_x^2 + \beta\tau_{xy}^2) = \Psi(x) = k^2$. Indeed, on the interval $0 \leq x \leq x_*$ of the beam's length, on which the height is distributed according to the rule $h = H(x)$, the first term in the braces has a maximum in (6.42), if we compute $\Psi(x) = k^2$. For $0 \leq x \leq x_*$, the maximum of the stress intensity, that is, of the quantity $\sigma_x^2 + \beta\tau_{xy}^2$, occurs on the outer beam surfaces $y = H/2$. On the remaining segment of the beam's length $x_* \leq x \leq l$, where the height of the beam is constant: $h = 3P\sqrt{\beta}/(2kb)$, the second term attains a maximum again equal to k^2. For this segment, while the beam is in the limiting allowed state, there exists a neutral surface ($y = 0$), such that at any point of that surface the stress intensity $\sigma_x^2 + \beta\tau_{xy}^2$ attains its maximum with respect to y. For $y = 0$ on $x_* \leq x \leq l$, $\sigma_x = 0$ and $\tau_{xy}^2 = k^2/\beta$. The function $H(x)$ can be computed by numerically integrating eq. (6.45), which may be written for convenience in dimensionless variables: $H' = kbH/(3P\sqrt{\beta})$ and $t = 2(l-x)kb/(3P\sqrt{\beta})$. In these new variables (with the primes omitted), these equations and initial conditions become

$$H_t = [H^4/t^2 - 1]^{1/2} \text{ and } H(t_1) = \sqrt{t_1}.$$

This is a one-parameter problem with parameter t_1. This equation can be integrated on the interval $t_0 \leq t \leq t_1$. To the left of t_1, this assumes the appearance of an initial-value problem $H(t_1) = \sqrt{t_1}$, and t_0 may be found from the condition $H(t_0) = 1/2$, which can be deduced from (6.46). The values of the function $H = H(t)$, computed for $t_1 = 0.3, 0.4, 0.5, 1$, and 2 are illustrated in Figure 6.5 by curves 1 to 5. The broken line in Figure 6.5 is the graph of the function $H = \sqrt{t}$.

Optimal shapes were obtained in [6.3] (also see [5], [6.2]) for cantilevered beams of circular cross-section loaded by a concentrated force applied to the free end.

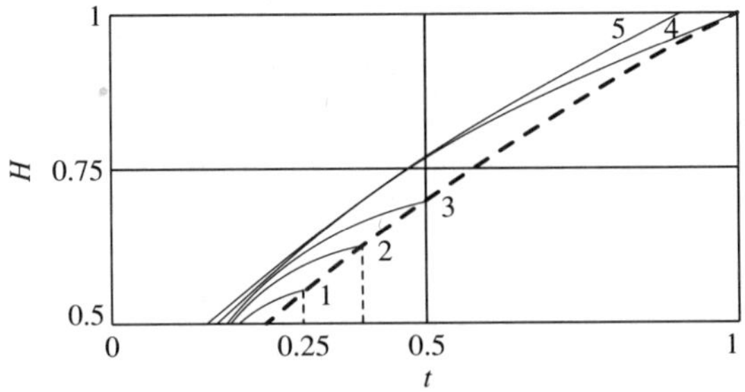

FIGURE 6.5. Optimal height function.

6.5 Optimal Shapes of Beam Cross-Sections with Unequal Strength

In the previous section, we discussed results concerning the optimal distribution of material along the length of a beam. Another basic problem that turns up in designs of systems of beams working in pure bending is the derivation of the optimal shape of the cross-section of a beam. This problem is considered in [6.8], [6.36], [6.89], and [6.91].

Let us consider the problem of finding the cross-sectional shape of a beam having least weight. We assume that the beam lies along the x-axis of a Cartesian coordinate system $Oxyz$. The cross-section of the beam in the yz-plane is characterized by the height $2H$ and is symmetric with respect to the origin of the coordinate system (the point O). The width of the cross-section is given by a functon $2Z(y)$, with $-a \leq y \leq +a$, and is equal to a constant value $2b$ on $a \leq y \leq H$ and $-H \leq y \leq -a$. Taking into account the symmetry, we can consider only a part of the cross-section, as indicated in Figure 6.6 (with $y \geq 0$ and $z \geq 0$).

The parameters H, a, and b are regarded as known. We also assume that we know the bending moment M and the shear load Q that act on a given cross-section, and the value of k, which characterizes the strength of the beam's material. The thickness of the web $Z(y)$ is the unknown "control" function. To describe the contour of the beam, we also need to examine the function $z(y) = Z(y)$ if $0 \leq y \leq a$, and $z(y) = b$ if $a \leq y \leq H$, which is defined on the interval $0 \leq y \leq H$.

The problem of finding an optimal design consists of finding a function $Z(y)$ $(0 \leq y \leq a)$ from the condition of minimizing the cross-sectional area:

$$J = 4b(H - a) + 4 \int_0^a Z(y)dy \to \min_Z, \tag{6.48}$$

while satisfying the strength constraint

$$\psi(y) = ((1 + \alpha)/2)\sqrt{\sigma_x^2 + \beta\tau_{xy}^2} + ((1 - \alpha)/2)|\sigma_x| \leq k, \tag{6.49}$$

where

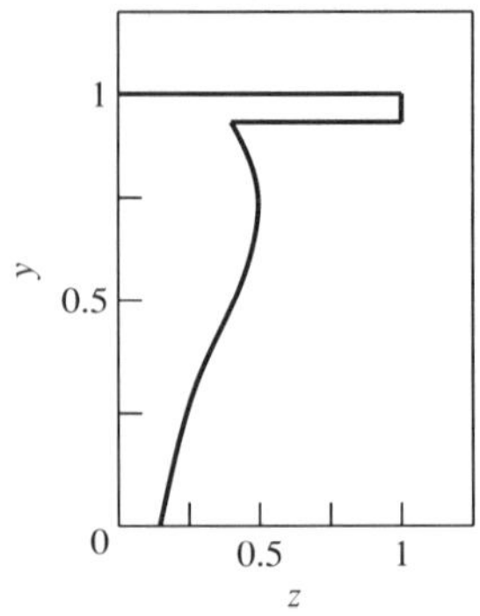

FIGURE 6.6. Half-section of the web of a beam.

$$\sigma_x = My/I, \qquad \tau_{xy} = QS/(2Iz), \qquad I = 4\int_0^a y^2 Z\,dy + 4b(H^3 - a^3)/3;$$

$$S(y) = 2\int_y^a \xi Z\,d\xi + S_0 \quad (0 \le y \le a), \quad S(y) = b(H^2 - y^2) \quad (a \le y \le H),$$

$$S_0 = b(H^2 - a^2). \tag{6.50}$$

Note the differences between the formulation of problem (6.48) to (6.50) and problems of beam optimization discussed in earlier sections. In previously discussed problems, we assumed that the shape of the cross-section was already determined, except for data supplied by the value of a single parameter (the width or height of a beam with rectangular cross-section, or a coefficient of similarity for similar cross-sections), which was an unknown function of the coordinate x. In determining the values of design variables from the condition of least mass and in studying shapes that were candidates for the optimal solution, one examines projects of uniform strength, whereby the distribution of the "control" functions ensures that at certain points of an arbitrary cross-section of a beam precisely the same critical value of the stress intensity is attained. Because of this feature, several articles have posed the "inverse" problems of determining the distribution of the thickness from the condition of attaining the limiting stress values in every cross-section of the beam.

The following feature is an essential difference of the problem of optimizing the shape of cross-section we are now considering and distinguishing it from problems of establishing a best distribution of load-bearing material along the span of the beam. In this postulation of the problem, there may not exist a uniformly stressed shape of a cross-section such that $\psi(y) = k$ for $0 \le y \le a$ (with $a \ne H$) (cf. [6.89] and [6.91]).

In determining the shape of a uniformly stressed part of a beam's cross-section for $0 \le y \le a$, that is, the shape $Z(y)$ of the web, we use the condition $\psi(y) = k$. Here we assume that the value of the parameter is chosen so that in the stressed state the material satisfies a strength constraint, that is, $\psi(y) \le k$

for $a \le y \le H$. The condition $\psi(y) = k$ for $0 \le y \le a$ can be written as

$$\left.\begin{aligned} &S(y) = 2\int_y^a \xi Z\,d\xi + S_0 = 2Z(y)/\chi(y),\\ &\chi(y) = (Q/M)\{(1/\beta)[(2/(1+\alpha))^2(c-(1-\alpha)y/2)^2 - y^2]\}^{-1/2},\\ &c = Ik/M. \end{aligned}\right\} \quad (6.51)$$

The function $Z(y)$ describing the shape of a panel with uniform strength satisfies an inhomogeneous Volterra's integral equation of the first kind. To solve it we can use a change of variable defined by the transformation $d\zeta/dy = -yZ(y)$, which reduces the integral equation (6.51) to a Cauchy problem for the ordinary first-order differential equation

$$d\zeta/dy = -(S_0 + 2\zeta)My\chi(y)/(2QT), \qquad \zeta(0) = 0.$$

Solving this equation and eliminating an arbitrary constant from the condition $\zeta(0) = 0$, and then substituting the original variable, we derive the following distribution of thickness for a uniformly stressed panel web:

$$Z_c = S_0\Phi(y), \qquad 0 \le y \le a, \qquad \Phi(y) = (\chi(y)/2y)\exp\left(\int_y^a \chi(\xi)\,d\xi\right). \quad (6.52)$$

This formula indicates that Z_c depends on the function $\chi(y)$ and on the parameter c, which is a linear function of the moment of inertia I of the cross-sectional area. To determine the value of c, we can use a relation obtained by substituting $Z = Z_c(y)$ from eq. (6.52), and substituting $I = Mc/k$ in (6.50) for the moment of inertia, and then completing the transformation:

$$\left.\begin{aligned} &4k\Psi(c)b = 3Mc\\ &\Psi(c) \equiv \frac{3}{2}(H^2 - a^2)\int_0^a \left\{y\chi(y)\exp\left[\int_y^a \chi(\xi)\,d\xi\right]\right\}dy + H^3 - a^3. \end{aligned}\right\} \quad (6.53)$$

Formula (6.53) allows us to determine the value of c, and then, applying (6.52), to find the shape of the panel.

It was shown in [6.8] that there exists a one-to-one relation between the parameter b and the value of the solution c of eq. (6.53), and also that $db/dc > 0$. This property can be utilized in finding the limits on the existence of an optimal solution. In addition, it was shown in [6.8] that there exists a unique value c of eq. (6.53) for arbitrary values of the parameters assigned to this problem. This value of c also appears in the formula $I = Mc/k$, which gives the magnitude of the moment of inertia. In turn, I and c must satisfy constraints on the strength of the panel web valid for all y, $0 \le y \le H$. The uniform strength condition $\psi(y) = k$ for $0 \le y \le H$, combined with the relation $c = Ik/M$, can be expressed in the form

$$\left.\begin{aligned} &c \ge c_{\min} = \max_{a \le y \le H}\,[(1-\alpha)y/2 + 1/4\sqrt{16y^2 + \beta^2 m^2(H^2 - y^2)^2}\,],\\ &m = 2Q/M. \end{aligned}\right\} \quad (6.54)$$

It can be shown that (6.54) is satisfied if the thickness of the web obeys the inequality

$$b \geq b_{\min} = 3/4(Mc_{\min})/(k\Psi(c_{\min})). \tag{6.55}$$

If the value of the parameter c computed from eq. (6.53) satisfies inequality (6.54), we also minimize the functional defining the area of the cross-section in constructing the optimal shape of the cross-section; that is, the corresponding value of Z_c turns out to be a solution of the optimization problem (6.48) to (6.50). The strength constraint on a beam with cross-section size Z is given by

$$\frac{(1-\alpha)}{2}\frac{My}{I} + \frac{(1+\alpha)}{2}\sqrt{\left(\frac{My}{I}\right)^2 + \beta\left(\frac{QS}{2I}\right)^2} \leq k, \qquad 0 \leq y \leq a,$$

which can be written as

$$\frac{S}{2} \equiv \frac{S_0}{2} + \int_y^a yZ\mathrm{d}y \leq \frac{yZ}{\chi}, \tag{6.56}$$

where χ is defined by (6.51). Now we introduce a variable $u = yZ/\chi$, and apply some elementary algebra to (6.56), obtaining

$$\chi \leq u\chi\left(\int_y^a u(t)\chi(t)\mathrm{d}t + S_0/2\right)^{-1}. \tag{6.57}$$

Integrating both sides of (6.57) between y and a, and making use of (6.56), we derive

$$\int_y^a \chi(y)\mathrm{d}y \leq \ln\left\{\int_y^a u\chi\mathrm{d}y + S_0/2\right\} - \ln(S_0/2)$$
$$\leq \ln(S_0/2) \leq \ln(Zy/\chi) - \ln(S_0/2). \tag{6.58}$$

From (6.58) we deduce the inequality

$$Z \geq \left(\frac{S_0}{2y}\right)\exp\left(\int_y^a \chi\mathrm{d}y\right) = Z_c, \qquad 0 \leq y \leq a.$$

Therefore, among all admissible cross-sections (i.e., those that satisfy the strength conditions) having identical moment of inertia I, the minimum of the cost functional is attained by a cross-section having a uniform stress distribution in the material of its web:

$$J = 4\int_0^a Z\mathrm{d}y + 4(H-a)b \geq 4\int_0^a Z_c\mathrm{d}y + 4(H-a)b = J_c.$$

We shall present computer output on the shapes of optimal cross-sections. In carrying out these computations, the quantities H, a, Q, M, b, and k were given. The value of the parameter $c = IkM$ was computed numerically by solving the transcendental equation (6.53) using Newton's method. The integral on the right-hand side of that equation was evaluated using Gaussian quadrature formulas. The boundary contour for the optimal solution was computed from (6.52). A sequence of cross-sections was determined for the values $H = 0.1$ m, $a = .095$ m, $Q = 10^4$ N, $k = .75 \times 10^9$ N/m^2, and $b = 0.006$ m, which were regarded as fixed, while the magnitude of M assumed various values.

Figure 6.7 shows graphs of the functions $z' = z'(y')$, $z' = 10^4 z$, and $y' = 10y$ (primes will be omitted) for values of the bending moment $M = 2 \times 10^5$, 4×10^4, and 6×10^4 (Nm). Curves corresponding to these values are numbered 1, 2, and 3, respectively. Comparison with a regular I-beam having the same strength, the same overall dimensions, and identical thickness of flanges reveals that the gain for the optimal beams amounts to 3%, 4%, and 6.2%, respectively.

The problem of optimizing the shape of a beam whose cross-section has an assigned moment of inertia may also be of some interest. As the strength constraint we can adopt condition (6.49). The constraint assigned to the value of the moment of inertia arises from a demand on the beam's bending rigidity.

If the width of the flanges is not fixed and is regarded as another unknown, together with the variable curve $Z(y)$, which describes the shape of the beam on $0 \leq y \leq a$, the optimization problem is reducible to the one discussed above.

The essential difference between that problem and a problem with fixed bending rigidity (i.e., fixed moment of inertia) consists in the list of quantities H, a, Q, M, and k that uniquely determine the shape of the optimal cross-section. The width of the web is determined by the formula $b = 3/4\, Mc/(k\Psi(c))$, where $c = Ik/M$. The optimization problem has a solution if

$$I \geq I_{\min} = (M/k) \max_{a \leq t \leq H} \left[((1-\alpha)/2)t + 1/4\sqrt{16t^2 + \beta m^2 (H^2 - t^2)^2}\right].$$

Figure 6.8 contains graphs of the functions $z' = z'(y)$, $z' = 10^4 z$, and $y' = 10y$ for values of the parameters $H = 0.1$ m, $a = 0.095$ m, $k = 7.5 \times 10^8$ N/m,

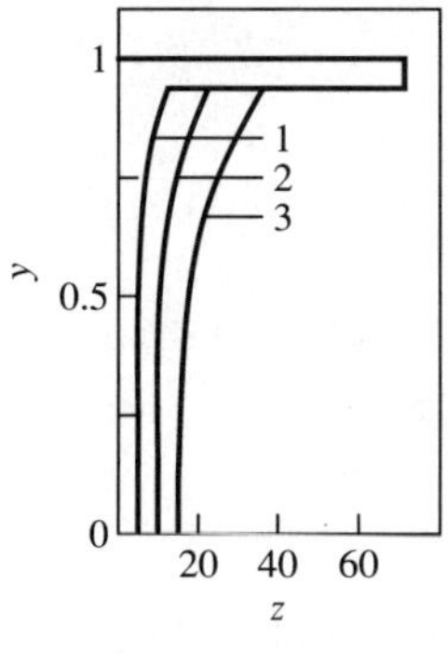

FIGURE 6.7. Optimal web shape.

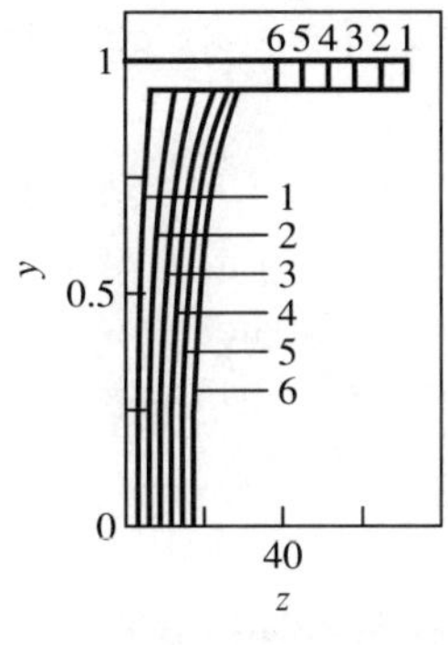

FIGURE 6.8. Optimal web shapes for different flange widths.

$M = 2 \times 10^5$ Nm, $I = 3.75 \times 10^{-4}$ m. The integers $1, 2, \ldots, 6$ are used as labels on curves corresponding to the values of the parameter Q, which is set equal to 10^4, 2.8×10^4, 4.6×10^4, 6.15×10^4, 8.23×10^4, and 10^5 H, respectively.

6.6 Design of Trusses Using Techniques of Mathematical Programming

A truss is a structure consisting of rods connected by swivel joints. The basic purpose of a truss is to "transmit" loads applied to some specific points of the truss to the locations of its support. Problems generally studied in optimal design theory consist in minimizing the quantity of the material (i.e., the weight of the truss) by choosing locations for the truss joints and the cross-sectional areas of its elements. Certain questions will be posed concerning the formulation of typical design projects, and examples will be offered of applications of mathematical programming techniques.

Let a set of admissible junction points be given. We shall consider all possible ways of connecting them by structural members. Designing along directions of forces in the members, we write the equilibrium equations at each joint point:

$$\sum_{j=1}^{n} a_{ij} T_j = F_i, \qquad i = 1, 2, \ldots, r, \tag{6.59}$$

where T_j are the forces acting on the jth rod, a_{ij} is the cosine of the angle between the direction of the jth member and the ith coordinate axis, and n is the number of structural members comprising the truss. F_j are the components of the total load applied to the joint ($F_j \neq 0$ only for joints to which a load is applied, or for points of support), and r is the number of scalar equations of equilibrium. We note that in (6.59) the only nonzero terms correspond to members conected to the joint in question. For all other members not connected to that joint, we set $a_{ij} = 0$. We denote by k the admissible values of the tensile (or, respectively, compressive) stresses, and by S_j the cross-section areas. We also state constraints assigned to forces T_j acting on the truss elements:

$$-kS_j \leq T_j \leq kS_j, \qquad j = 1, \ldots, n. \tag{6.60}$$

The optimization problem consists in finding the values of S_j and T_j (with $j = 1, 2, \ldots, n$) such that (6.59) and (6.60) are satisfied, and such that they minimize the total weight of the truss:

$$J = \rho \sum_{j=1}^{n} S_j l_j. \tag{6.61}$$

Here ρ is the weight per unit volume, and l_j is the length of the jth rod.

In an optimal design of a truss, all stress constraints must be active; that is, the inequalities in (6.60) must become strict equalities. If for some truss member it is true that $|T_j| < kS_j$, it is possible to reduce the magnitude of the cross-sectional area S_j without violating constraints (6.59) and (6.60). Therefore, for an optimal truss we must have

$$|T_j| = kS_j, \qquad j = 1, 2, \ldots, n. \tag{6.62}$$

Using (6.62), the size of S_j can be evaluated from the stress values in the structural members, and thus the optimization problem (6.59) and (6.60) can be restated in the form

$$\left.\begin{aligned} J = (\rho/k) \sum_{j=1}^{n} l_j |T_j| \to \min, \\ \sum_{j=1}^{n} a_{ij} T_j = F_i, \qquad i = 1, 2, \ldots, r. \end{aligned}\right\} \tag{6.63}$$

The current design variables S_j do not appear explicitly in (6.63). They can be determined from (6.62) after the problem posed by (6.63) has been solved. If in the solution of (6.63) we obtain $T_j = 0$ for some value of j, we can assign $S_j = 0$ to the corresponding member, which implies that in the design process we have eliminated that particular member. Thus, we have changed the basic configuration of the truss, and the truss obtained as a result of the weight optimization procedure may contain only a subset of the initially assigned junction points.

It is possible to reduce (6.63) to a problem of linear programming (see [6.16]). Indeed, we can introduce the quantities

$$T_j^+ = \begin{cases} T_j, & \text{if } T_j \geq 0, \\ 0, & \text{if } T_j \leq 0, \end{cases} \qquad T_j^- = \begin{cases} 0, & \text{if } T_j \geq 0, \\ -T_j, & \text{if } T_j \leq 0 \end{cases}$$

and represent the stresses in the members and their absolute values by

$$T_j = T_j^+ - T_j^- \quad \text{and} \quad |T_j| = T_j^+ + T_j^-. \tag{6.64}$$

We note that both T_j^+ and T_j^- are nonnegative; that is, $T_j^+ \geq 0$ and $T_j^- \geq 0$. We also define a matrix $\mathbf{L}$ and vectors $\mathbf{T}$, $\mathbf{F}$, $\mathbf{l}$ by the formulas

$$\left.\begin{aligned} &\mathbf{L} = \begin{bmatrix} a_{11} \ldots a_{1n} & -a_{11} \ldots -a_{1n} \\ \text{------} & \text{----------} \\ \text{------} & \text{----------} \\ a_{m1} \ldots a_{mn} & -a_{m1} \ldots -a_{mn} \end{bmatrix}, \\ &\mathbf{F} = \{F_1, \ldots, F_m\}, \qquad l = \{l_1, \ldots, l_n; l_1, \ldots, l_n\}, \\ &\mathbf{T} = \{T_1^+, \ldots, T_n^+, T_1^-, \ldots, T_n^-\}. \end{aligned}\right\} \tag{6.65}$$

Using these new variables in the notation of (6.64) and (6.65), we can reduce this optimization problem to an equivalent problem in linear programming:

$$\mathbf{lT} \to \min, \qquad \mathbf{LT} = \mathbf{F}. \tag{6.66}$$

The cost functional in (6.66), which may be written as the scalar product $\mathbf{lS}$, is related to the functional J, that is, $\mathbf{lT} = kJ/\rho$. In several instances it is of interest to consider the linear programming problem dual to (6.66):

$$\mathbf{FN} \to \max, \qquad \mathbf{NL} \leq \mathbf{l}. \tag{6.67}$$

There is a difference, since unlike the components of the vector $\mathbf{T}$, the components of $\mathbf{N} = \{N_1, \ldots, N_m\}$ do not have to be nonnegative. According to the basic theorem of linear programming, the solution of the original and of the dual problem are related by the equality

$$\min \mathbf{lT} = \max \mathbf{FN}.$$

We shall now consider the problem given by (6.59) to (6.61) with an additional constraint on the size of the cross-sectional areas:

$$S_i \geq S_{i\,\min}, \qquad i = 1, 2, \ldots, n, \tag{6.68}$$

where $S_{i\,\min} > 0$ is the minimal admissible value for the design variable S_i. In statically determinate cases, when for a given set of loads, the forces in members can be uniquely determined from the equations of equilibrium and are not functions of the deformed state, material properties, or cross-sectional area parameters of the members, the optimization process is greatly simplified. It is clear that in this case the optimal solution is given by

$$S_i = \max\{S_{i\,\min}, |T_i|/k\}, \tag{6.69}$$

which implies that for each member of the truss, the equality sign must replace the inequality sign in either (6.68) or (6.60). This means that at least one of these constraints must be active. In statically indeterminate cases, generally speaking, the presence of the limiting stress values does not assure optimality, and to solve such design problems, one must apply other mathematical optimization techniques (such as those given in Chapter 5).

Next, we shall present some facts related to an algorithm for iterative optimization, which uses the finite elements methods and gradient projection techniques. (We outline here results contained in the dissertation of E.V. Makeev.) This is an iterative algorithm that consists in successively approximating the optimal solution. At each step of this algorithm, computation of the stress-strain state of the structure is accomplished by finite-element methods. Following that, active, or ε-active, constraints are identified, and once these constraints are shown to hold, one uses the gradient projection technique to improve the design variations and make corrections in the design variables. This algorithm is applied until such time as the necessary conditions for optimality are almost satisfied, within a sufficiently small error. The algorithm was applied to a truss with 72 members (as shown in Figure 6.9). The design was carried out with the following values of the basic parameters: Young's modulus $E = 2 \times 10^6$ kg/cm^2, material density $\rho = 78 \times 10^{-4}$ g/cm^3, and admissible stress $k = 1.5 \times 10^4$ kg/cm^2. In one particular case, the loads were applied to joints 17, 18, 19, and 20 only, and were of the form

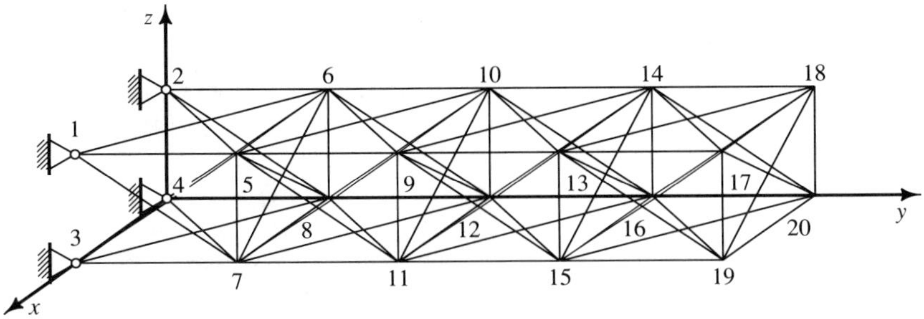

FIGURE 6.9. Optimization of a 72-member truss.

$$(F)_{17} = (\beta, \beta, -\beta), \qquad (F)_{18} = (\beta, -\beta, -\beta),$$

$$(F)_{19} = (-\beta, \beta, -\beta), \qquad (F)_{20} = (-\beta, -\beta, -\beta),$$

where $\beta = 3 \times 10^4$ kg. Initially, member 1-5 was assigned the maximum admissible stress value. The weight J of the truss was 29.94 kg ($S_{i0} = 2$ cm^2, $S_{i\,\min} = 0.4$ cm^2). For the computed optimal design the total weight was $J = 10.18$ kg and the limiting stress value was attained in members 1-5, 2-6, 4-6, 3-7, 3-5, 6-11, 5-9, 6-9, 6-10, 8-12, 7-11, 9-13, 10-14, 12-16, 11-15, 11-13, 15-16, 13-15, 14-17, and 16-19.

A more general optimization problem consists in finding not only the cross-sectional areas of the membrs, but also the locations of the joints for a truss. Again we minimize the weight, obeying conditions (6.60), and also the additional constraint

$$S_{i\,\min} \leq S_i \leq S_{i\,\max}, \tag{6.70}$$

where $S_{i\,\min} \leq S_{i\,\max}$ are assigned positive constants. In many cases, in order to make this problem regular, it is necessary to introduce additional constraints. Here we shall consider as the additional constraints the inequalities

$$|x_j - x_j^0| \leq \delta_j, \qquad |y_j - y_j^0| \leq \delta_j, \qquad |z_j - z_j^0| \leq \delta_j,$$

where x_j, y_j, and z_j are the current design coordinates of the jth joint, while x_j^0, y_j^0, and z_j^0 are the initial design coordinates of the jth joint, and δ_j is a regularization parameter. As a specific example, we shall consider optimization of a truss with 13 members (such as shown on Figure 6.10).

This problem is solved using the same successive optimization algorithm. The following (fixed) parameters for this problem are given by $E = 10^6$ kg/cm^2, $\rho = 8 \times 10^{-3}$ kg/cm^3, $k = 9 \times 10^3$ kg/cm^2, $S_{i0} = 2.5$ cm^2, and $S_{i\,\min} = 1$ cm^2. External forces $F = 3 \times 10^3$ kg are applied at the joints connecting the first and sixth member, to the ninth member, respectively. In the initial design the weight of the truss is 31.43 kg.

The optimal solution for the weight minimization problem with cross-

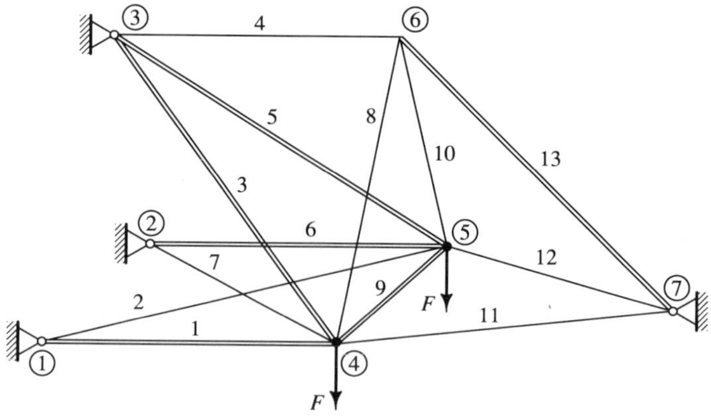

FIGURE 6.10. Optimization of a 13-member truss.

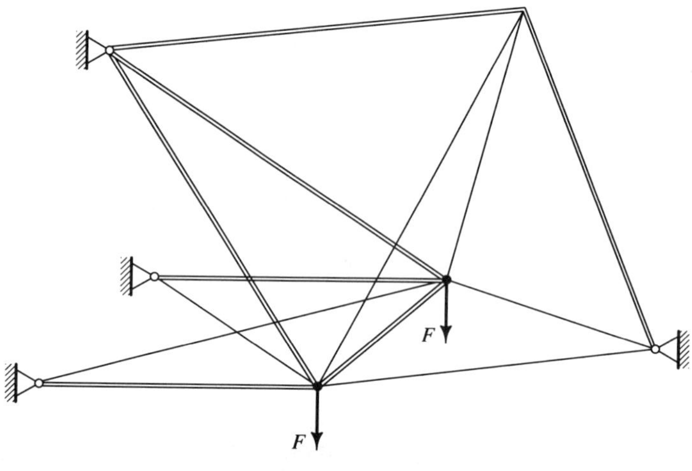

FIGURE 6.11. Optimization of the 13-member truss with location of the joint 6 regarded as unknown.

sectional areas of members representing the only design variables is illustrated on Figure 6.10. The solution illustrated in Figure 6.11 represents a more complex case, in which the design variables include not only the cross-sectional areas of the members, but also the coordinates of the sixth joint. For the projects shown in Figures 6.10 and 6.11, we have, respectively, $J = 22.85$ kg and $J = 21.04$ kg. Double lines indicate truss members carrying the maximum allowable stress. The values of cross-sectional areas for individual members are displayed in Table 6.1.

Here i denotes the number of the structural member, and $(S_i)_1$ and $(S_i)_2$ denote the cross-sectional areas for the variants illustrated on Figures 6.10

TABLE 6.1. Comparison of optimal cross-sectional areas for optimization problems on Figures 6.10 and 6.11, respectively.

i	$(S_i)_1$	$(S_i)_2$	i	$(S_i)_1$	$(S_i)_2$	i	$(S_i)_1$	$(S_i)_2$
1	1,836	1,340	6	1,836	1,340	10	1	1
2	1	1	7	1	1	11	1	1
3	4,354	3,972	8	1	1	12	1	1
4	1	1,076	9	1,409	1,346	13	1,362	1,357
5	4,354	3,972						

and 6.11, respectively. Thus, a considerable reduction in the weight of this truss has been achieved. The variation in the location of only a single joint permitted an additional reduction of weight by about 4%.

6.7 Maxwell–Michell Trusses

Our discussion of the example in the preceding section should make it clear that the location of the joints has a decisive influence on the weight of a truss. However, to discover optimal truss configurations and some basic laws governing the designs of optimal structures, it makes sense to use a continuum model, instead of discretizing this problem, as was done in Section 6.6 above. A fundamental advance based on a continuum approach was accomplished by Maxwell and Michell (see [6.57] and [6.60], and also [6.9], which will be now studied in our presentation).

The problem first studied by Maxwell consists of finding a truss of minimum weight capable of transferring loads applied to specified locations to the truss supports. Let us recall some classical results concerning estimates of the minimal quantity of the necessary material. We assume that the material has unequal resistance to tension and compression, and we denote by k^t and k^c the maximum admissible stresses for the tension and compression, respectively.

Let the locations of the junction points in the truss be given by the radius vectors $\mathbf{R}_i$. The lengths and cross-sectional areas of the members acting in tension and compression, respectively, are denoted by l_i^t, S_i^t, and l_i^c, S_i^c. The tensile and compressive loads are denoted by T_i^t and T_i^c, respectively. Strength constraints are assigned by limiting the maximum values of stresses in all the members:

$$-k^c S_i^c \leq -T_i^c, \qquad T_i^t \leq k^t S_i^t. \tag{6.71}$$

To establish some basic relations in the space in which the truss is located, we apply an infinitesimal virtual deformation, represented by a homogeneous dilation of magnitude $3e$ with its center coinciding with the origin of the coordinate system. As a result the truss members acquire a strain of magnitude e, while the points of applications of the external loads and the changes in the lengths of the members are $e\mathbf{R}_i$ and el_i, respectively. According to the virtual work principle, the change in the strain energy of the tensioned and compressed members is equal to the work of the external loads:

$$e\left(\sum_{i=1}^{n_t} T_i^t l_i^t - \sum_{i=1}^{n_c} T_i^c l_i^c\right) = e\sum_{i=1}^{n_f} \mathbf{F}_i\mathbf{R}_i = \text{const.} \tag{6.72}$$

In (6.72) n_t and n_c denote the number of structural members in tension and compression, respectively ($n_t + n_c = n$), and n_f denotes the total number of points to which external loads are applied. We use the strength condition (6.71) to select the minimum admissible cross-sectional area for each structural member. Thus,

$$S_i^c = T_i^c/k^c, \qquad S_i^t = T_i^t/k^t. \tag{6.73}$$

Substituting formulas (6.73) for T_i^c and T_i^t in (6.72), we obtain a fundamental relation in the theory of minimizing the weight of trusses first derived by James Clerk Maxwell (see [6.57]):

$$k^t V^t - k^c V^c = \sum_{i=1}^{n_f} \mathbf{F}_i\mathbf{R}_i. \tag{6.74}$$

Here V^t and V^c denote the total volumes of the structural members in tension and compression, respectively:

$$V = V^t + V^c, \qquad V^t = \sum_{i=1}^{n_t} l_i^t S_i^t, \qquad V^c = \sum_{i=1}^{n_c} l_i^c S_i^c.$$

Using (6.74), we obtain

$$V = V^c[1 + k^c/k^t] + (1/k^t)\sum_{i=1}^{n_t} \mathbf{F}_i\mathbf{R}_i = V^t[1 - k^t/k^c] + (1/k^c)\sum_{i=1}^{n_c} \mathbf{F}_i\mathbf{R}_i, \tag{6.75}$$

or

$$V = [(k^c + k^t)/(2k^t k^c)]\left(\sum_{i=1}^{n_t} T_i^t l_i^t + \sum_{i=1}^{n_c} T_i^c l_i^c\right) - [(k^t - k^c)/(2k^t k^c)]\left(\sum_{i=1}^{n_f} \mathbf{F}_i\mathbf{R}_i\right). \tag{6.76}$$

In many cases formulas (6.75) and (6.76) allow us to obtain an efficient estimate for the total weight of an optimal truss design. These relations have also a very important corollary, which may be formulated as a lemma: A uniformly stressed structure (or one of uniform strength), all of whose members are subjected to either pure tension or compression (if such a structure exists), is also a structure having minimal weight (or volume).

In general, there do not exist structures with uniform strength, all of whose members work only in pure tension or compression, and a direct application of Maxwell's theory is rather difficult. A generalization of this theory for designing lightest (or minimal volume) structures was given by Michell. It can be seen from (6.76) that a minimal volume (or weight) is attained when the expression inside the parentheses in (6.76) is minimized. A basic lemma asserting it can be stated as follows (without a proof): Among all trusses subjected to a given system of loads distributed inside a region Ω, the lightest truss, if it exists, must satisfy the following conditions:

1. Stresses in all members working in tension or compression must be equal to the limiting admissible values k^t and k^c, respectively.
2. There exists a virtual strain of the entire region Ω equal to e along the length of each member of the truss, which does not exceed in absolute value the strain e computed in any other direction. Moreover, the corresponding virtual displacements are equal to 0 at all support points.

If we apply this type of virtual strain to an optimal truss, then the change in the value of the strain energy is given by the formula

$$\delta U^* = \left(\sum_{i=1}^{n_t^*} T_i^{t*} l_i^{t*} + \sum_{i=1}^{n_c^*} T_i^{c*} l_i^{c*} \right) e. \tag{6.77}$$

If we apply this virtual strain to some other nonoptimal truss that is also fully stressed, then the virtual work of the external loads is exactly the same as before, and the changes in the strain energy of an optimal truss and of a fully stressed truss are identical. Aside from that, since, roughly speaking, the location of members of a fully stressed truss discussed here does not agree with the arrangement of these members for an optimal truss, at least some of these members of the nonoptimal truss will be subjected to strains smaller (in absolute value) than e. Denoting by e_i^t and e_i^c the mean values of the strains in the tensile and compression members of the truss, respectively, we compute the value of the change in strain energy:

$$\delta U = \sum_{i=1}^{n_t} T_i^t l_i^t e_i^t - \sum_{i=1}^{n_c} T_i^c l_i^c e_i^c. \tag{6.78}$$

Comparing the values of δU and δU^* given by (6.78) and (6.77), and recalling that $e_i^t \leq e$ and $e_i^c \leq e$, we derive the inequality

$$\begin{aligned} \left(\sum_{i=1}^{n_t^*} T_i^{t*} l_i^{t*} + \sum_{i=1}^{n_c^*} T_i^{c*} l_i^{c*} \right) e &= \sum_{i=1}^{n_t} T_i^t l_i^t e_i^t - \sum_{i=1}^{n_c} T_i^c l_i^c e_i^c \\ &\leq \left(\sum_{i=1}^{n_t} T_i^t l_i^t + \sum_{i=1}^{n_c} T_i^c l_i^c \right) e. \end{aligned} \tag{6.79}$$

Comparing (6.79) with (6.76) we conclude that

$$V^* \leq V. \tag{6.80}$$

The proof of inequality (6.80) uses fairly routine arguments, which include the principle of virtual work.

It is possible to obtain more detailed information concerning an optimal truss in the two-dimensional case. In this case the members of the truss lie along the lines of principal strain. Otherwise, at some point of the truss it would be possible to find a direction along which the stress exceeds the value e, contradicting our assumptions concerning the optimal truss design (in particular, it violates condition 2).

Another obvious condition concerns the orthogonality of the lines whose direction at the junction points of the structure indicates the orientation of the tension and compression members, respectively. Therefore, a constraint on the shape of the curves that coincide with the directions of a layout for the structural members (of the "Michell curves of principal strain") is identical to the constraint that is offered by the Hencky theory for the slip lines in a two-dimensional plastic flow for a continuous medium (see [6.26], [6.27], [6.76], or [6.90]).

Some sample designs of optimal trusses are shown in Figures 6.12 to 6.15. The total volume of material utilized in these trusses is given by the formulas

$$V_1 = (Fl/2)(1/k^t + 1/k^c)(1/2 + \pi/4), \qquad V_2 = (Fl/2)(1/k^t + 1/k^c)\pi/2,$$

$$V_3 = Fl(1/k^t + 1/k^c)\log(r/2), \qquad V_4 = 2M(1/k^t + 1/k^c)\log\tan(\pi/4 + r/2),$$

where F is the load, M the applied torque, and r the cross-sectional radius of the load-bearing struts.

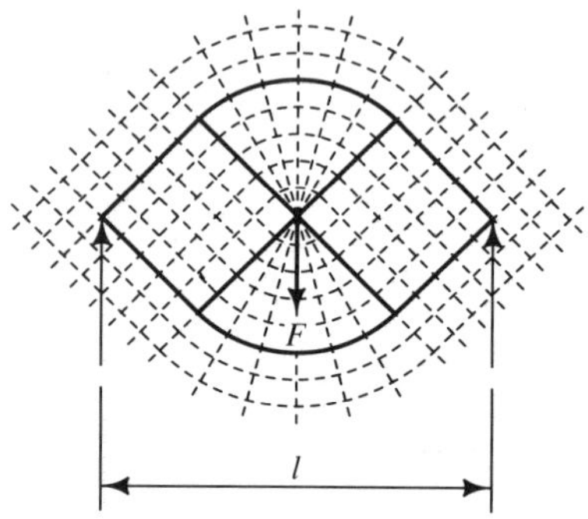

FIGURE 6.12. Optimal truss shape layout (Michell curves of principal strain coincide with positioning of members).

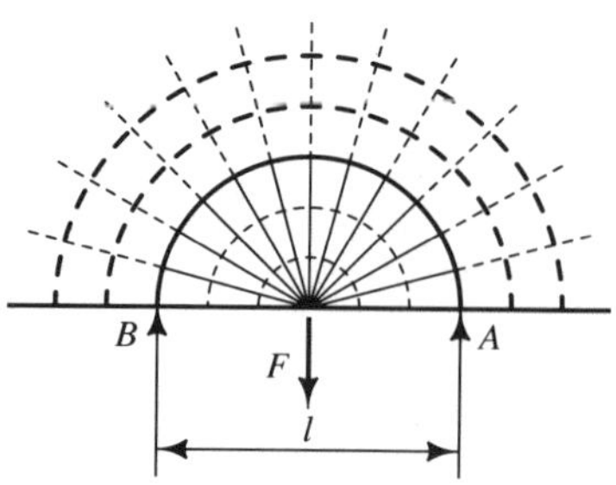

FIGURE 6.13. Optimal truss shape layout (Michell curves of principal strain coincide with positioning of members).

FIGURE 6.14. Optimal truss shape layout (Michell curves of principal strain coincide with positioning of members).

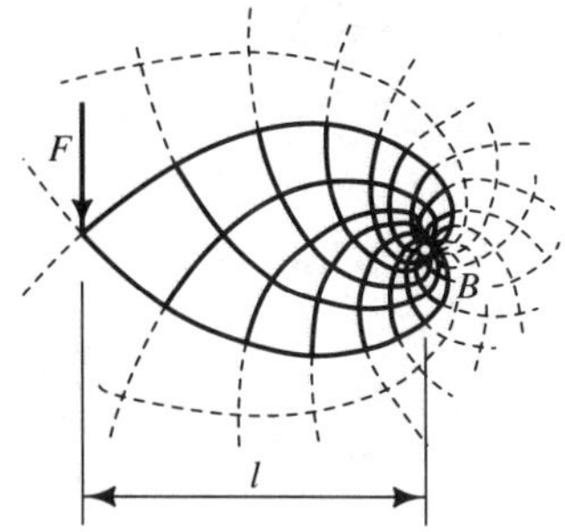

FIGURE 6.15. Optimal truss shape layout (Michell curves of principal strain coincide with positioning of members).

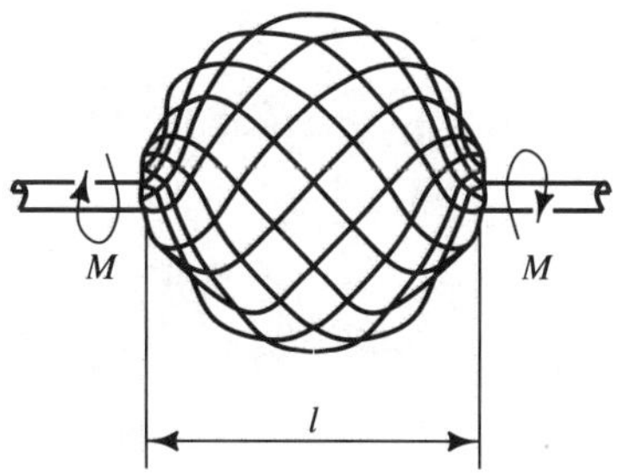

In conclusion, we remark that although the Maxwell–Michell theory is highly idealized, and the specific solutions developed in its application are difficult to construct due to the necessity of having a large number of members and junctions, this theory is valuable, as it reveals certain essential design principles for all truss-type structures.

6.8 Shells of Minimal Mass with Constraints on Strength

In several articles on the subject of optimal design theory, the authors investigated the problem of weight reduction for a shell with certain conditions specified regarding its strength. Some of the best studied cases involve moment-transmitting, or moment-free shells of revolution (see [6.7], [6.20], [6.23], [6.31], [6.38], [6.48], [6.55], [6.59], [6.78], [6.87], and [6.102]). Following the fundamental work [6.7], we shall consider a shell of revolution subjected to axially symmetric loading. Location of a point on the middle surface of the shell is uniquely determined by the values of the two angles θ and φ, where $0 \leq \theta \leq \pi$ is the angle between the direction normal to the median surface and the axis of revolution, while φ is the angle determining the position of the point on the meridian. The principal radia of curvature and the radius of a circle formed by intersecting the shell by a plane normal to the axis of revolution are denoted by R_1, R_2, and R, respectively.

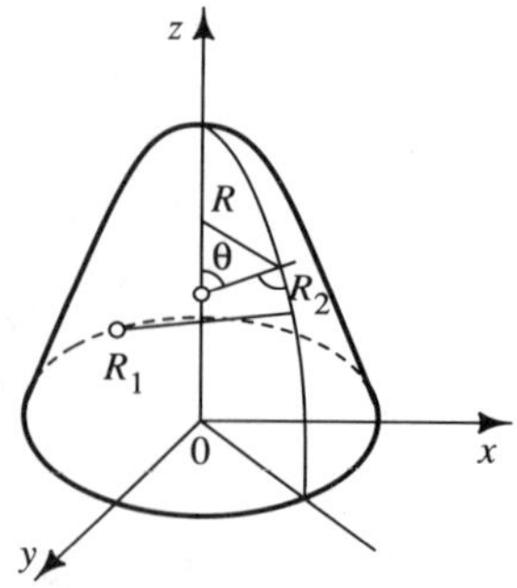

FIGURE 6.16. Illustration of notation used for shells of revolution.

Let $p = p(\theta)$ stand for the exterior pressure (i.e., force per unit surface area), and let $Q = Q(\theta)$ be the total force acting on the part of the shell below the circle θ (with radius R) (see Figure 6.16). From the conditions of equilibrium for that part of the shell, we can deduce the following equations for the stresses:

$$\left.\begin{aligned} \sigma_\theta &= \varphi_1(\theta)/h, \qquad \varphi_1(\theta) \equiv (-Q/2\pi R)\sin\theta, \\ \sigma_\varphi &= \varphi_2(\theta)/h, \qquad \varphi_2(\theta) \equiv -pR_1 - (\varphi_1 R_1/R_2), \end{aligned}\right\} \tag{6.81}$$

where $h = h(\theta)$ is a function describing the distribution of the thickness along the meridian. The magnitudes of the stresses σ_θ and σ_φ completely characterize the state of the shell.

The admissibility of stresses is given by the strength requirement

$$g(\sigma_\theta, \sigma_\varphi) \le k^2, \tag{6.82}$$

where g is a function determining the strength properties and k^2 a given constant.

The mass of a shell element lying between the parallel circles θ_0 and θ is given by the formula

$$J = \rho \int_{\alpha=\theta_0}^{\theta} hR_1R_2 \sin\alpha\, d\alpha. \tag{6.83}$$

The optimization problem consists in finding the thickness of shell function $h(\theta)$ that minimizes the functional (6.83) (the weight) such that stresses computed from (6.81) satisfy the strength condition (6.82).

The unknown design variable $h(\theta)$ must be regarded as a parameter determining in (6.81) the location of a point on an infinite ray L emanating from the origin of the coordinate system. The segment of this ray belonging to the region bounded by the critical (or limiting) curve (see Section 6.1) corresponds to a stress condition complying with the criterion (6.82).

The minimal admissible thickness of the shell corresponds to the intersection of this ray with the curve ($g = k^2$). Shells with thickness satisfying this condition are said to have uniform strength. It follows from the arguments

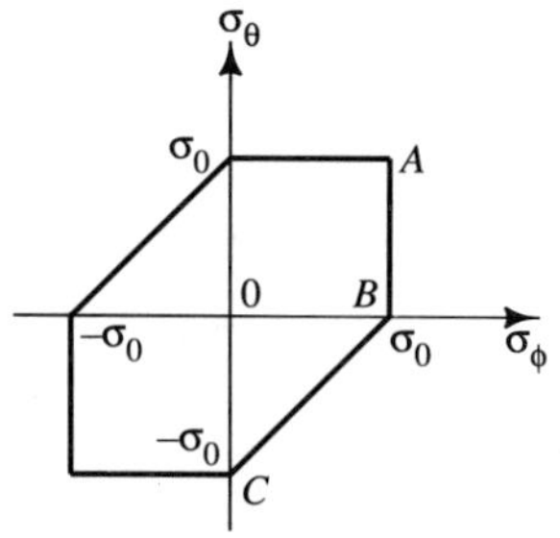

FIGURE 6.17. Treska's region.

presented in Section 6.3 that shells of revolution having uniform strength also minimize the total mass functional (6.83). Note that a shell of uniform strength is optimal also in the case when torques applied to it produce shear stresses.

As an example let us consider the project of optimizing a shell subjected to a hydrostatic pressure. A liquid with specific weight μ fills the entire volume of an inverted semispherical bowl of radius R. The stress components are

$$\sigma_\theta = (\mu R^2/3h)(1 - \cos^3\theta)/\sin^2\theta,$$

$$\sigma_\varphi = (\mu R^2/3h)[3\cos\theta - (1 - \cos^3\theta)/\sin^2\theta]. \tag{6.84}$$

We adopt Treska's criterion as our strength constraint:

$$\max(|\sigma_\theta|, |\sigma_\varphi|, |\sigma_\theta - \sigma_\varphi|) \le \sigma_0. \tag{6.85}$$

In the plane spanned by the coordinate system of the principal stresses σ_θ, σ_φ, the limiting polygonal curve is illustrated in Figure 6.17. After examining (6.84) and (6.85), it should be clear that for the values $0 < \theta < \operatorname{arc}\cos((\sqrt{3} - 1)/2) \approx 76°$, both σ_θ and σ_φ are positive and that $|\sigma_\theta| < |\sigma_\varphi|$. A point in the plane of $(\sigma_\theta, \sigma_\varphi)$ describing the stressed state for an optimal shell having uniform strength must lie on line segment AB. Within this range of angles, the shell's thickness can be computed from the formula

$$h = \mu R^2(1 - \cos^2\theta)/(3\sigma_0 \sin^2\theta).$$

The stress σ_φ becomes negative, and the stress state belongs to the line BC, defined by the equation $\sigma_\theta - \sigma_\varphi = \sigma_0$ for $76° \le \theta \le 90°$. The thickness distribution is

$$h = (\mu R^2/3\sigma_0)[2(1 - \cos^3\theta)/\sin^2\theta - 3\cos\theta].$$

A graphical representation of the thickness distribution is shown in Figure 6.18.

Now, let us carry out an optimization of a toroidal shell subjected to internal pressure p. Let the median surface of this shell be formed by revolving a circle of radius a in a circular path with radius R. The stress components are

$$\sigma_\theta = (pR/2h)(2a + R\sin\theta)/(a + R\sin\theta), \qquad \sigma_\varphi = pR/2h.$$

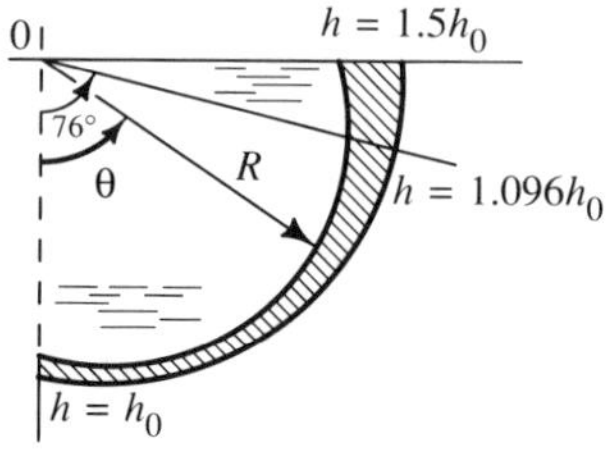

FIGURE 6.18. Distribution of thickness for a uniformly stressed shell.

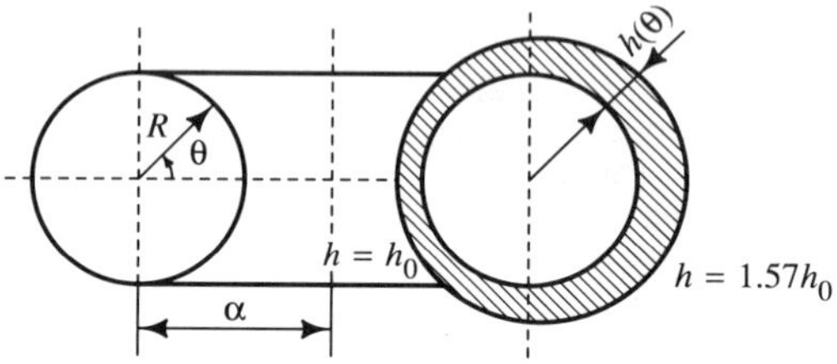

FIGURE 6.19. Distribution of thickness generated by the von Mises criterion in a toroidal shell.

We assume that the shell's material satisfies the von Mises criterion

$$\sigma_\theta^2 + \sigma_\varphi^2 + \sigma_\theta\sigma_\varphi \le \sigma_0^2.$$

Then the optimal distribution of thickness is given by the formula

$$h(\theta) = (pR/2\sigma_0)\sqrt{(3a^2 + 3aR\sin\theta + R^2\sin^2\theta)}/(a + R\sin\theta).$$

A graph of this functon is given in Figure 6.19.

6.9 Lowering Stress Concentrations to Acceptable Limits for Elastic Plates with Variable Thickness

In Section 6.2 we discussed problems of strength maximization (or minimization of stress concentrations) for elastic structural members. Here we shall formulate a similar problem concerning the bending of a continuous plate (without holes) with variable thickness (see [5], and [2.3]). The plate is situated in the xy plane and is supported along the contour of its boundary Γ. A transverse load $q(x, y)$ is applied to the plate. On segment Γ_1 of Γ the plate is simply supported, while on the remaining part Γ_2 it is rigidly built-in ($\Gamma = \Gamma_1 + \Gamma_2$). The area of the region Ω bounded by the contour Γ and the volume of the plate are given and will be denoted, respectively, by S and V. Let $\sigma_x, \sigma_y, \ldots, \tau_{xy}$ denote the usual components of the stress tensor, which uniquely determine the state of the material at each point (x, y, ζ), with $(x, y) \in \Omega$ and $-h/2 \le \zeta \le h/2$.

The strength criterion is defined by the magnitude of the second invariant of the stress deviator tensor:

$$g \equiv 1/3[\sigma_x^2 + \sigma_y^2 - \sigma_x\sigma_y + 3\tau_{xy}^2 + 3(\tau_{xz}^2 + \tau_{yz}^2)]. \tag{6.86}$$

For the sake of convenience, we shall introduce the dimensionless variables

$$\left.\begin{aligned} &x' = x/\sqrt{S}, \qquad y' = y/\sqrt{S}, \qquad w' = w/\sqrt{S}, \qquad h' = hS/V,\\ &q' = (12(1 - \nu^2)S^{3/2}/EV^3)q, \qquad \sigma_x' = \sigma_x/E, \ldots, \tau_{yz}' = \tau_{yz}/E,\\ &g' = (12(1 - \nu^2)S^{3/2}g)/VE^2, \end{aligned}\right\} \tag{6.87}$$

where, as before, E is Young's modulus, ν is Poisson's ratio, and w is the transverse displacement function. The equation of state and the boundary conditions written in terms of dimensionless variables become

$$L(h)w \equiv [h^3(w_{xx} + \nu w_{yy})]_{xx} + [h^3(w_{yy} + \nu w_{xx})]_{yy} + 2(1 - \nu)(h^3 w_{xy})_{xy} = q, \tag{6.88}$$

$$(W)_\Gamma = 0, \qquad (\partial w/\partial n)_{\Gamma_2} = 0, \qquad \{h^3[\Delta w - ((1 - \nu)/R)\partial w/\partial n]\}_{\Gamma_1} = 0. \tag{6.89}$$

As before, the subscripts x, y, etc., denote partial differentiation, and $\partial w/\partial n$, R, and Δ denote differentiation in the direction of the normal to the contour Γ, the radius of curvature of Γ, and the Laplacian operator, respectively.

Formulas for the stress components produced by the plate bending are written in the form

$$\left.\begin{aligned} \sigma_x &= [\zeta V(w_{xx} + \nu w_{yy})]/[(1 - \nu^2)S^{3/2}],\\ \sigma_y &= [\zeta V(w_{yy} + \nu w_{xx})]/[(1 - \nu^2)S^{3/2}],\\ \tau_{xy} &= -\zeta V w_{xy}/[(1 + \nu)S^{3/2}],\\ \tau_{xz} &= -\{V^2(h^2 - 4\zeta^2)/[8S^3(1 - \nu^2)h^3]\}\\ &\quad \times \{[(1 - \nu)(h^3 w_{xy})_y + (h^3 w_{xx})_y + \nu(h^3 w_{yy})_x]\},\\ \tau_{yz} &= -\{V^2(h^2 - 4\zeta^2)/[8S^3(1 - \nu^2)h^3]\}\\ &\quad \times \{[(1 - \nu)(h^3 w_{xy})_x + (h^3 w_{yy})_x + \nu(h^3 w_{xx})_y]\}. \end{aligned}\right\} \tag{6.90}$$

After substituting (6.90) in (6.86), we obtain an explicit formula for g as a function of the derivatives of the displacement function:

$$\left.\begin{aligned} g &= 4\zeta^2 g_1 + \{V^2(h^2 - 4\zeta^2)^2/(S^3h^6)\}g_2,\\ g_1 &= (w_{xx} + \nu w_{yy})^2 + (w_{yy} + \nu w_{xx})^2 - (w_{xx} + \nu w_{yy})(w_{yy} + \nu w_{xx})\\ &\quad + 3(1 - \nu)^2(w_{xy})^2,\\ g_2 &= 3/16\{[(1 - \nu)(h^3 w_{xy})_y + (h^3(w_{xx} + \nu w_{yy}))_x]^2 + [(1 - \nu)(h^3 w_{xy})_x\\ &\quad + (h^3(w_{yy} + \nu w_{xx}))_y]^2\}. \end{aligned}\right\} \tag{6.91}$$

As our quality functional we take the quantity

$$J = \max_{xy\zeta} g, \qquad (x, y) \in \Omega, \qquad \zeta \in [-h/2, +h/2], \tag{6.92}$$

and study the problem of optimizing the function J given by (6.91) and (6.92). We need to find a distribution of thickness that minimizes the maximum of the quantity g in the domain $\Omega_h = \Omega \times [-h/2, +h/2]$, that is

$$J_* = \min_h J = \min_h \max_{xy\zeta} g. \tag{6.93}$$

The minimum with respect to h is sought under an isoperimetric constraint on the constant volume of the plate and bounds on the thickness:

$$\int_\Omega h\mathrm{d}x\mathrm{d}y = 1, \tag{6.94}$$

$$h_{\min} \le h \le h_{\max}. \tag{6.95}$$

Here $h_{\min}$ and $h_{\max}$ are given constants.

We denote by $\Omega_h^0 \subset \Omega_h$ the set of points on which g attains its maximum. Thus,

$$J = (g)_{|\Omega_h^0} = \max_{xy\zeta} g.$$

The plastic state, according to the von Mises criterion, will be reached first on the set of points Ω_h^0 as the loads are proportionally increased. We recall that the von Mises criterion stipulates that a plastic state is reached when $g = k^2$. It is clear that the smaller is J, the bigger are the loads that initiate plastic deformation in the plate. Consequently, minimizing J leads to enlargement of the range of loads that will not cause the appearance of a plastic zone.

Observing that g depends explicitly on ζ, we can compute the interior maximum with respect to ζ in (6.93). It is easy to show that

$$\max_\zeta g = \max\{h^2 g_1, V^2 g_2/S^3 h^2\}.$$

By following the techniques of Section 2.3, we replace the functional J by a functional J_p defined by the formula

$$J_p = \left[\int_\Omega (h^2 g_1)^p \mathrm{d}\Omega\right]^{1/p},$$

and carry out some routine computations, deriving an expression for the variation of this functional,

$$\left.\begin{aligned} \delta J_p &= \int_\Omega \Lambda \delta h \mathrm{d}\Omega, \\ \Lambda &= 2h\chi[(1 - \nu + \nu^2)(w_{xx}^2 + w_{yy}^2) + (4\nu - 1 - \nu^2)w_{xx}w_{yy} \\ &\quad + 3(1 - \nu^2)w_{xy}^2] - [w_{xx}v_{xx} + w_{yy}v_{yy} + \nu(w_{xx}v_{yy} + w_{yy}v_{xx}) \\ &\quad + 2(1 - \nu)w_{xy}v_{xy}], \end{aligned}\right\} \tag{6.96}$$

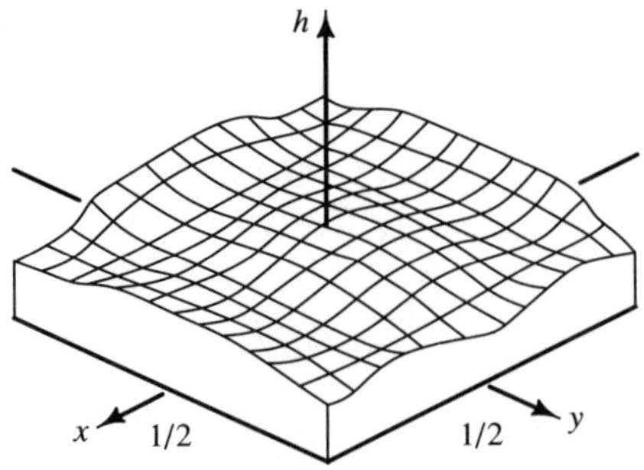

FIGURE 6.20. Optimal shape of a plate (in an L^p approximation).

where $\chi = \{h^2 g_1/\|h^2 g_1\|_{L_p}\}^{p-1}$, and v is the adjoint variable that solves the boundary-value problem

$$\left.\begin{aligned} &L(h)v = \Phi_p, \qquad (v)_\Gamma = 0, \qquad (\partial v/\partial n)_{\Gamma_2} = 0,\\ &(h^3[\Delta v - ((1-\nu)/R)\partial v/\partial n])_{\Gamma_1} = 0,\\ &\Phi_p = 2(1-\nu+\nu^2)[(w_{xx}\chi h^2)_{xx} + (\chi h^2 w_{yy})_{yy}]\\ &\qquad + (4\nu - 1 - \nu^2)[(\chi h^2 w_{xx})_{yy} + (\chi h^2 w_{yy})_{xx}]\\ &\qquad + 6(1-\nu^2)(\chi h^2 w_{xy})_{xy}. \end{aligned}\right\} \tag{6.97}$$

By using eqs. (6.96) and (6.97) we are able to apply a sequential optimization algorithm (see Chapter 4) to the problem of strength optimization. As an example, we offer computational results for the optimal shape of a square plate ($-1/2 \le x \le +1/2$, $-1/2 \le y \le +1/2$) freely supported along the edge $y = -1/2$ (for all $x \in [-1/2, +1/2]$), and rigidly built-in along the three remaining edges. The plate parameters listed in (6.88) and (6.95) were assumed to be $h_{\min} = 0.8$, $h_{\max} = 1.2$, and $q = 1$.

The computed variation of thickness is illustrated on Figure 6.20. We note that this distribution of thickness differs very little from the distribution of thickness for the strongest plate computed for exactly the same loads, boundary conditions, and values of plate parameters.

6.10 Minimizing Stress Concentrations in Elastic Plates with Holes

First, we shall introduce the basic equations describing the bending of an infinite elastic plate of constant thickness with a single hole, and formulate an optimization problem. Let Ω denote a doubly connected region in the xy plane that is occupied by the plate material and Γ the boundary of the hole. We shall use the xy Cartesian coordinates, and also orthogonal coordinates ns, related to the contour Γ and pointing in the directions normal and tangential to the boundary of the hole, respectively. It is assumed that this contour does not have any sharp corner points (i.e., it is "smooth"). The components of the bending moments and shear forces, expressed in the coordinates xy and ns, shall be denoted by M_x, M_y, M_{xy}, Q_x, Q_y, and M_n, M_s, M_{ns}, Q_n, Q_s, respectively. The plate is subjected to the bending moments

$$M_x = M_1 \geq 0, \qquad M_y = M_2 \geq 0, \qquad M_{xy} = 0 \tag{6.98}$$

applied at infinity, while no forces are applied to the contour of the hole.

In the absence of transverse loads, the displacement function $w(x, y)$ satisfies the biharmonic equation $\Delta^2 w = 0$. In the absence of loads on the boundary of the hole, the boundary conditions are given by

$$M_n = 0, \qquad M_{ns} = 0, \qquad Q_n = 0, \qquad (x, y) \in \Gamma. \tag{6.99}$$

The bending moments and shear forces are related to the displacements by the well-known formulas (see [6.81])

$$\left.\begin{array}{lll} M_x = -D(w_{xx} + \nu w_{yy}), & M_y = -D(w_{yy} + \nu w_{xx}), & \\ M_{xy} = -D(1 + \nu)w_{xy}, & Q_x = -D(\Delta w)_x, & Q_y = -D(\Delta w)_y, \end{array}\right\} \tag{6.100}$$

where $D = Eh^3/(12(1 - \nu^2))$ is the cylindrical rigidity modulus of the plate, E is Young's modulus, ν is Poisson's ratio, and h is the thickness of the plate.

As is well known (see [6.94]), one of the conditions in (6.99) is superfluous. Here we must acknowledge the fact that for a hole of arbitrary shape, a solution of the biharmonic equation cannot simultaneously satisfy all of the three conditions listed above. For this reason an approximate formulation of boundary conditions utilizes the two equalities $M_n = 0$ and $Q_n + \partial M_{ns}/\partial s = 0$. The solution of the boundary-value problem with these boundary conditions leads to a well-known distortion of the stress field at the boundary of the hole. Note that for holes of optimal shape, all three conditions given in (6.99) are somehow satisfied. In investigating this optimization problem, we shall make use only of the conditions $M_n = 0$ and $M_{ns} = 0$. The condition $Q = 0$ will be automatically satisfied for an optimal shape of the hole.

If the deflection function $w(x, y)$ is found as a solution of the boundary-value problem, with the additional conditions (6.98) satisfied at infinity, the moments and shear forces are given by (6.100), and all nonzero components of the stress tensor may be computed (see [6.81] and [6.82]) from the formulas

$$\left.\begin{array}{lll} \sigma_x = 12\zeta h^{-3} M_x, & \sigma_y = 12\zeta h^{-3} M_y, & \tau_{xy} = 12\zeta h^{-3} M_{xy}, \\ \tau_{xz} = \frac{3}{2}(h^2 - 4\zeta^2)h^{-3} Q_x, & \tau_{yz} = \frac{3}{2}(h^2 - 4\zeta^2)h^{-3} Q_y. & \end{array}\right\} \tag{6.101}$$

The real variable ζ in (6.101) measures the distance from the middle surface along a normal direction and varies in the interval $-h/2 \leq \zeta \leq h/2$. As our function g, which characterizes the stress concentrations and determines the transition to the plastic state, we choose the magnitude of the second invariant of the stress tensor deviator:

$$\left.\begin{aligned} g &\equiv \sigma_x^2 + \sigma_y^2 - \sigma_x\sigma_y + 3(\tau_{xy}^2 + \tau_{xz}^2 + \tau_{yz}^2) \\ &= 144\zeta^2 h^{-6}[(M_x + M_y)^2 + 3(M_{xy}^2 - M_x M_y)] \\ &\quad + \tfrac{27}{4}h^{-6}(h^2 - 4\zeta^2)^2[Q_x^2 + Q_y^2], \end{aligned}\right\} \tag{6.102}$$

where $(x, y) \in (\Omega + \Gamma)$ and $-h/2 \le \zeta \le h/2$. Note that g depends explicitly on ζ, but is only an implicit function of x and y in terms of M_x, M_y, M_{xy}, Q_x, and Q_y. Now the optimization problem consists in finding an admissible shape of the contour Γ such that in the region $\Omega + \Gamma$ occupied by the material, the maximum value of g is minimized:

$$J_* = \min_{\Gamma} J = \min_{\Gamma} \max_{xy} \max_{\zeta} g. \tag{6.103}$$

To find the minimum of (6.103) with respect to Γ, it is assumed that the unknown contour cannot shrink to a single point, that is, the area of the hole cannot vanish.

Let us now represent the function g as a sum of two terms:

$$\left.\begin{aligned} g &= g_1 + g_2, \\ g_1 &= 144\zeta^2 h^{-6}[(M_x + M_y)^2 + 3(M_{xy}^2 - M_x M_y)], \\ g_2 &= \tfrac{27}{4}h^{-6}(h^2 - 4\zeta^2)^2[Q_x^2 + Q_y^2], \end{aligned}\right\} \tag{6.104}$$

and consider first the problem of minimizing the maximum value of g_1 with respect to Γ,

$$J_{1*} = \min_{\Gamma} \max_{xy} \max_{\zeta} g_1, \tag{6.105}$$

where $(x, y) \in \Gamma$, $\zeta \in [-h/2, h/2]$. We note that the expression inside the brackets in (6.104) for g_1 is positive. This assures that g will be positive-definite. To show that this is true, let us assume the contrary, that is, that $|\zeta| = h/2$ and $g = g_1 < 0$. Using the previously discussed property, we conclude that the maximum of the functional in (6.105) with respect to ζ occurs on the boundary $|\zeta| = h/2$. We also note that the functional in (6.104) is invariant under a change from the xy to the ns coordinate system, that is,

$$(M_x + M_y)^2 + 3(M_{xy}^2 - M_x M_y) = (M_n + M_s)^2 + 3(M_{ns}^2 - M_n M_s).$$

Also, using the boundary conditions $M_n = 0$ and $M_{ns} = 0$, for all $(x, y) \in \Gamma$, we arrive at an expression for g_1 that is valid at all points of the boundary:

$$(g_1)_\Gamma = aM_s^2, \quad \text{where } a = 144\zeta^2 h^{-6}. \tag{6.106}$$

We introduce an auxiliary function $\chi = M_x + M_y$. From (6.100) we conclude that $\chi = D(1 + \nu)\Delta w$. Since the deflection function w obeys a biharmonic equation, it is clear that χ is a harmonic function. Using the invariance equality $(M_x + M_y) = (M_n + M_s)$, plus conditions assigned at Γ and at infinity, we derive relations for χ:

$$\Delta\chi = 0, \qquad (\chi)_\Gamma = M_s, \qquad (\chi)_\infty = M_1 + M_2. \tag{6.107}$$

Recall the following property of harmonic functions, which will be used in the subsequent presentation (for a more detailed presentation, see [5] and [6.4]). Suppose that in the xy plane we have n holes bounded by simple closed curves Γ_i, $i = 1, 2, \ldots, n$. In the region Ω containing the point at infinity and bounded by $\Gamma = \Sigma\Gamma_i$ (Ω is the region exterior to the holes), we consider a family of harmonic functions that are continuous in $(\Omega + \Gamma)$ and have (constant) positive limit A at infinity. Let g be the value attained on Γ by the member of this family χ, that is, $(\chi)_\Gamma = g$. Among members of this family of functions, the minimum of the functional $\max_{\xi\eta} |g|$ ($(\xi, \eta) \in \Gamma$) with respect to g is attained uniquely by the function $g(\xi, \eta) = A$, that is, $\chi(x, y) \equiv A$, and this minimum is equal to A. That is,

$$\min_{g} \max_{\xi\eta} |g| = A. \tag{6.108}$$

Examining further properties of (6.106) and (6.107) and making use of (6.108), we come to the conclusion that the minimum of the maximum value of g on the boundary Γ is attained if and only if

$$(M_s)_\Gamma = M_1 + M_2. \tag{6.109}$$

Therefore, (6.109) is the necessary and sufficient condition for optimality in the auxiliary problem (6.105). Now the formula $J_{1*} = 36h^{-4}(M_1 + M_2)^2$ can be deduced from (6.106) and (6.109).

We shall now study the problem of finding the minimum for a maximum value of g in the region $\Omega + \Gamma$, and we shall prove that the same equality (6.109) again constitutes an optimality condition. To prove it we shall study the behavior of g in the region $\Omega + \Gamma$, and show that for contours satisfying condition (6.109) the function g_1 attains its maximum value on the boundary of a hole.

We shall use a complex representation of bending moments and shears (see [6.81]), expressing M_x, M_y, M_{xy}, Q_x, and Q_y in terms of two analytic functions $\Phi(z)$ and $\Psi(z)$ as follows:

$$\left.\begin{aligned} M_x + M_y &= -2D(1+\nu)[\Phi(z) + \bar{\Phi}(z)], \\ M_y - M_x + 2iM_{xy} &= 2D(1-\nu)[\bar{z}\Phi'(z) + \Psi(z)] \\ Q_x - iQ_y &= -4D\Phi'(z), \end{aligned}\right\} \tag{6.110}$$

where $z = x + iy$ and $\bar{z} = x - iy$, and i is the imaginary unit. As usual, the bar over Φ ($\bar{\Phi}$) denotes the complex conjugate.

Let us assume that condition (6.109) is satisfied. In that case the analytic function $\Phi(z)$ must be constant in its entire domain of definition ($\Phi'(z) = 0$), and (6.110) becomes

$$\left.\begin{aligned} M_x + M_y &= M_1 + M_2, \\ M_y - M_x + 2iM_{xy} &= 2D(1-\nu)\Psi(z) \end{aligned}\right\} \tag{6.111}$$

The second equality in (6.111) and its complex conjugate equality $M_y - M_x - 2iM_{xy} = 2D(1 - \nu)\bar{\Psi}(z)$ can be multiplied by each other, side by side.

Then we obtain

$$g_0 \equiv (M_x + M_y)^2 + 4(M_{xy}^2 - M_x M_y) = b\Psi\bar{\Psi}, \tag{6.112}$$

where $b = 4D^2(1 - \nu)^2$. We rewrite Ψ and $\bar{\Psi}$ in the form $\Psi = \psi_1 + i\psi_2$ and $\bar{\Psi} = \psi_1 - i\psi_2$, where ψ_1 and ψ_2 are real-valued functions. Then (6.112) becomes $g_0 = b(\psi_1^2 + \psi_2^2)$, which is a very convenient form for estimating. The functions ψ_1 and ψ_2 satisfy the Cauchy–Riemann equations $\psi_{1,x} = \psi_{2,y}$ and $\psi_{1,y} = -\psi_{2,x}$, and are harmonic ($\Delta\psi_1 = \Delta\psi_2 \equiv 0$). Applying the Laplacian operator to the function g_0, and performing some routine substitutions using the properties of ψ_1 and ψ_2, we derive the inequality $g_0 = 4b(\Delta\psi_1)^2 \geq 0$. We express g_1 as a function of g_0 and the constants assigned to this problem, $g_1 = \frac{1}{4}a[M_1 + M_2]^2 + 3g_0$. Again, applying the Laplacian operator to this expression, we obtain the inequality $\Delta g_1 = 3ab(\Delta\psi_1)^2 \geq 0$. Consequently, the function g_1 cannot attain its maximal value at any interior point of its domain $(\Omega + \Gamma)$. After comparing the values of g_1 which are assumed by it on Γ and at infinity respectively, that is, $a[M_1 + M_2]^2$ and $a([M_1 + M_2]^2 - 3M_1 M_2)$, we see that the maximum of g_1 must be reached on the boundary Γ.

Therefore, if (6.109) is satisfied, the maximum of g_1 over the domain $\Omega + \Gamma$ must occur at some points of Γ. Thus, (6.109) is also a sufficient and necessary condition for the minimization of the maximum value of g_1 over the domain $\Omega + \Gamma$.

Let us return to the initial problem of minimizing the maximum value of g. We shall show that (6.109) is again a necessary and sufficient condition for this problem. Indeed, for any holes for which (6.109) is true, $Q_x - iQ_y = -4\Phi'(z) = 0$, and consequently

$$Q_x = Q_y = 0. \tag{6.113}$$

In that case $g_2 = 0$ at all points $(x, y) \in \Omega + \Gamma$, and

$$g = g_1, \qquad (x, y) \in \Omega + \Gamma. \tag{6.114}$$

For holes of arbitrary shape,

$$g = g_1 + g_2 \geq g_1, \qquad (x, y) \in \Omega + \Gamma. \tag{6.115}$$

Comparing (6.114) and (6.115) and recalling that the minimum for the maximum value of g_1 in the domain $\Omega + \Gamma$ is reached only if (6.109) is true, we arrive at the conclusion that this equality is a necessary and sufficient condition for optimality. The proof of the optimality of the shapes of holes that satisfy condition (6.109) never mentions the connectivity of Ω. Consequently, eq. (6.109) constitutes also a necessary and sufficient condition for optimality if there are n holes instead of 1.

Our arguments can be generalized to the case when constant moments $M_n = -M_0$ are applied to the boundary of the holes (here M_0 is a known constant). It is easy to show that in the absence of distributed moments on the boundary, the optimality condition becomes $(M_{ns})_\Gamma = M_1 + M_2 + M_0$.

Finding an optimal shape with the help of eq. (6.109) reduces to a well-posed boundary-value problem for a biharmonic equation with boundary conditions $M_n = 0$ and $M_{ns} = 0$, and given conditions at infinity. Equation (6.109) is used to determine the unknown boundary Γ.

Using eqs. $M_x + M_y = -D(1 + \nu)/\Delta w$ and $M_x + M_y = M_1 + M_2$, we convert this problem into the inverse problem for the Poisson equation

$$\Delta w = c, \qquad c = -(M_1 + M_2)/D(1 + \nu), \tag{6.116}$$

which can be studied using techniques from the theory of functions of complex variables (see [6.11]). Following [6.11], we derive the following result: For a plate in pure bending, holes having an elliptical shape,

$$x^2(M_1)^{-2} + y^2(M_2)^{-2} = \lambda^2, \tag{6.117}$$

are optimal, where λ^2 denotes an arbitrary positive constant.

Since the problem of finding the shapes of optimal holes for plates subjected to simultaneous bending and tension leads to an almost identical mathematical formulation, we refer the reader to [6.11], [6.32], [6.39], [6.40], [6.95], and [6.96] for details of the techniques applied in determining the contours of a system of holes.

It turns out that for an optimal plate, the "ignored" boundary condition $Q_n = 0$ is satisfied, as evidenced by (6.113). Another consequence of (6.113) is that all the stress tensor components and the functional g vanish on the median surface ($\zeta = 0$).

Note that for a bent plate with an optimally shaped hole, the mean curvature is constant everywhere,

$$1/r_x + 1/r_y = c, \tag{6.118}$$

where c is a constant given by (6.116), and r_x and r_y denote the curvatures in the directions of the axes x and y, respectively.

Note, too, that the plastic state will be first reached for such a plate on the boundary $\Gamma = \sum_i \Gamma_i$, and it will be reached uniformly on the entire boundary.

Problems somewhat similar to those discussed above also arise in investigating the optimal shapes of elastic bodies of differing geometry, particularly in designing optimal "connections" between regions with sharply different geometries and high stress concentrations (see [6.66], [6.67], [6.95], and [6.99]). Problems of reinforcing thin-walled structures are discussed in [6.46], [6.47], [6.54], [6.56], and [6.61].

6.11 Optimal Design Taking into Account the Plastic Properties of the Material

In preceding sections of this chapter, the entire discussion presumed that we are dealing with the ideal elastic behavior of the material right down to the point when it fails, according to some strength criterion. However, for many real structures, long before its load-bearing capacity is destroyed the response of the material becomes essentially nonlinear. If we acknowledge these nonlinear characteristics of materials in our design theory, we may accomplish substantial savings in the quantity of material (weight of the structure), as compared with results obtained from computions using the elastic body hypotheses. However, modern research into the theory of structural design, assuming an adequate failure mechanism, has encountered serious mathematical difficulties.

One possible explanation concerns our present knowledge of structural optimization using materiels other than elastic materials, which is a far from satisfactory solution. Recent research in this area is directed toward finding new ways of formulating various types of nonlinear behavior of materials and of complex structural components (such as combined plates, shells, and three-dimensional objects), where various techniques have been developed for sensitivity analysis and other effective numerical approaches to optimization.

Here we shall mention certain works on the theory of optimal design, using the idea of exhausting the bearing capacity of a structure. Problems of weight minimization for structures with a given critical load are discussed in [6.14], [6.17], [6.18], [6.29], [6.49], [6.51], [6.69]–[6.73], [6.80], [6.84], [6.85], [6.87], [6.88], and [6.100], from the viewpoint of elastoplastic material behavior. Some optimal design problems that place constraints on the admissibility of variables, in particular, on the cyclic nature of the applied loads, are discussed in [6.43], [6.58], [6.72], [6.74], and [6.75]. A variety of problems related to plastic properties of materials is discussed in [13], [38], [43], [6.12], [6.13], [6.19], and [6.49], and the related equal strength criterion in [6.4], [6.5], [6.7], [6.8], [6.36], and [6.79].

Let us consider the equilibrium of a deformed body occupying a region Ω and subjected to body forces q_i and exterior loads T_i applied to parts of the body surface Γ_i $(i = 1, 2, \ldots, n)$. On the remainder of its surface Γ_u the body is rigidly restrained $(\Gamma_\sigma + \Gamma_u = \Gamma)$, $\Gamma_\sigma = \bigcup \Gamma_i$. This condition can be restated as $(u)_{\Gamma_u} = 0$. We assume that the material is elastoplastic. A plastic flow condition is realized at the point when there is strict equality in the condition $g(\sigma_{ij}, k) \leq 0$. On the other hand, if strict inequality is true, the material is elastic. Here k denotes the constant of plasticity, σ_{ij} are the components of the stress tensor, and g is a given function. The equation $g(\sigma_{ij}, k) = 0$ in the stress space defines a family of convex surfaces corresponding to different values of k and containing the origin of the coordinate system in the interior. These surfaces shrink to the origin as $k \to 0$.

In all arguments given below, it is assumed that for the loads applied to certain parts of the body's surface, a plastic flow will occur somehwere. The occurrence of plastic zones is admissible; however, the existence of plastic deformation must not exhaust the load-carrying capacity and cause failure. By exhaustion of load-carrying capacity, we understand the condition when the magnitude of strain becomes unbounded while the load remains constant (see [6.41]. In what follows we assume that strains are "small" prior to failure.

An analysis of the stress-strain state for an elastoplastic body may be conducted with the help of various well-researched numerical techniques. Such computations can be completed only by using modern (mainframe-type) computer facilities. An extremely large volume of computations is necessary in optimizing elastoplastic structures to compare the numerous different variants. Apparently this explains the absence of solutions to problems in optimal design that make full use of the elastoplastic analysis.

In several elastoplastic analysis problems, it is possible to achieve significant simplifications if one estimates only the load-carrying capacity and is not concerned with detailed information on the stress-strain state and strength characteristics. In such cases we can apply a fundamental theorem on the critical static equilibrium (see [6.28] and [6.41]). According to this theorem on the state of a static critical equilibrium, a body may withstand a given load if there exists a safe, statically admissible stress field σ_{ij}, that is, a stress distribution satisfying the equilibrium conditions

$$\sigma_{ij,j} + q_i = 0, \tag{6.119}$$

$$(\sigma_{ij} n_j)_{\Gamma_\sigma} = T_i, \tag{6.120}$$

and such that

$$g(\sigma_{ij}, k) < 0, \tag{6.121}$$

where n_j is the cosine between the unit outward normal vector and the surface of the body. As usual, lower indices following a comma indicate the operation of partial differentiation with respect to the indicated coordinates.

To make use of this theorem, we can formulate the following problem of minimizing the volume of material:

$$\int_\Omega d\Omega \underset{\Gamma_v}{\to \min}, \tag{6.122}$$

with eqs. (6.119) and (6.120) acting as conditions on preservation of the load-carrying ability. Minimization is accomplished by specifying the part of the body surface $\Gamma_v \subset \Gamma_\sigma$ on which the loads are assigned.

The set of safe stresses obeying (6.119) to (6.121) is closed. This property implies that optimization problems of the type (6.119) to (6.122) are well posed.

We can apply the following device to regularize it: Let us consider two plastic flow surfaces that are close to each other and correspond to the parameters k and k_f, respectively, where k and k_f differ by a small amount.

Here $k > k_f$ and $k = k_f + \varepsilon$, where $\varepsilon > 0$ can be chosen to be an arbitrarily small number. Recalling the geometric formulas for plastic flow surfaces, it is easy to see that the strict inequality (6.121) is satisfied "automatically" for any statically admissible flow fields that satisfy both (6.119) and (6.122), and the strength inequality

$$g(\sigma_{ij}, k_f) \leq 0. \tag{6.123}$$

Therefore, such flow fields are also safe, statically admissible flow fields for the initially postulated plastic flow surface.

A solution of the optimization problem may now be derived within arbitrary accuracy by selecting a small ε and substituting inequality (6.121) and condition (6.123) in (6.119) to (6.122).

This device, which consist of modifying the flow surface to account for inaccuracies in the values of the constants k for materials used in the design, does not cause any additional noticeable errors in practice.

To investigate this formulation of our optimization problem, we use an approach based on sensitivity analysis. We introduce a slack (auxiliary) variable μ, and transform inequality (6.123) into an equality constraint:

$$g(\sigma_{ij}, k_f) + \mu^2 = 0. \tag{6.124}$$

Now we construct a generalized Lagrangian functional

$$J = \int_\Omega \mathrm{d}\Omega + \int_\Omega \psi_i(\sigma_{ij,j} + q_i)\mathrm{d}\Omega + \int_\Omega \lambda(g + \mu^2)\mathrm{d}\Omega, \tag{6.125}$$

where ψ_i $(i = 1, 2, \ldots, n)$ and λ are adjoint variables. Varying Γ_v and substituting the boundary conditions, we obtain an expression for δJ. Omitting all intermediate formulas, we display the final result:

$$\left.\begin{aligned} \delta J = &\int_\Omega (\lambda \partial g/\partial\sigma_{ij} - \psi_{i,j})(\delta\sigma_{ij} - \sigma_{ij,k}\delta x_k)\mathrm{d}\Omega + \int_\Omega 2\lambda\mu \mathrm{d}\Omega \\ &+ \int_\Gamma \psi_i n_j(\delta\sigma_{ij} - \sigma_{ij}\delta x_k)\mathrm{d}\Gamma + \int_{\Gamma_v} \{1 + \lambda(g + \mu^2)\} n_k \delta x_k \mathrm{d}\Gamma. \end{aligned}\right\} \tag{6.126}$$

To relate δJ to the variation of the surface Γ_v, we use the symmetry of the stress tensor $(\sigma_{ij} = \sigma_{ji})$ and define the adjoint variable ψ_i, which solves the boundary-value problem

$$1/2(\psi_{i,j} + \psi_{j,i}) = \lambda(\partial g/\partial\sigma_{ij}), \tag{6.127}$$

$$(\psi_i)_{\Gamma_u} = 0. \tag{6.128}$$

We also assume that the functions λ and μ satisfy the relation

$$\lambda\mu = 0 \tag{6.129}$$

for all $x \in \Omega$.

If assumptions (6.127) to (6.129) are true, then the first and second integral on the right-hand side of (6.126) must be equal to 0.

Also we regard the loads $q_i(x_k)$ and $T_i(x_k)$ as known functions of the spatial coordinates. Therefore, as we vary the shape of the domain, the corresponding variation of the boundary conditions assumes the form

$$\delta(\sigma_{ij}n_j) = n_j\delta\sigma_{ij} + \sigma_{ij}\delta n_j = T_{i,j}\delta x_j.$$

The variation of the normal vector δn_j can be expressed in terms of the variation of the vector l tangent to the surface and the coefficients of the first quadratic form for the shape of the varied surface. Therefore, the term containing n_j in the surface integral can be integrated by parts. In the resultant expression we can apply Gauss's formula and identify the term containing the curvature H of the surface. Finally, we derive a formula for δJ:

$$\delta J = \int_{\Gamma_v} \{1 + \lambda g - 2H\psi_i T_i - (\psi_i \sigma_{ij})_{,j} + (\psi_i T_i)_{,k} n_k\} \delta l d\Gamma. \tag{6.130}$$

Equation (6.130) defines the sensitivity of the optimized functional with respect to changes in the design variable (i.e., changes in the shape of the domain) and permits us to employ numerical techniques of iterative optimization (see [2]). We use (6.130) to obtain a necessary condition for optimality:

$$1 + \lambda g - 2H\psi_i T_i - (\psi_i \sigma_{ij})_{,j} + (\psi_i T_i)_{,k} n_k = 0 \quad \text{on } \Gamma_v. \tag{6.131}$$

We shall examine a special case when there are no body forces ($q_i = 0$) and the varied boundary segment is load-free ($T_i = 0$). We also assume that $g(\sigma_{ij}, k_f) = g_0(\sigma_{ij}) - k_f^2$, where $g_0(\sigma_{ij})$ is a homogeneous function of order 2 of the stress components. Then the optimality condition is considerably simplified:

$$\lambda[2k_f^2 + g] = 1 \quad \text{on } \Gamma_v. \tag{6.132}$$

In this case the optimality condition is not a function of the adjoint variables ψ_i, and the solution of the optimal design problem can be derived by using (6.119), (6.120), (6.122), and (6.129), and the following formula for δJ:

$$\delta J = \int_{\Gamma_v} \{1 - \lambda[2k_f^2 + g]\} \delta l d\Gamma, \tag{6.133}$$

corresponding to condition (6.132). In the general case, it is necessary to use the full system of relations (6.119), (6.120), (6.124), and (6.127) to (6.130) to derive an optimal solution. It is easy to see the similarity between eqs. (6.119), (6.120), (6.127), and (6.128) and the equations for a model of a rigid-plastic body. In this comparison, we establish a correspondence between the adjoint variables ψ_i and the velocities v_i, and between the term $1/2(\psi_{i,j} + \psi_{j,i})$, which occurs on the left-hand side of eq. (6.127), and the velocity of deformation for the associated law of plastic flow.

We use the notation $\dot{\varepsilon} = 1/2(\psi_{i,j} + \psi_{j,i})$. Recalling the homogeneity of the function g_0, we use (6.127) to derive

$$\sigma_{ij}\dot{\varepsilon}_{ij} = 2g_0\lambda. \tag{6.134}$$

This equality permits us to express the optimality condition (6.132) and eq. (6.133) for δJ in a more convenient form. We shall consider two cases, when $g \neq 0$ and when $g = 0$. If on the varied part of the boundary $g \neq 0$ ($g_0 < k_f^2$), then $\mu \neq 0$. Recalling condition (6.129), we see that $\lambda = 0$. But this equality implies that our assumption stating that the varied part of the boundary was not fully stressed contradicts the optimality relation (6.132). Therefore, let us assume that in (6.132) $g = 0$ (with $g_0 = k_f^2$). Because of (6.124) and (6.129), we must have $\mu = 0$ and $\lambda \neq 0$. But according to (6.134), λ satisfies the relation $\lambda = \sigma_{ji}\dot{\varepsilon}_{ij}/2k_f^2$. Substituting this in (6.132) and recalling that $g = 0$, we obtain the optimality condition in the form

$$\sigma_{ij}\dot{\varepsilon}_{ij} = 1 \quad \text{on } \Gamma_v. \tag{6.135}$$

Therefore, if the plasticity criterion is given by a homogeneous function of second order of homogeneity with respect to the components of the stress tensor, in the absence of body forces and with no external loads applied to Γ_v, we arrive at the classical condition given in [6.17]. Completely analogous simplifications are obtained in (6.133) for δJ.

Condition (6.135) implies that on the unknown segment of the boundary, which is free from external loads, energy is dissipated at a constant rate. Note that along Γ_v, the stress intensity $g = k_f^2$ is constant.

Consequently, if in solving our optimization problem the sequence of boundary surfaces converges, the limit surface Γ_v is uniformly strong.

Numerical solution of such optimization problem may be accomplished by an iterative technique of sequential optimization (see [2]). It permits us to perform computations of the following two types: The first type is used in solving problems of rigid-plastic analysis. As a result of numerically solving such problems for a given shape of a body, one finds the values of σ_{ij}, ψ_i, λ, and μ. The second type of computation relates to constructions that vary and improve the shape of the body. These computations are carried out with the aid of previously derived formulas of sensitivity analysis using the current values of σ_{ij}, ψ_i, λ, and μ. Several techniques of both the calculus of variations and of mathematical programming can be utilized in sensitivity analysis. Note, too, that many well-known techniques are used within the framework of an algorithm for successive approximation. Gradient-type techniques turn out to be most suitable for practical applications (see [5], [8], [21], and [24]).

An approach based on the application of a successive optimization algorithm has been developed for solving problems of minimizing the weight of elastoplastic structures with constraints on load-bearing capacity. Results of computations for certain elastoplastic members are illustrated on Figures 6.21 to 6.24. The critical behavior of these members is determined by the two-dimensional equations of equilibrium and the von Mises criterion. Figure 6.21 illustrates an improved shape of a support member. The symmetry of the external loads, boundary conditions, and geometric shape of the member with respect to the x axis was taken into account both in the computations and in the graphic display. Therefore, only one-half of the member is shown in Figures 6.21 and 6.22. A uniformly distributed compressive load is applied to the boundary segment CD. The unknown part of the boundary BC is load-free. The shape of BC is regarded as a design variable and is improved during the iterative process. Along the line AB, this compressed member contacts an ideally smooth and perfectly rigid foundation. Results of a design of an

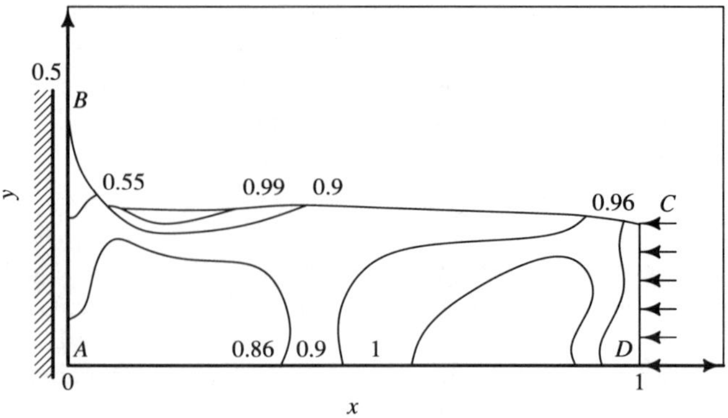

FIGURE 6.21. Elasto-plastic design.

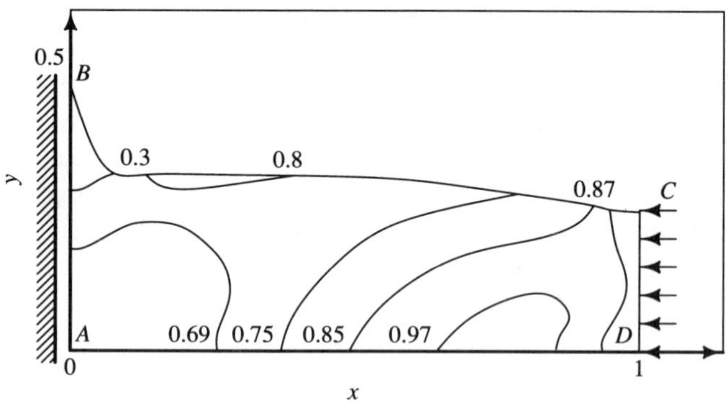

FIGURE 6.22. Elastic design.

elastoplastic element are then compared with a design of a perfectly elastic element. With this in mind, we also solve a problem of designing an elastic member that minimizes its weight, with constraints assigned to the stress intensity functional $g_0(\sigma_{ij}) \leq 1$. The solution shown in Figure 6.22 was obtained using an algorithm for optimization of elastic members proposed in [3.2] and [4.6].

In the elastic design (6.22) all the computations were carried out for a well-posed system of equations of elasticity using exactly the same values of the geometric, physical, and load parameters that were used in the plastic design.

Comparing the numerical results of the elastic and plastic designs, we see that, allowing for the existence of a plastic failure mechanism permits an additional decrease in weight of more than 6.6%. At the same time, the stress field becomes noticeably smoother. The level curves for a constant stress intensity are indicated in Figures 6.21 and 6.22 by solid lines.

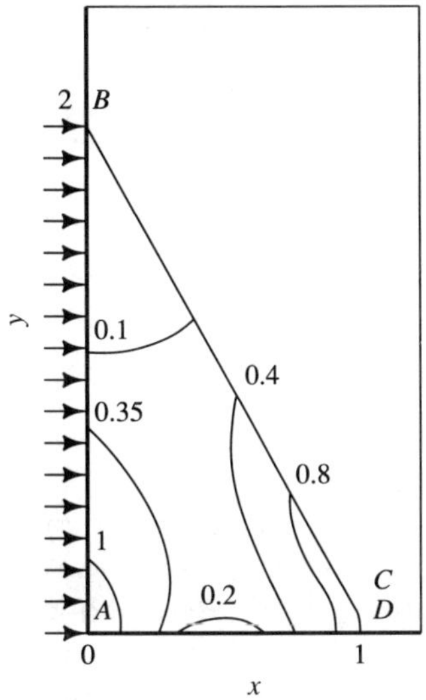

FIGURE 6.23. Initially assumed shape.

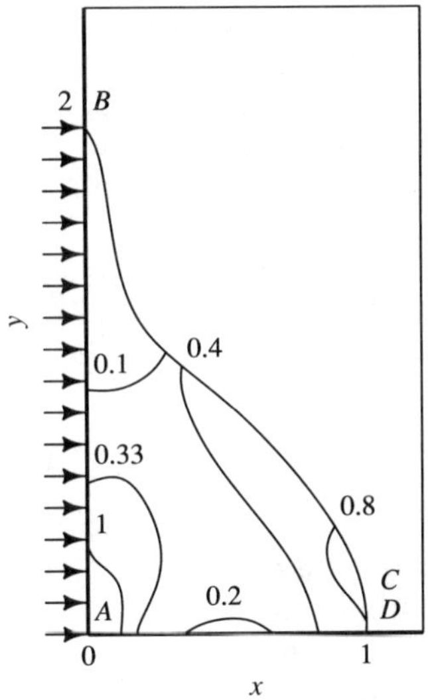

FIGURE 6.24. The final shape.

Another sample design of a two-dimensional structural member is shown in Figures 6.23 and 6.24. Loads acting on this element are uniformly distributed along the boundary AB and point in the direction of the x axis. The AD segment of the boundary is rigidly held. In the optimization process, we can vary the BC part of the surface, which is load-free. The initial approximation to the shape of the BC segment of the boundary consists of a line segment (see Figure 6.23). The improved shape of the boundary resulting from our computations is illustrated in Figure 6.24. As in the preceding example, the solid lines in Figures 6.23 and 6.24 indicate the level curves for values of stress intensity functional. The total mass of the material has been decreased by about 10%.

6.12 Design of Elastoplastic Structures with Minimal Weight and Adaptability to Variable Loads

The theory offered in the preceding section concerning optimal designs with constraints on the limiting load-carrying ability suffices to ensure a safe response of the structure to known (fixed) systems of applied loads. However, if different types of loads may be acting on that structure, even if none of them is a critical load (in the sense of the theory of critical equilibria), but nevertheless may initiate a plastic deformation, it is possible for the structure to fail. A basic reason for the failure is the presence of cyclic loads that cause the structure to undergo cyclic velocities of plastic deformation. In that case failure is caused by cyclic plastic deformations that accumulate to produce a total (large) deformation.

Structures may adapt to smaller loads (i.e., survive them) if the residual stresses following a loading program are such that in a subsequent loading program the deflections are fully elastic. Below we shall utilize a basic theorem (Melan's theorem) of shakedown theory (see [6.41]). If it is possible to find time-independent residual stresses σ_{ij}^r, such that at every point inside the structure the sum of these stresses with the elastic streses σ_{ij}^e produces a safe stress; that is, if

$$g(\sigma_{ij}^e + \sigma_{ij}^r, k) < 0,$$

for all admissible combinations of loads within the assigned bounds, the structure will adjust to a time-independent distribution of residual stresses, and for all subsequent variations of loads within the admissible bounds, the response of the structure is fully elastic (shakedown will be reached by the structure). Here σ_{ij}^e denotes the stress distribution caused by any admissible realizable load satisfying a full system of classical equations of the theory of elasticity.

We shall formulate an optimization problem with skakedown constraints. Let a deformed body occupy a region Ω and be subjected to body forces q_i and external loads T_i that act on a segment of the boundary denoted by Γ_σ. The remaining part of the boundary Γ_u is rigidly held ($\Gamma = \Gamma_\sigma + \Gamma_u$). As before, the material is assumed to be ideally elastoplastic. We need to determine the shape of a segment of the boundary $\Gamma_v \in \Gamma_\sigma$ in order to minimize the functional

$$\int_\Omega d\Omega \to \min_{\Gamma_v}, \tag{6.136}$$

with a condition assigned to the stresses σ_{ij}^e and σ_{ij}^r, which are solutions of the boundary-value problem

$$\left.\begin{aligned} (\sigma_{ij}^e)_{,j} + q_i = 0, \qquad \sigma_{ij}^e &= 1/2[C_{ijkl}(u_{k,l}^e + u_{l,k}^e)], \\ (\sigma_{ij}^e n_j)_{\Gamma_\sigma} = T_i, \qquad & (u_i^e)_{\Gamma_u} = 0, \end{aligned}\right\} \tag{6.137}$$

$$(\sigma_{ij}^r)_{,j} = 0, \qquad \dot{\sigma}_{ij}^r = 0, \qquad (\sigma_{ij}^r n_j)_{\Gamma_\sigma} = 0. \tag{6.138}$$

The strength condition is given by the inequality

$$g(\sigma_{ij}^e + \sigma_{ij}^r, k_f) \le 0, \tag{6.139}$$

for all active loads T_i and q_i. The dot over a variable in (6.138) denotes differentiation with respect to time. Other symbolism used in (6.137) to (6.139) is commonly accepted. In analogy with the previous section, the plasticity constant k has been replaced by $k_f = k - \varepsilon$, with $\varepsilon > 0$. We presume that the loads acting on the body obey some loading sequence $T_i = T_i(x, t), q_i = q_i(x, t)$, $t \in [0, t_k]$, where t_k is some fixed finite time for completion of the loading program. If loading is periodic, then we set t_k to be equal to the period.

To formulate our optimization problem, we use relations from sensitivity analysis, and necessary optimality conditions. Thus, the minimized functional representing the volume of the body is rewritten as

$$(1/t_k)\int_0^{t_k}\int_\Omega \mathrm{d}\Omega \mathrm{d}t, \tag{6.140}$$

and inequality (6.139) is transformed into the equality $g + \mu^2 = 0$ after introduction of a multiplier (or slack variable) μ^2. Combining this equality with (6.136) to (6.138) and designating the functional (6.140) as a cost functional representing the volume, we construct the generalized Lagrangian functional:

$$J = \int_0^{t_k}\int_\Omega f^l \mathrm{d}\Omega \mathrm{d}t, \qquad f^l = (1/t_k)[1 + \lambda(g + \mu^2) + \psi_i^e(\sigma_{ij}^e + q_i)$$
$$+ \chi_{ij}^e(\sigma_{ij}^e - C_{ijkl}u_{k,l}^e) + \psi_i^r(\sigma_{ij}^r)_{,j}]. \tag{6.141}$$

Here ψ_i^e, ψ_i^r, χ_{ij}^e, and λ denote adjoint variables. To obtain a formula for δJ, we vary the boundary surface Γ_v while obeying the boundary conditions that are assigned to the state function. We can assume that the variation of σ_{ij}^e is time-independent and compute the values of the adjoint variables from the system of relations:

$$\tfrac{1}{2}(\psi_{i,j}^e + \psi_{j,i}^e) = \chi_{ij}^e + \lambda\partial g/\partial\sigma_{ij}^e \qquad \text{in } \Omega, \tag{6.142}$$

$$(\psi_i^e)_{\Gamma_u} = 0, \tag{6.143}$$

$$(C_{ijkl}\chi_{kl}^e)_{,j} = 0 \qquad \text{in } \Omega, \tag{6.144}$$

$$(C_{ijkl}\chi_{kl}^e n_j)_{\Gamma_\sigma} = 0, \tag{6.145}$$

$$\tfrac{1}{2}(\psi_{i,j}^r + \psi_{j,i}^r) = (1/t_k)\int_0^{t_k}\lambda(\partial g/\partial\sigma_{ij}^r)\mathrm{d}t \quad \text{in } \Omega, \tag{6.146}$$

$$(\psi_i^r)_{\Gamma_u} = 0, \tag{6.147}$$

$$\lambda\mu = 0 \qquad \text{in } \Omega. \tag{6.148}$$

If (6.142) to (6.148) are satisfied, the expression for δJ can be written as follows:

$$\delta J = \int_{\Gamma_v}\{(1/t_k)\int_0^{t_k}[1 + \lambda g - 2H\psi_i^e T_i - (\psi_i^e\sigma_{ij}^e)_{,j}$$
$$+ (\psi_i^r\sigma_{ij}^r)_{,j} + (\psi_i^e T_i)_{,k}n_k]\mathrm{d}t\}\delta l \mathrm{d}\Gamma, \tag{6.149}$$

where H denotes (as before) the mean curvature of the surface Γ_v and δl stands for the variation of the normal to that surface. Equation (6.149), which is a basic sensitivity formula, relates the variation of the boundary surface to the variation of the cost functional, and indicates that to compute a variation that improves the cost functional by this formula, one must have complete information concerning the values of the elastic stresses σ_{ij}^e, of the adjoint variables ψ_i^e, ψ_i^r, χ_{ij}^e, and λ for all values of $t \in [0, t_k]$. Some routine algebra applied to

(6.149) results in the necessary optimality condition

$$(\psi_i^r \sigma_{ij}^r)_{,j} - 1 = (1/t_k) \int_0^{t_k} [\lambda g - 2H\psi_i^e T_i - (\psi_i^e \sigma_{ij}^e)_{,j} + (\psi_i^e T_i)_{,k} n_k)] \mathrm{d}t. \quad (6.150)$$

In the case when no body forces or external loads are applied to the surface segment Γ_v and the plasticity criterion is given by a homogeneous function of order 2 of homogeneity with respect to the components of the stress tensor $g_0(\sigma_{ij})$, that is, if we have $g(\sigma_{ij}, k) = g_0(\sigma_{ij}) - k_f^2$, then the optimality condition (6.150) simplifies to the formula

$$(\psi_i^r \sigma_{ij}^r)_{,j} - 1 = (1/t_k) \int_0^{t_k} [\lambda g - (\psi_i^e \sigma_{ij}^e)_{,j}] \mathrm{d}t. \quad (6.151)$$

References for Chapter 6

6.1. Aleksandrov, M.A., Korshinin, M.S., and Smoliarov, N.N., Computation of elastic plates and shells that are close to uniform strength, *Prikl. Mekh.*, 1978, *14*, No. 10, pp. 41–46.

6.2. Banichuk, N.V., On a game theoretic approach to optimization problems for elastic bodies, *Prikl. Mat. Mekh.*, 1973, *37*, No. 6, pp. 1098–1108.

6.3. Banichuk, N.V., Optimal design for a one-dimensional bending problem for fixed and moving loads, *Izv. Akad. Nauk SSSR. MTT*, 1974, No. 5, pp. 113–123.

6.4. Banichuk, N.V., The problem of optimizing the shape of a hole in a plate subjected to bending, *Izv. Akad. Nauk SSSR. MTT*, 1977, No. 3, pp. 81–88.

6.5. Banichuk, N.V., Conditions for optimality in problem of finding the shapes of holes in elastic bodies, *Prikl. Mat. Mekh.*, 1977, *41*, No. 5, pp. 920–925.

6.6. Banichuk, N.V., Designing of plates for minimum stress and deflection, In: *Optimization of Distributed Parameter Structures*, Iowa, 1980, E.J. Haug and J.J. Cea (eds.), Alphen aan den Rijn, Sijthoff-Noordhoff, 1981, pp. 333–361.

6.7. Banichuk, N.V., and Kobelev, V.V., Some problems of optimal design of shells of revolution, *Izv. Arm. SSR, Mekhanika*, 1983, No. 2, pp. 10–17.

6.8. Banichuk, N.V., and Kobelev, V.V., On optimal but not uniformly strong shapes of cross-sections of beams, *Izv. Akad. Nauk SSSR. MTT*, 1983, No. 5, pp. 162–167.

6.9. Barnett, R.L., Survey of optimum structural design, *Exp. Mech.*, 1966, *6*, No. 12, pp. 19–26.

6.10. Beliaev, N.V., *Strength of Materials*, Gostekhizdat., Moscow-Leningrad 1951, 856 pp.

6.11. Cherepanov, G.P., Inverse problems in the two-dimensional theory of elasticity, *Prikl. Mat. Mekhanika*, 1974, *38*, No. 6, pp. 963–979.

6.12. Chiras, A.A., Barkauskas, A.E., and Karkauskas, R.P., *Theory and Technique in Optimizing Elastoplastic Systems*, Leningrad, Stroiizdat., 1974.

6.13. Cinquini, C., and Sacchi, G., Problems of optimal design for elastic and plastic structures, *J. Mech. Appl.*, 1980, *4*, No. 1, pp. 1–29.

6.14. Cinquini, C., Limit analysis and optimal plastic design of circular cylindrical shells. In: *Optimization Methods in Structural Design*, H. Eschenauer and N. Olhoff (eds.), Mannheim Bibliogr. Institute, 1983, pp. 309–318.

6.15. Cinquini, C., Structural optimization of plates of general shape by finite elements, *J. Structural Mechanics*, 1981, *9*, No. 4, pp. 465–481.

6.16. Dorn, W.S., Gomory, R.E., and Greenberg, H.G., Automatic design of optimal structures, *J. Mech.*, *3*, No. 1, 1964.

6.17. Drucker, D.C., and Shields, R.T., Design for minimum weight. In: *Proc. 9th International Congr. Applied Mechanics*, Brussels, 1957, Vol. 5, pp. 212–222.

6.18. Drucker, D.C., and Shields, R.T., Bounds on minimum weight design. *Quart. Appl. Math.*, 1957, *15*, No. 3, pp. 269–281.

6.19. Erkhov, M.I., On the theory of optimal design for dynamically loaded rigid-plastic bodies. In: *Mechanics of deformable solid bodies*, Novosibirsk, Institute of Hydrodynamics, 1979, pp. 44–47.

6.20. Flügge, W., *Statics and Dynamics of Shells*, Moscow, Gostekhizdat., 1961, 306 pp.

6.21. Francavilla, A., Ramakrishnan, C.V., and Zienkiewicz, O.C., Optimization of shape to minimize stress concentration, *J. Strain Analysis*, 1975, *10*, pp. 63–70.

6.22. Galilei, Galileo, Discorsus et demonstrationes mathematicae, circa duas novas scientias, Leyden, 1699, a Latin translation from the 1638 edition by Mathias Bernegger, including Galileo's direction to Elzeviers. New translation into English with introduction and notes by Stillman Drake, University of Wisconsin Press, Madison, 1974.

6.23. Ganeeva, M.S., Kornishin, M.S., and Malakhov, V.G., Uniformly stressed elastic shells of revolution. In: *Seminar on Theory of Shells*, Kazan', Phys.-Tekh. In., 1975, Vol. 3, pp. 92–106.

6.24. Gol'denblat, I.I., and Kozhnov, V.A., *Strength and Plasticity Criteria for Structural Materials*, Moscow, Mashinostroenie, 1968, 192 pp.

6.25. Haug, E.J., and Kwak, B.M., Contact stress minimization by contour design, *Intern. J. Numer. Meth. Eng.*, 1978, *12*, pp. 917–930.

6.26. Hegemeier, G.A., and Prager W., On Michell trusses, *Intern. J. Mech. Science*, 1969, *11*, p. 209.

6.27. Hemp, W., On Michell framework for uniform load between fixed supports, *Eng. Optim.*, 1974, *1*, No. 1, pp. 61–69.

6.28. Hodge, F.G., *Computation of Structures with Plastic Deformations*, Moscow GNTI, 1963, 380 pp.

6.29. Hopkins, H.G., and Prager, W., Limits on economy of material in plates, *J. Appl. Mech.* 1955, *22*, pp. 372–374.

6.31. Ivanov, G.B., Computation of an optimal variable thickness of a shell. In: *Problems of Mekhanics of a Solid Deformable Body*, Leningrad, Sudpromgiz, 1970, pp. 171–177.

6.32. Ivanov, G.B., and Kosmodem'ianskii, A.S., Inverse problem of bending for thin isotropic plates, *Izv. Akad. Nauk SSSR. MTT*, 1974, No. 5, pp. 53–56.

6.33. Il'iushin, A.A., A theory of elastoplastic strains, *Prikl. Mat. Mekh.*, 1946, *10*, No. 3, pp. 347–356.

6.34. Il'iushin, A.A., *Plasticity*, Moscow and Leningrad, Gostekhizdat., 1948, 308 pp.

6.35. Ishlinskii, A.Iu., *Mechanics, Ideas, Problems and Applications*, Moscow, Nauka, 1985, 624 pp.

6.36. Ishlinskii, A.Iu., On a uniformly stressed cross-section of a beam, Nauchn. Zap, Mekh. Moscow State University, 1940, *39*, pp. 87–90.

6.37. Ishlinskii, A.Iu., Strength hypothesis for a change of shape, *Nauchn. Zap. Mekh.* Moscow State University *Scientific Notes (Mechanics) Moscow State University* 1940, *16*, pp. 111–124.

6.38. Khuberian, K.M., *Rational Shapes of Pipes, Reservoirs, and Pressure Vessels*, Moscow, Gosstroiizdat, 1956.

6.39. Khutorianskii, N.M., Certain inverse and optimization problems in a two-dimensional theory of elasticity. In: *Applied Problems of Strength and Plasticity*, Gorkii, Gorkii University, 1978, *6*, pp. 81–87.

6.40. Khutorianskii, N.M., On solution of certain two-dimensional and three-dimensional shape optimization problems in elastic bodies. In: *Applied Problems of Strength and Plasticity*, Gorkii, Gorkii University, 1978, *8*, pp. 66–74.

6.41. Koiter, V.T., *A General Theory of Elastoplastic Media*, Moscow, Innostr. Liter. (Russian translation) 1961, 80 pp.

6.42. Komarov, V.A., On a rational distribution of material in structures, *Izv. Akad. Nauk SSSR (Mekhanika)*, 1965, No. 5, pp. 85–87.

6.43. Konig, J.A., On optimum shakedown design. In: *Proc. IUTAM Symposium on Optimal Design*, Warsaw, 1973, Springer-Verlag, Berlin, 1975, pp. 405–414.

6.44. Kornishin, M.S., Aleksandrov, M.A., and Smoliarov, N.N., Computation of close to uniformly stressed deformable plates and shallow shells using numerical techniques. In: *Proc. All-Union Conf. on Numerical Solutions of Problems in Elasticity and Plasticity*; Part II, Novosibirsk, V.Ts. SO Akad. Nauk SSSR, 1976, pp. 69–76.

6.45. Kristenson, E.S., and Madson, N.F., On optimum shape of a fillet in plates subjected to multiple in-plane loading cases, *Intern. J. Numer. Meth. Eng.*, 1976, *10*, pp. 1007–1019.

6.46. Kurshin, L.M., and Rastorguyev, G.I., On the problem of reinforcing the boundary of a hole in a plate, *Izv. Akad. Nauk SSSR. MTT*, 1979, No. 6, pp. 94–102.

6.47. Kurshin, L.M., and Rastorguyev, G.I., On the problem of reinforcing the boundary of a hole in a plate, using a moment-free, elastic rod, *Prikl. Mat. Mekh.*, 1980, *44*, No. 5, pp. 905–915.

6.48. Lepik, Iu.R., Application of Pontryagin's maximality principle for an optimal design of cylindrical shells made from rigid-plastic materials. In: *Advances in Mechanics of Deformable Media*, Moscow, Nauka, 1975, pp. 340–349.

6.49. Lepik, U. [Iu.], Application of Pontryagin's maximum principle for minimum weight design of rigid-plastic circular plates. *Intern. J. Solids Structures*, 1973, *9*, pp. 615–624.

6.50. Litvinov, V.G., and Panteleev, A.D., The problem of optimization of plates with variable thickness, *Izv. Akad. Nauk SSSR. MTT*, 1980, No. 2, pp. 174–181.

6.51. Maier, G., Zavelani-Rossi, A., and Benedetti, D., A finite element approach to optimal design of plastic structures in plane stress, *Intern. J. Numer. Meth. Eng.*, 1972, *4*, pp. 455–473.

6.52. Malkov, V.P., and Strongin, R.G., Optimization of structures with respect to weight and strength criteria. In: Solution techniques for problems in elasticity and plasticity. Gorkii, Gorkii University, 1971, *4*, pp. 138–149.

6.53. Malkov, V.P., and Salganskaya, E.A., Optimal distribution of material in rotating discs from a strength criterion, *Izv. Vuzov, Aviats. Tekhnika*, 1976, No. 3.

6.54. Malkov, V.P., Equivalent reinforcement for a boundary of a cutout in a thin-walled member. In: *Applied Problems of Strength and Plasticity*. Gorkii, Gorkii University, 1979, *10*, pp. 96–113.

6.55. Mansfield, E.H., An optimum surface of revolution for pressured shells, *Intern. J. Mech. Science*, 1981, *23*, pp. 57–62.
6.56. Mansfield, E.H., Neutral holes in plane sheet-reinforced plates which are elastically equivalent to the uncut sheet, *Quart. J. Mech. Appl. Math.*, 1953, *6*, Part 3, pp. 370–378.
6.57. Maxwell, C., *Scientific Papers*, Cambridge University Press, 1880, Vol. 2, pp. 175–177.
6.58. Mazzarella, C., and Polizzotto C., Optimum design of rigid-workhardening structures with constraints on deformation, *Eng. Struct.* 1980, *2*, No. 3, pp. 138–146.
6.59. Medvedev, N.G., and Totskii, N.P., Optimization of cylindrical shells with variable thickness and axially symmetric loads, *Prikl. Mat. Mekh.* 1984, *20*, No. pp. 53–57.
6.60. Michell, A.G., and Melbourne, M.C.S., The limits of economy of material, *Phil. Mag.*, Series 6, 1904, *8*, pp. 589–597.
6.61. Mikhailovskii, E.I., Optimal reinforcement for the edges of a shell, *Izv. Akad. Nauk SSSR-MTT*, 1975, No. 1, pp. 42–51.
6.62. Mushtari, Kh.A., On the bending theory for a rectangular plate with variable thickness, *Inzh. Zh.*, 1964, *4*, No. 1, pp. 45–49.
6.63. Mushtari, Kh.A., On the bending theory for a plate with minimal weight and made of a composite material, *Prikl. Mekh.*, 1967, *3*, No. 4, pp. 1–7.
6.64. Nemirovskii, Iu.V., Estimates of the weight of optimal plastic structures, *Inzh. Zhurnal, Mekh. Tve. Tela*, 1968, No. 4, pp. 159–162.
6.65. Nemirovskii, Iu.V., and Reznikov, B.S., Uniform strength in conditions of creep for beams and plates, *Mashinovedenie*, 1969, No. 2, pp. 58–64.
6.66. Neuber, H., Der Zugbeanspruchte Flachstab mit optimalen Querschnittsubergang, *Forsch. Ingenieurwiss.* 1969, *35*, pp. 29–30.
6.67. Neuber, H., Zur Optimierung der Spannungskonzentration. In: *Mechanics of Continuous Media and Related Problems of Analysis*, Moscow, Nauka, 1972, pp. 375–380.
6.68. Oda, J., On a technique to obtain an optimum strength shape by the finite element method, *Bull. JSME*, 1977, *20*, pp. 160–167.
6.69. Onat, E.T., and Prager, W., Limit of economy of material in shells, *Ingenieur*, 1955, *67*, pp. 46–49.
6.70. Onat, E.T., Shumann, W., and Shield, R.T., Design of circular plates for minimum weight, *ZAMP*, 1957, *8*, No. 6, pp. 485–499.
6.71. Pisarenko, G.S., and Lebedev, A.A., *Deformation and Strength of Materials in a Complex Stressed State*. Kiev, Naukova Dumka, 1976, 415 pp.
6.72. Pochtman, Iu.M., and Piatigorskii, Z.I., *Computation and Optimal Design of a Structure with Consideration of Adaptability*. Moscow, Nauka, 1978, 208 pp.
6.73. Polizzotto, C., Optimal design for multiple sets of loads, *Meccanica*, 1974, *9*, pp. 206–219.
6.74. Polizzotto, C., Mazzarella, C., and Panzeca, T., Optimal design for workhardening adaptation, *Comp. Meth. Appl. Mech. Eng.*, 1977, *12*, No. 2, pp. 129–144.
6.75. Polizzotto, C., and Panzeca, T., Optimal design of beams for workhardening adaptation. In: *Optimization of Distributed Parameter Structures*, Iowa, 1980, E.J. Haug and J. Cea (eds.), Alphen aan den Rijn, Sijthoff-Noordhoff, 1981, pp. 910–927.

6.76. Prandtl, L., Anwendungsbeispiele zu einem Henkyschen Satz über das plastische Gleichgewicht, *Ztschrft angew. Math. Mech.*, 1923, *3*, No. 6.
6.77. Rabinovich, I.M., Beam systems of minimal weight. In: *Proc. II-nd All-Union Conf. Theoretical and Applied Mechanics*, Moscow, Nauka, 1966, Vol. 3, pp. 265–275.
6.78. Runge, C., and Konig, H., *Vorlesungen uber numerischen Rechnen*, Berlin, 1924.
6.79. Save, M.A., Some aspects of minimum weight design. In: *Engineering Plasticity*, J. Heymann and F.A. Leckie (eds.), Cambridge Univ. Press, 1968, pp. 611–6626.
6.80. Save, M.A., A unified formulation of the theory of optimal plastic design with convex cost functions, *J. Struct. Mech.*, 1972, *1*, No. 2, pp. 257–276.
6.81. Savin, G.N., *Stress Distribution around Openings*, Kiev, Naukova Dumka, 1968, 887 pp.
6.82. Savin, G.N., *Stress Concentration around Openings*, Moscow and Leningrad, Gostekhizdat, 1951, 496 pp.
6.83. Schnack, E., An optimization procedure for stress concentrations by the finite element technique, *Intern. J. Numer. Meth. Eng.*, 1979, *14*, pp. 115–124.
6.84. Shamiev, F.G., On designing ring-shaped plates of minimal weight, *Izv. Akad. Nauk. Az. SSR, Fiz-Mat. Tekh. Nauk*, 1963, No. 3, pp. 13–20.
6.85. Shamiev, F.G., On designing elliptically shaped plates of minimal weight, *Izv. Akad. Nauk, Az. SSR, Fiz-Mat Tekh. Nauk*, 1966, No. 4, pp. 23–27.
6.86. Shefer, G., and Demkowicz, L., Optimal design of elastic nonlinear plates by means of the mini-max theory. In: *Optimization Methods in Structural Design*, H. Eschenauer and N. Olhoff (eds.), Mannheim Bibliogr. Institute, 1983, pp. 409–414.
6.87. Shirko, I.V., Axially symmetric bending of a cylindrical shell with uniform strength. *Prikl. Mekh.*, 1969, *5*, No. 4, pp. 45–53.
6.88. Shirko, I.V., The shape of a plate with uniform strength, *Inzhin. Zh.*, 1965, *5*, No. 2, pp. 293–298.
6.89. Sirazutdinov, Iu.K., Beam cross-sections of uniform strength, *Proc. Kazan Aviation Institute*, 1974, No. 168, pp. 11–18.
6.90. Strang, G., and Kohn, R.V., Hencky-Prandtl nets and constrained Michell trusses. In: *Proc. Intern. Symp. Optimum Structural Design*, University of Arizona, Tucson, Arizona, 1981, pp. 4.17–4.22.
6.91. Sirazutdinov, Iu.K., Designs of optimal systems of beams, *Archivum Inzynierji Lądowej*, 1979, *25*, No. 2, pp. 245–264.
6.92. Sofronov, Iu.D., Designs of uniformly stressed and of uniform strength trusses, beams and frames subjected to cyclic loads, with consideration of hysteresis losses in materials. In: *Loss of Energy in Vibrations of Mechanical Systems*. Kiev, Naukova Dumka, 1968, pp. 76–82.
6.93. Timoshenko, S.P., *Strength of Materials*, Moscow, Fizmatgiz, 1960, Vol. 1, 380 pp.
6.94. Timoshenko, S.P., and Voinovski-Krieger, S., *Plates and Shells*, Moscow, Nauka, 1966, 636 pp.
6.95. Tveergand, V., On the optimum shape of a fillet bar with restrictions. In: *IUTAM Symposium on Optimal Design*, Warsaw, 1973. Springer Verlag, Berlin, 1975, pp. 181–195.
6.95. Vigderhauz, S.B., An integral equation for a problem in two-dimensional theory of elasticity, *Prikl. Mat. Mekh.*, 1976, *40*, No. 3, pp. 566–569.

6.96. Vigderhauz, S.B., On a case of the inverse problem of two-dimensional theory of elasticity, *Prikl. Mat. Mekh.*, 1977, *41*, No. 5, pp. 902–908.

6.97. Vigderhauz, S.B., Optimality conditions in axially symmetric problems of the theory of elasticity, *Prikl. Mat. Mekh.*, 1982, *46*, No. 2, pp. 278–282.

6.98. Vigderhauz, S.B., An inverse problem in a three-dimensional theory of elasticity, *Izv. Akad. Nauk SSSR. MTT*, 1983, No. 2, pp. 90–93.

6.99. Wheeler, L., On the role of constant stress surfaces in the problem of minimizing elastic stress concentrations, *Intern. J. Solids Structures*, 1976, *12*, No. 11, pp. 779–789.

6.100. Zavelani, A., Maier, G., and Binda, L., Shape optimization of plastic structures. In: *IUTAM Symposium on Optimal Design*, Warsaw, 1973. Springer Verlag, Berlin, 1975, pp. 181–195.

6.101. Zienkiewicz, O.C., and Campbell, J.S., Shape optimization and sequential linear programming. In: *Optimum Structural Design*, R.H. Gallagher and O.C. Zienkiewicz (eds.), Wiley, New York, 1973, pp. 109–126.

6.102. Ziegler, N., Dome with uniform strength, *Mechanika*, 1952, pp. 127–132.

7
Rigidity and Weight Criteria

7.1 Rigidity Criteria and Their Use in Problems of Optimal Design

All real structures are characterized by some degree of deformability. As external loads are applied and the weight of the structure itself is acting on, the deformations may be large, and some parts of the structure may have displacements that are not admissible and inconsistent with a proper functioning of the structure. Therefore, making certain that the rigidity of the structure is maintained and reducing the weight, while observing constraints on the rigidity, is one of the most important design objectives.

Structural rigidity may be defined in various ways. One possible measure of rigidity is the magnitude of the deflection at certain crucial points within the entire structure or in certain structural members. Other criteria may involve the total work performed by elastic displacements, or the magnitudes of strains, or the maximal deflection for a thin-walled structure.

The problem of minimizing the weight of a structure with constraints assigned to its rigidity, and the closely related problem of maximizing the rigidity of a body with a constant weight, has been the main topic of a large number of publications (see [5], [24] and [48]). The distribution of material has been found either numerically or analytically in these monographs for several types of structural members. This is particularly true for a class of problems with constraints assigned to the deflection function and for problems of minimizing the maximum deflection of a thin-walled structure, where minimax techniques have been most effective (see [4.7], and [7.33] to [7.36]). Problems in which the criteria of rigidity are given by integral functionals have been investigated in minute detail (see [7.2], [7.40], [7.46], [7.47], [7.52] to [7.56], [7.58], [7.61], [7.82], [7.83], [7.97], [7.78] to [7.84], and [7.99] to [7.101]).

We shall consider a deformed body occupying a region Ω that is bounded by a surface $\Gamma = \Gamma_\sigma + \Gamma_u$ (see Figure 7.1). The displacements $\mathbf{u}$ are prescribed on the part Γ_u, and loads $\mathbf{q}$ are given on Γ_σ:

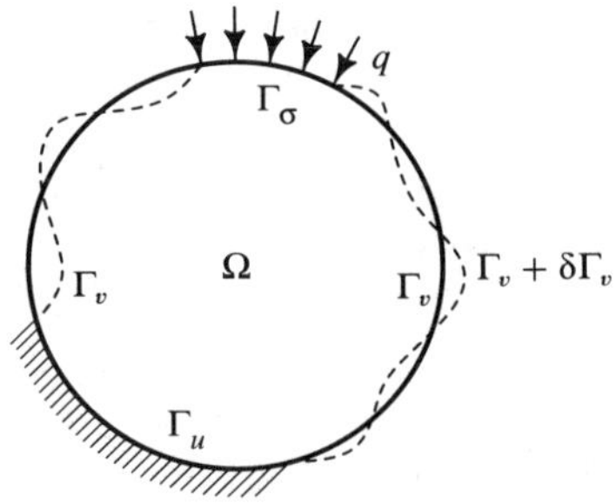

FIGURE 7.1. Shape of a deformed body.

$$(u_i)_{\Gamma_u} = U_i, \qquad (\sigma_{ij} n_j)_{\Gamma_\sigma} = q_i, \tag{7.1}$$

where U_i and q_i are given functions.

Let Γ_v denote a segment of the surface Γ_σ that is load-free ($\mathbf{q} = 0$). The optimization problem consists in finding the shape of the boundary segment Γ_v such that the compliance functional for the structure attains a minimum, that is

$$J = \frac{1}{2} \int_{\Gamma_\sigma} \mathbf{qu} d\Gamma_\sigma \to \min_{\Gamma_\sigma} \tag{7.2}$$

and such that the isoperimetric condition of constant volume is maintained,

$$\text{meas}\, \Omega = V, \tag{7.3}$$

and the basic equations defining the stressed state of the body are obeyed,

$$\sigma_{ij,j} + q_i^v = 0, \qquad \sigma_{ij} = A_{ijkl}\varepsilon_{kl}, \qquad \varepsilon_{kl} = 1/2(u_{k,l} + u_{l,k}). \tag{7.4}$$

Here A_{ijkl} are elastic moduli and q_i^v are components of the vector of body forces. For the optimization problem given by (7.1), (7.3), and (7.4) with integral quality functional (7.2), the optimality condition may be derived directly by varying the definite integrals with a variable domain of integration. It assumes the form

$$\Pi(\sigma_{ij}, \varepsilon_{ij}) \equiv \tfrac{1}{2}(\sigma_{ij} \cdot \varepsilon_{ij})_{\Gamma_v} = \lambda, \tag{7.5}$$

where Π is the strain energy density and λ is an unknown constant.

Condition (7.5) implies that along an optimal boundary surface, the strain energy density is constant.

Now, using Hooke's law we may represent the expression on the left-hand side of (7.5) as a quadratic form of the components of the stress tensor. In two-dimensional problems of the theory of elasticity, only one stress component σ_s is nonzero along any curve in Γ_v, where s denotes the direction tangential to the boundary and σ_s is a normal stress component acting in the direction orthogonal to s. Therefore, the optimality condition assumes the form $(\sigma_s)_{\Gamma_v} = \text{const}$, which means that the optimal boundary is uniformly stressed, or uniformly strong.

As we have already stated in Section 1.3, one of the reasons why functionals of the form (7.2) have found such wide application in the theory optimal design is the ease of obtaining optimality conditions and solutions to problems. One shortcoming that must be mentioned in the use of the functional (7.2) as the quality criterion is that just because the absolute value of the integral J is small does not mean that the magnitudes of the displacements at certain locations inside the elastic body are small. We shall discuss some examples that illustrate this shortcoming.

Let the load $\mathbf{q}$ that bends a beam be represented by a single concentrated force P acting at the point x_0, that is, $q = P\delta(x - x_0)$, where δ is the Dirac delta function. Then the compliance of the beam is given by

$$J = 1/2Pu(x_0).$$

To reduce the compliance for a constant value of P, we need to reduce the magnitude of the deflection at the point of application of the load.

Similarly, if the load acting on the beam consists of a single concentrated moment applied to the free end, that causes rotation of this end by an angle φ, then the compliance J is given by the formula

$$J = 1/2M\varphi.$$

In this case the minimization of the compliance (therefore, maximization of rigidity) consists in minimizing the angle of rotation of the free end of the beam.

Let an elastic body be rigidly supported along segment Γ_u of the boundary, and a perfectly rigid die be applied (with Γ_σ denoting the region of contact).

Let u denote the displacement of the die and P the force applied to the die. Then we have

$$J = \frac{1}{2} u \int_{\Gamma_\sigma} q \, d\Gamma_\sigma = \frac{1}{2} uP.$$

To reduce the compliance J for a given value of P, we need to reduce the depth of penetration of the die.

For a pipe with circular cross-section subjected to internal pressure q, the compliance is proportional to the radial displacement of points on the inner contour of its cross-section, that is,

$$J = \frac{1}{2} \int_0^{2\pi} qur \, d\varphi = \pi rqu,$$

where r is the radius of the inner contour.

Note certain properties of the compliance functional. Let us suppose that displacements are assigned on the boundary segment Γ_u, that is $U_i = 0$, and let λ and μ denote the elastic (Lamé) constants of the material. Let us also assume that the material is isotropic, but in general inhomogeneous. The inhomogeneity is understood in the sense that the elastic constants for this material have different values at different points inside the material. The following assertion is known to be true (see reference (7.27)). If we decrease (increase) the values of the elastic constants λ and μ of our material in an arbitrary region inside the body, then the compliance does not decrease (or, respectively, increase). To prove this assertion, let us look at the potential energy of our system in a state of equilibrium (see [7.51])

$$W(u) = \int_{\Omega} \Pi \, d\Omega - \int_{\Gamma_\sigma} qu \, d\Gamma_\sigma. \tag{7.6}$$

Here u denotes the displacement field corresponding to the equilibrium state, and Π is the potential energy density for the deformed elastic body:

$$\Pi = \tfrac{1}{2}\lambda(\varepsilon_{11} + \varepsilon_{22} + \varepsilon_{33})^2 + 2\mu(\varepsilon_{11}^2 + \varepsilon_{22}^2 + \varepsilon_{33}^2 + 2\varepsilon^2{}_{12} + 2\varepsilon_{23}^2 + 2\varepsilon_{13}^2)].$$

We reduce the values of the elastic constants λ and μ in some region Ω_1 inside the body to some new values λ_1 and μ_1, respectively, without altering the displacement field. Consequently the value of the first term in (7.6) is decreased, and

$$W(u) \geq W^1(u), \tag{7.7}$$

where $W^1(u)$ is the potential energy of the altered body corresponding to the same displacement field u. We note that this displacement field is admissible for the altered body (see [7.57]), since on the boundary Γ_u it assumes the same 0 values as before, and the displacements u^1 satisfy the conditions of equilibrium for identical applied loads. Recalling the principle of minimum potential energy for the system (see [7.57] and [7.92]), we conclude that

$$W^1(u) \geq W^1(u^1).$$

Using Clapeyron's theorem we can assert that in the state of equilibrium the following equalities are true for the original and altered bodies:

$$\left.\begin{aligned} W(u) &= -\frac{1}{2}\int_{\Gamma_\sigma} qu \, d\Gamma_\sigma = -J, \\ W^1(u^1) &= \frac{1}{2}\int_{\Gamma_\sigma} qu^1 \, d\Gamma_\sigma = -J^1. \end{aligned}\right\} \tag{7.8}$$

Using (7.8), with (7.6) and (7.7), we derive

$$J = \frac{1}{2}\int_{\Gamma_\sigma} qu \, d\Gamma_\sigma \leq \frac{1}{2}\int_{\Gamma_\sigma} qu^1 \, d\Gamma_\sigma = J^1. \tag{7.9}$$

It is clear that if we interchange the original and the altered body, the functional J will not increase with the increase in the constants λ and μ.

We observe that in the more general case when displacements on the boundary segment Γ_u are not identically equal to 0 ($U_i \neq 0$), the assertion proved above is still true if the characteristic (quality) integral functional is replaced by the quantity

$$J = \frac{1}{2}\int_{\Gamma_\sigma} qu\mathrm{d}\Gamma_\sigma - \frac{1}{2}\int_{\Gamma_u} \sigma_n U\mathrm{d}\Gamma_u$$

(see [7.27]). Here σ_n is the vector of forces acting in the plane of the boundary Γ_u with a normal direction n.

The assertion proved above is also true if instead of forces moments are applied to the segment of the body surface Γ_σ or if the body consists of several parts in contact with each other with friction absent on the contact surfaces.

7.2 Local Rigidity Criteria

A very important property of designs for many types of structures or structural members is the magnitude of the largest deflection

$$J = \max_{x \in \Omega} |u(x)|, \tag{7.10}$$

where $u(x)$ is the vector of displacements and Ω is the region occupied by the structural material. Typical difficulties encountered in solving problems with functional of the form (7.10) can be described as follows: If, for example, we wish to solve the problem of minimizing the maximal deflection for a plate by finding an optimal distribution of its thickness (see [5], [6.3], [7.7], [7.38], [7.39], and [7.83]), then the derivation of the necessary optimality conditions is complicated by the fact that we do not know in advance at what point within the structure the maximum deflection occurs. But the location of that point does not depend only on the loads applied to the structure, but also on the distribution of the thickness (i.e., the design function). If on the other hand we consider an optimization problem with inequality-type constraints applied to local properties of the form (7.10), then similar difficulties arise in locating the points or regions in which strict equality is realized. Essential simplifications can be effected in those cases in which the location of the points at which the values of such local properties are to be evaluated are assigned in advance (see [7.37]), or when the location of the greatest value of the cost functional is known (e.g., it may be deduced from the symmetry of the problem).

To estimate the rigidity of structures (this includes curved beams, shells, etc.), in many instances one can use a technique that consists of applying a concentrated force to a fixed location and estimating (generally at some other location) the resulting displacement. Now the applied load is permitted to point in all possible directions, that is, the end of the load vector describes the surface of a sphere. The corresponding displacement vector describes an ellipsoidal surface. (This assertion is a direct consequence of linear elasticity hypotheses concerning the properties of the structural material.)

Estimating the rigidity consists in finding the spatial orientation of this ellipsoid and computing the magnitudes of its semiaxes. As a measure of rigidity, we can take the size of the major axis of the ellipsoid.

This distribution of displacements of various orientations of the load vector indicates a minimal number of boundary-value problems whose solutions permit us to estimate the structural rigidity.

Let us describe another technique for estimating the rigidity of thin-walled structures, which reduces to a computation of displacements that are caused by concentrated loads (see [5] and [7.7]). Let w denote the displacement function for a thin-walled structure obtained by solving the differential equation

$$L(h)w = q, \tag{7.11}$$

with some assigned boundary conditions. It corresponds to a variational principle according to which w minimizes the value of the functional

$$\int_{\Omega} \{\Pi(w, h) - 2qw\} d\Omega \tag{7.12}$$

(among admissible functions, that is, once continuously differentiable functions that may satisfy certain additional conditions).

Here Ω, h, q, L, and Π denote in (7.11) and (7.12) the median surface of a structural member, the distribution of thickness, the external loads, a positive-definite differential operator, and the strain energy density function, respectively. We shall consider the class of admissible loads

$$q(x) \geq 0, \quad \text{and} \quad \text{such that} \int_{\Omega} q(x) d\Omega \leq P. \tag{7.13}$$

As a measure of the rigidity of our structure, we shall take the quantity

$$J = J(h) = \max_{q} \max_{x \in \Omega} w(x, h, q). \tag{7.14}$$

Here P denotes a given constant value. Let us prove a property of the criterion (7.14) with (7.13) that permits us to simplify computations of rigidity estimates.

Property 1. Independently of the distribution of thickness (i.e., of the design), the load that causes a maximum deflection is a concentrated load.

To prove this property, it suffices to consider two equivalent displacement functions w and w^c that correspond, respectively, to a distributed load in (7.13) and to a concentrated load P, applied to the point x^1 of maximum value for the deflection function w. The values of the functions w and w^c computed at the point x^1 are denoted w^1 and w^{c_1}, respectively. Using (7.13) we obtain an estimate of the deflection function,

$$\int_{\Omega} qw d\Omega \leq \left(\max_{x} w\right) \int_{\Omega} q d\Omega \leq w^1 P. \tag{7.15}$$

Now using (7.15) and the minimum principle for the total energy, we arrive at a triple inequality

$$\int_\Omega \Pi(w^c, h)d\Omega - 2Pw^{c_1} \leq \int_\Omega \Pi(w, h)d\Omega - 2Pw^1$$
$$\leq \int_\Omega \{\Pi(w, h) - 2qw\}d\Omega. \qquad (7.16)$$

The energy relations

$$Pw^{c_1} = \int_\Omega \Pi(w^c, h)d\Omega, \qquad \int_\Omega qw d\Omega = \int_\Omega \Pi(w, h)d\Omega,$$

expressing the equality between work performed by external loads and the strain energy stored permit us to simplify inequality (7.16):

$$Pw^{c_1} \geq 2Pw^1 - \int_\Omega qw\,d\Omega \geq \int_\Omega qw d\Omega. \qquad (7.17)$$

It follows from inequality (7.17) that $w^{c1} \geq w^1$. Therefore, we have demonstrated that, for an arbitrary admissible load q satisfying (7.13), it is possible to indicate a point of application of a concentrated load P that causes a maximal deflection of the beam not smaller than the maximal deflection caused by the distributed load $q(x)$. Thus, property 1 has been proved.

Next, let us examine (7.14) if $J = J(h)$ stands for the maximal deflection of a plate. Let the maximum with respect to x in (7.14) occur at $x = x^1$ and the maximum with respect to q occur for $q = P\delta(x - x^2)$, that is, for a concentrated load applied at some point $x = x^2$.

Property 2. The maximum deflection $J = J(h)$ (where the maximum is taken with respect to both q and all $x \in \Omega$) is attained at the point of application of the concentrated load, that is, for $x^1 = x^2$.

We shall prove this assertion by contradiction. Let us assume that, to the contrary, the maximum deflection is attained at some point x^1, while the concentrated load is applied at a different point x^2, where the deflection is equal to w^2. We also consider the deflection function w^a produced by applying the load P at the point x^1. Let w^{a1} denote the magnitude of deflection w^a evaluated at the point x^1. Using a well-known variational principle,

$$\int_\Omega \Pi(w^a, h)d\Omega - 2Pw^{a_1} \leq \int_\Omega \Pi(w, h)d\Omega - 2Pw^1$$
$$\leq \int_\Omega \{\Pi(w, h)d\Omega - 2Pw^2. \qquad (7.18)$$

Using again the energy relations

$$Pw^2 = \int_\Omega \Pi(w, h)d\Omega, \qquad Pw^{a_1} = \int_\Omega \Pi(w^a, h)d\Omega,$$

we transform inequality (7.18) into the form

$$Pw^{a_1} \geq 2Pw^1 - \int_\Omega \Pi(w, h)\mathrm{d}\Omega \geq \int_\Omega \Pi(w, h)\mathrm{d}\Omega. \tag{7.19}$$

Looking at the right-hand side of inequality (7.19), we see that

$$Pw^1 - \int_\Omega \Pi(w, h)\mathrm{d}\Omega \geq 0.$$

This estimate, combined with the left inequality in (7.19), implies that $w^{a1} \geq w^1$. This contradiction completes the proof of Property 2.

Using these properties of criteria (7.14) with (7.13), we may restrict our discussion in estimating rigidity to applications of concentrated loads of magnitude P and to computations of the deflections only at the points of application of these loads.

7.3 Optimization of Rigidity for Beams

Research into optimization of structures originated with the classical work of Galileo devoted to the design of shapes of beams. Afterwards many important problems have been solved related to optimization of beams subjected to bending. Nevertheless, a majority of modern works on optimal design still explores the beam model. The equations modelling the bending of beams are one of the simplest equations encountered in strength of materials theory and are very convenient for stating new problems and for making comparisons between various algorithms and techniques. In this and the preceding section, we have examined some typical questions arising in the optimization of rigidity characteristics within the framework of a simple beam model.

Let a beam be freely supported at the points $|x| = l$, of the x axis. It is bent due to the action of a transverse concentrated load applied at the point $x = 0$ (i.e., $q = P\delta(x)$). The length of the beam ($2l$) and its volume V are given. This imposes a constraint on the distribution of transverse cross-sectional areas $S(x)$.

We postulate the problem of minimizing the magnitude of deflection at the point of application of the load ($x = 0$) by finding the best shape of the beam, that is, finding a control function $h(x)$ that uniquely determines the magnitude of $S(x)$ and of $D(x) = EI(x)$. (Here we take $S = B_\alpha h$ and $D = EI = A_\alpha h^\alpha$; see Section 1.4 in Chapter 1.)

As before, E is Young's modulus, I is the moment of inertia of the beam's cross-sectional area about the neutral axis, A_α and B_α are constants that depend on the type of transverse cross-section, and α is a parameter that may take values 1, 2, or 3. We display again the basic equations for our problem:

$$\left.\begin{aligned} &(Dw_{xx})_{xx} = q, \qquad (w)_{|x|=l} = (Dw_{xx})_{|x|=l} = 0, \\ &\int_{-l}^{l} S\mathrm{d}x = V, \qquad J = w(0) \to \min_h \end{aligned}\right\} \tag{7.20}$$

Note that the problem dual to (7.20) consists in minimizing the total volume of the beam with a constraint assigned to the magnitude of the deflection $w(0)$. The necessary optimality condition for both the direct and the dual problem (see Section 3.1) assumes the form

$$h^{\alpha-1}(w_{xx})^2 = \lambda^2, \tag{7.21}$$

where λ^2 is an unknown constant. To determine the value of λ^2, we use the equality between the work performed in deflecting the beam and the strain energy:

$$P = (A_\alpha/w(0)) \int_{-l}^{l} h^\alpha (w_{xx})^2 dx = (A_\alpha \lambda^2 V)/(B_\alpha w(0)). \tag{7.22}$$

We shall consider the case $\alpha = 1$ corresponding to a trilayer beam with a variable thickness $h/2$ of the reinforcing layers. In this case S is the area of the transverse cross-section of the reinforcing layers; that is, $S = bh$, where b is the width of a rectangular beam. Problems of optimal design of trilayer beams are investigated in [3.17], where a very interesting property of trilayered beams was found. For trilayered structures (beams, plates, etc.) the optimality condition does not explicitly depend on the control function. Therefore, if a sufficient number of constraints is assigned to the state function w, independent of the control h, the optimality condition and some boundary conditions can be used to determine the state function w. To determine the control function h (the thickness of the reinforcing layers), one needs to make use of the basic equations and of the other boundary conditions, which may depend on both w and h. This scheme for solving problems of optimizing trilayered structures was proposed in [3.17]. Let us apply it to the case of a simply supported trilayered beam.

Using the optimality condition $(w_{xx})^2 = \lambda^2$ and the boundary conditions $w(-l) = w(+l) = 0$, we obtain $w = \frac{1}{2}\lambda(l^2 - x^2)$. Substituting this solution in (7.20), we have $l^2 = Pbw(0)/(A_1 V) = [(Pbl^2)/(2A_1 V)]^2$. Now, looking at the equilibrium equation in (7.20) with the boundary conditions $hw_{xx} = 0$ at $x = -l$ and $x = l$, we formulate the boundary value problem $h_{xx} = 0$ $(-l < x < 0$ and $0 < x < l)$ with $h(-l) = h(+l) = 0$. Using the isoperimetric condition in (7.20), we can solve this problem for $h(x)$. After some elementary steps we obtain

$$\left.\begin{aligned} &h = V(l + x)/(bl^2), \quad \text{if } -l \le x \le 0, \qquad h = V(l - x)/(bl^2), \quad \text{if } 0 \le x \le l, \\ &w = [(Pbl^2)/(4A_1 V)](l^2 - x^2), \qquad J_* = w(0) = Pbl^4/(4A_1 V). \end{aligned}\right\} \tag{7.23}$$

Figure 7.2 indicates the distribution of thickness and the deflection of an optimal beam ($l = 1$, $V/b = 0.1$, $Pb/(4A_1 V) = 0.3$).

As was shown in [7.46] and can be seen in Figure 7.3, the thickness of the optimal beam reaches its maximum at the point of application of the load, and decreases toward the ends, becoming equal to 0 at the free support points. Thus, an optimal beam differs from beams having a constant cross-section, or

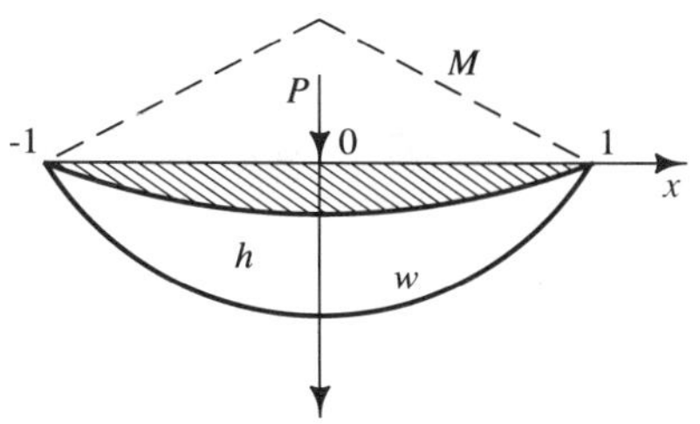

FIGURE 7.2. Thickness and deflection graphs.

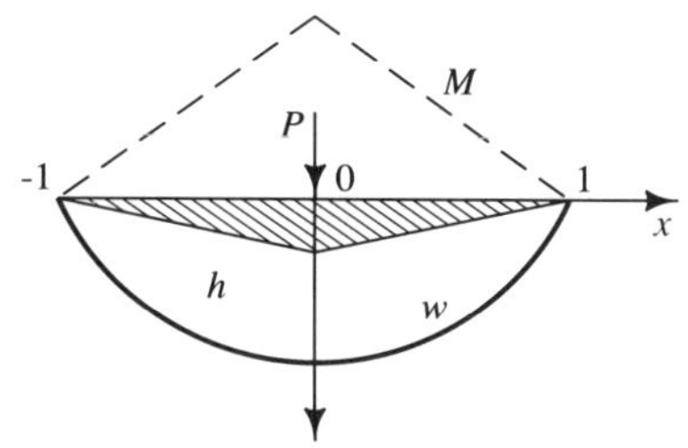

FIGURE 7.3. Thickness and deflection graphs.

in general from beams with $h \neq 0$. For such beams, the boundary condition $hw_{xx} = 0$ is satisfied by setting the curvature of the neutral axis equal to 0 (i.e., $w_{xx} = 0$ at $x = -l$ or $x = +l$), while for the optimal beam this condition is satisfied by setting the thickness of the beam at the endpoints equal to 0. The curvature of the optimal beam is constant throughout its span $-l \leq x \leq l$. The bending moment M (shown by the broken line in Figure 7.2) and the strain energy density stored by the beam segment with coordinate x vary as functions of x in a manner similar to $h(x)$; that is, they attain a maximum at $x = 0$ and decrease to 0 at the endpoints $x = -l$ and $x = +l$.

We shall now evaluate the effectiveness of the optimal solution, comparing it with a beam of constant thickness. For a beam of constant thickness having exactly the same volume, we have $J = w(0) = Pl^4/(3A_1V)$, and consequently, the relative gain in the value of the cost functional for the optimal beam is given by $(J - J_*)/J = \frac{1}{4}$, that is, 25%.

Under the same initial assumptions, let us now consider a uniform beam having a variable height of the cross-section. Setting $\alpha = 3$ (see Section 1.4) and using the optimality condition (7.21), we transform the state equation and the boundary conditions into the form $((w_{xx})^{-2})_{xx} = 0$ for $-l < x < 0$ and for $0 < x < l$, with $w(-l) = w(+l) = w_{xx}^{-2}(-l) = w_{xx}^{-2}(+l) = 0$. Applying the condition $w_x = 0$, we obtain a well-posed boundary-value system for the deflection function w. The condition $w_x(0) = 0$ is a consequence of symmetry with respect to the point $x = 0$, and the assumption that $h(0) \neq 0$. This perceived symmetry permits us to determine all unknown quantities on the interval $0 < x < l$. Integrating the equation $((w_{xx})^{-2})_{xx} = 0$, and computing the three constants of integration from the conditions $w_x(0) = 0$, $w(l) = (w_{xx})^{-2}(l) = 0$, we obtain an expression for w containing one unknown constant. To find the values of that constant and of λ, we need to use the optimality condition, eq. (7.22), and the isoperimetric condition (7.20). We obtain

$$\left.\begin{aligned} h &= 3V/(4bl)(1 - x/l)^{1/2}, \qquad J_* = w(0) = 64Pl^6b^3/(81A_3V^3), \\ w &= [64Pl^5b^3/(81A_3V^3)](1 - x/l)(3 - 2(1 - x/l)^{1/2}) \end{aligned}\right\} \tag{7.24}$$

for $0 \leq x \leq l$.

The functions $h(x)$ and $w(x)$ are illustrated in Figure 7.3 (with $l = 1$, $3V/(4b) = 0.1$, and $64Pb^3/(81A_3V^3) = 0.3$).

As was the case with the trilayered beam, the thickness of a uniform beam attains its maximum in the middle of the beam (at $x = 0$) and reduces to 0 at both ends. However, in the case $\alpha = 1$ the thickness decreased proportionally to the distance from the end, but for $\alpha = 3$, the function $h(x)$ behaves like $t^{1/2}$, where $t = l - x$. There is also an essential difference in the behavior of the deflection function for $\alpha = 1$ and $\alpha = 3$. Now the curvature of the beam's axis varies along the span and becomes unbounded as we approach one of the freely supported ends.

Comparing the deflection $w(0)$ from (7.24) with the corresponding value of $w(0) = 4Pl^6b^3/(3A_3V^3)$ computed for a beam having the same volume but of constant thickness, we find that the gain due to optimization amounts to 40.7%.

7.4 Design of Compound Beams with Additional Supports

In designing a compound structure composed of separate structural elements, as one redistributes different materials between these elements the stresses change, and interactions between the members and the mode of securing them change also. Therefore, in an optimization of an entire structure, there is the problem of utilizing preliminary results concerning the optimization of each separate member. In what follows we offer an example of the design of compound, reinforced beams and draw conclusions concerning the effectiveness of using optimal elements and additional supports (see [7.15]).

We shall examine a structure consisting of $N + 1$ freely supported beams with variable rigidity. These beams are freely supported at the endpoints $x_0 = 0$ and $x_{N+1} = l$, and at the intermediate points x_i ($i = 1, 2, \ldots, N$) on the x axis. The free support at the intermediate locations is accomplished by perfectly rigid supports that do not allow any vertical displacement (see Figure 7.4).

Therefore, boundary conditions at the ends and at the intermediate support points are $w = EIw_{xx} = 0$. As before, E stands for Young's modulus, and I is the moment of inertia of the cross-sectional area about the neutral axis. We

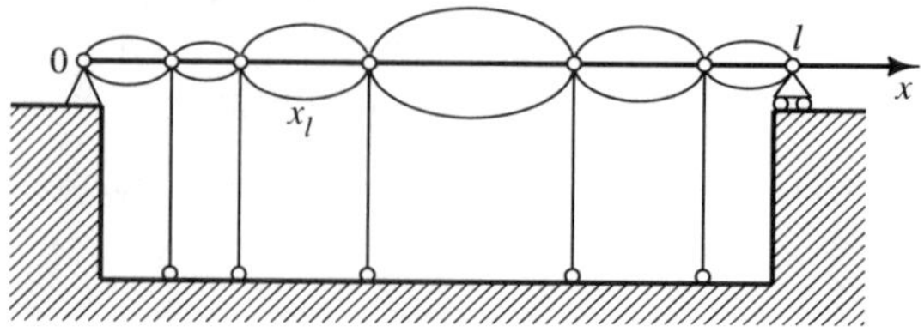

FIGURE 7.4.

shall consider different types of beams that satisfy the dependence relation

$$I(x) = C_k S^k(x), \qquad k = 1, 2, 3,$$

between the moment of inertia I and the magnitude of the beam's cross-sectional area S (see Section 1.4). The coefficient C_k in this formula depends on the shape of the cross-section but is not a function of x. The distribution of rigidity (we mean by it the distribution of cross-sectional areas) and also the number and location of the beam supports are regarded as the design variables. The cost functional depending on these variables is the quantity

$$J = \mu N + \rho \sum_{i=1}^{N+1} V_i. \tag{7.25}$$

The parameters μ and ρ are regarded as given constants and may have either a financial or weight interpretation. For example, let ρ represent the cost per unit volume of beam's material and μ the cost of a single support, while V_i stands for the volume of the ith beam, as given by the formula

$$V_i = \int_{x_i}^{x_{i+1}} S(x)\mathrm{d}x.$$

We assume that a quite arbitrary load $q(x)$ can be applied to our structure, provided the total load does not exceed some given value P, that is,

$$q(x) \geq 0, \qquad \int_0^l q(x)\mathrm{d}x \leq P. \tag{7.26}$$

In particular, in each span of the beam we can assign a concentrated load P. If the load is moving, with a quasi-static change of location, we assume that in the working environment of the structure that force will be applied successively to each point $x \in [0, l]$, and in particular it will be applied to the middle of the span of each beam.

The design problem consists in finding the distribution of rigidity in each beam and finding the optimal number of supports and their location such that the quality functional (7.25) is minimized for any loads that satisfy (7.26), while a rigidity constraint is maintained (i.e., the maximum deflection does not exceed some a priori prescribed magnitude):

$$w(x) \leq w_0. \tag{7.27}$$

Let us examine an arbitrary beam supported at the points x_i and x_{i+1}. A proof can be found in [6.3] that the cross-sectional area of a minimal volume beam that satisfies condition (7.27) and is subjected to external loads of the form (7.26) is given by the formula

$$\left.\begin{aligned} &S(x) = ([(k+3)V_i]/[(k+1)(x_{i+1}-x_i)])\{(16\Psi)/(x_{i+1}-x_i)^2\}^{1/(k+1)}, \\ &\qquad\qquad\qquad\qquad\qquad\qquad\qquad\qquad x_i \leq x \leq x_{i+1}, \\ &V_i = \gamma(x_{i+1}-x_i)^{(k+3)/k}, \qquad \gamma = [P/(16EC_k w_0)]^{1/k}[(k+1)/(k+3)]^{(k+1)/k}, \end{aligned}\right\} \tag{7.28}$$

where Ψ is a function defined by the formula

$$\left.\begin{aligned}\Psi &= (x - x_i)^2/4 \quad \text{if } x_i \le x \le (x_{i+1} + x_i)/2,\\ \Psi &= (x - x_{i+1})^2/4 \quad \text{if } (x_{i+1} + x_i)/2 \le x \le x_{i+1}.\end{aligned}\right\} \tag{7.29}$$

For structures composed of optimal beams satisfying (7.28) and (7.29), if we apply the condition that the functional (7.25) attains a minimum, it is easy to show that $x_i = (x_{i-1} + x_{i+1})/2$. Consequently, in an optimal structure all supports are spaced at an equal distance from each other and all beams have exactly the same length:

$$x_{i+1} - x_i = l/(N + 1), \qquad i = 0, \dots, N. \tag{7.30}$$

Taking into account (7.28) to (7.30), we derive a formula for the optimality of the cost functional,

$$J = \mu N + \rho(\gamma l^{(k+3)/k})/(N + 1)^{3/k}. \tag{7.31}$$

Thus, the initial optimization problem has been reduced to the problem of finding an optimal number of intermediate supports, or equivalently to finding the integer N. From the condition $\partial J/\partial N = 0$, we obtain, after some elementary algebra that include rounding off the value of N to the nearest integer,

$$\left.\begin{aligned}&N = a - 1 \quad \text{if } \alpha \ge \beta, \quad \text{and} \quad N = a \quad \text{if } \alpha < \beta, \quad \text{where}\\ &a = \operatorname{entier}(3/k\beta)^{k/(k+3)}, \qquad \alpha = [(a + 1)^{3/k} - a^{3/k}]/[(a^2 + a)^{3/k}],\\ &\beta = (\mu/\rho\gamma)l^{-(k+3)/k}.\end{aligned}\right\} \tag{7.32}$$

The operation entier denotes taking the nearest integer.

Formulas (7.28) to (7.32) fully determine the solution of our problem. For the case $k = 1$, $\rho l^3/\mu = 10^5$, $3Pl/(64EC_k w_0) = 2.56 \cdot 10^{-3}$ the optimal compound beam is illustrated in Figure 7.5.

This description of an optimization synthesizing procedure, using only optimal members, involved the simplest possible case. It allows an immediate generalization and can be applied to computations of more complex optimal trusses, systems of beams supported on elastic foundations, bridges, etc.

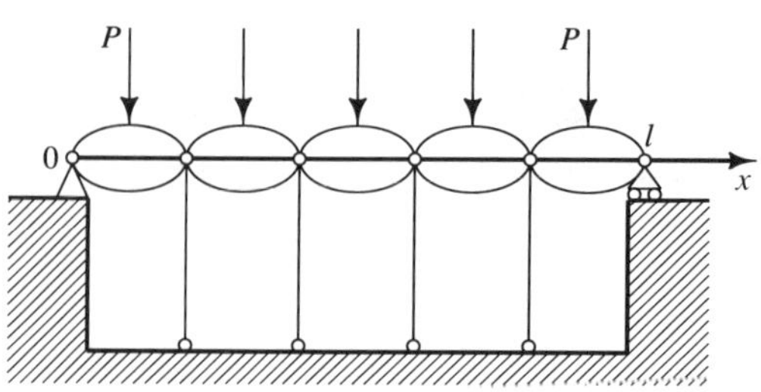

FIGURE 7.5.

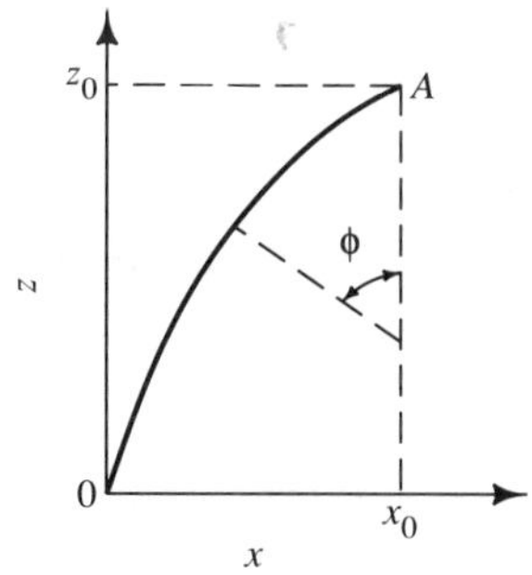

FIGURE 7.6.

7.5 Design of Curved Beams for a Given Shape of the Axis

We shall consider the bending of an elastic curved beam that is rigidly held at the endpoint O (see Figure 7.6) and is subjected to static loads only. We limit our discussion to the case when the axis of the beam is a planar curve defined by the parametric equations $x = x(s)$ and $z = z(s)$, where s is a parameter identified with length measured along the axis from the point O. We assume that one of the principal axes of inertia of the beam's cross-section lies in the plane xz, and the loads acting on it also lie in that plane. Therefore, the axis of the beam remains to be a planar curve after deformation. Strains in the beam are assumed to be "small."

The shape of the curved beam is given by the geometry of the axis, that is, by the functions $x(s)$ and $z(s)$, and by the magnitude of $h = h(s)$, which shall be called "the thickness" (according to our convention). In fact, this parameter may be interpreted as the height of a rectangular cross-section in analogy with straight beams or it may the radius for beams of circular cross-section.

The concept of rigidity utilized in this analysis and various geometric properties of the beam, such as the first moment of the area d, the radius of inertia I, and the magnitude of the area of cross-section S, are all functions of h. Now, denoting by u, w, M, N, and Q, respectively, the components of the deflection vector measured along the axes x and z, the bending moment, and the compression and shear forces, and noting that $u = w = 0$ at the point O, we can use a formula (which can be found in [7.93] or [7.94]) to compute the displacement at the free end of the beam:

$$\left.\begin{aligned} u(l) &= -(1/E)\int_0^l \{(N + M/R)\cos\varphi + [(MS)/(Rd) + N/R](z_0 - z) \\ &\quad - \kappa Q \sin\varphi\}\,\mathrm{d}s/S, \\ w(l) &= -(1/E)\int_0^l \{(N + M/R)\sin\varphi - [(MS)/(Rd) + N/R](x_0 - x) \\ &\quad + \kappa Q \cos\varphi\}\,\mathrm{d}s/S, \end{aligned}\right\} \tag{7.33}$$

where x_0 and z_0 are the coordinates of the free end of the beam in the undeformed state and l is its length. R, E, κ, and φ denote, respectively, the radius of curvature of the axis, Young's modulus, a coefficient that depends on the material properties of the cross-section ($\kappa = 3(1 + \nu)$, where ν is Poisson's ratio), and the angle between the normal to the beam's axis and the coordinate axis z (see Figure 7.6).

The location of the beam's free end (x_0, z_0) is assumed to be given. The quantities M, N, Q, and R in the integrands of (7.33) are assumed to be known functions of the coordinates x and z and of their derivatives with respect to s along the length of the beam. For R and φ, these are familiar geometric relations, while for M, N, and Q these relations may be derived in each specific case of loading applied to the beam from the equations of equilibrium, since the problems discussed here are statically determinate.

We shall assume that the shape of the undeformed axis, that is, the formula for functions $x(s)$ and $z(s)$ is known, as is the total volume of the beam. This assumption implies a constraint assigned to the distribution of the cross-sectional area

$$\int_0^l S(h)\mathrm{d}S = V. \tag{7.34}$$

We adopt as our cost functional some function F of the displacements at the free end of the beam

$$J = F(u(l), w(l)) \to \min. \tag{7.35}$$

We need to find a function $h(s)$ that represents the distribution of "thickness" along its length for a beam with rectangular cross-section, or thickness of the load-bearing layers in a trilayered beam, where $EI = A_\alpha h^\alpha$ and $S = bh$.

We shall investigate the case of slender beams, when

$$h_{\max}/l \ll 1 \qquad \text{and} \qquad h_{\max}/R_{\min} \ll 1. \tag{7.36}$$

Here $h_{\max} = \max_s h(s)$ and $R_{\min} = \min_s R(s)$. Under assumptions (7.36), the displacements at the free end of the beam may be written as

$$\begin{aligned} u(l) &= -(1/A_\alpha)\int_0^l [(z - z_0)M/h^\alpha]\mathrm{d}s, \\ w(l) &= -(1/A_\alpha)\int_0^l [(x - x_0)M/h^\alpha]\mathrm{d}s. \end{aligned} \tag{7.37}$$

For the nonadditive (in general) functionals (7.35) and (7.37), with the isoperimetric constant volume condition (7.34), the Euler equation may be written in the form (see Section 3.2)

$$Mh^{-(\alpha+1)}[(z_0 - z)F_u - (x_0 - x)F_w] = \lambda,$$

where λ is the Lagrangian multiplier corresponding to the isoperimetric condition. The partial derivatives $F_u = F_u(u, w)$ and $F_w = F_w(u, w)$ are computed for the values of the functionals $u(l)$ and $w(l)$ corresponding to the extremal solutions of the variational problem.

To solve Euler's equation we can use the following device: Let the values of F_u and F_w corresponding to the extremal solution of the variational problem be denoted by γ_1 and γ_2, respectively. The optimality condition is written as

$$h = (M/\lambda)[(z_0 - z)\gamma_1 - (x_0 - x)\gamma_2]^{1/(\alpha+1)}. \tag{7.38}$$

The constants γ_1 and γ_2 may be found from the system of equations

$$F_u \equiv f_1(\gamma_1, \gamma_2, \lambda) = \gamma_1, \qquad F_w = f_2(\gamma_1, \gamma_2, \lambda) = \gamma_2,$$

which are derived by substituting eq. (7.38) for h followed by computation of F_u and F_w. After finding the values of the constants γ_1 and γ_2, the multipler λ can be computed from the formula

$$\lambda = (b/V)^{1/(\alpha+1)}\left[\int_0^l \{M[(z_0 - z)\gamma_1 - (x_0 - x)\gamma_2]\}^{1/(\alpha+1)} ds\right]^{\alpha+1}. \tag{7.39}$$

In the specific case when $F = \beta_1 u(l) + \beta_2 w(l)$ and β_1 and β_2 are known constants, the above formulas can be used to derive

$$h = (V/b)\{M[(z_0 - z)\beta_1 - (x_0 - x)\beta_2]\}^{1/(\alpha+1)}$$
$$\times\left[\int_0^l \{M[(z_0 - z)\beta_1 - (x_0 - x)\beta_2]\}^{1/(\alpha+1)} ds\right]^{-1}. \tag{7.40}$$

In an example of a trilayered beam, we can estimate the gain in rigidity of the optimal beam by comparing it with a beam having a uniform thickness of the load-bearing layers. In this example, the axis of the beam has the shape of a segment of a circle: $z = r \sin\varphi$, $x = r(1 - \cos\varphi)$, $x_0 = r$, $z_0 = r$, $0 \le \varphi \le \pi/2$, with r denoting the radius. The exterior forces consist of a single concentrated load applied at the point (x_0, z_0) and acting in the direction of the negative x axis. As the minimized cost functional, we take the magnitude of the vertical displacement w of the free end ($\beta_2 = 1$, $\beta_1 = 0$). The magnitude of the displacement w for a trilayered beam with constant cross-section, as determined from (7.34) and (7.37), is equal to $w = \pi^2 r^4 Pb/(8A_1 EV)$. For a beam having exactly the same volume (of the reinforcing layers) with an optimal distribution of thickness given by $h = V\cos\varphi/(rb)$, the magnitude of the corresponding maximal displacement is $w_* = \mathrm{P}r^4 b/(EA_1 V)$.

Comparing these displacements, we see that the relative gain in rigidity of the optimal beam, as compared with a beam having constant thickness, is approximately 18%.

Let us now discuss a more general case, without making any assumptions concerning the thickness of the beam. We consider uniform beams with rectangular cross-section (therefore, $I = bh^3/12$ and $S = bh$). We rewrite (7.37) in the form

$$u(l) = \int_0^l (\Phi_1/h + \Phi_2/h^3) ds, \qquad w(l) = \int_0^l ((\Phi_3/h + \Phi_4/h^3) ds, \tag{7.41}$$

where the quantities $\Phi_1, \ldots, \Phi_4$ are not functions of h and are given as

$$\left.\begin{aligned}\Phi_1 &= -[1/(Eb)]\{(N + M/R)\cos\varphi + N/R(z_0 - z) - \kappa Q \sin\varphi\},\\ \Phi_3 &= -[1/(Eb)]\{(N + M/R)\sin\varphi - N/R(x_0 - x) + \kappa Q \cos\varphi\},\\ \Phi_2 &= -[12M/(Eb)](z_0 - z), \qquad \Phi_4 = [12M/(Eb)](x_0 - x).\end{aligned}\right\}$$

Necessary conditions for a minimum of the functional (7.35) and (7.41) with the integral constraint (7.34) assigned to the function h have the form (see Section 3.2)

$$h^{-2}(\Phi_1 F_u + \Phi_3 F_w) + 3h^{-4}(\Phi_2 F_u + \Phi_4 F_w) = \lambda. \tag{7.42}$$

Here λ is a Lagrangian multiplier. The functions F_u and F_w have the same meaning as in the previous discussion of thin beams, and in solving (7.42) we can use the same techniques as before. If the minimized functional is given by $F = \beta_1 u(l) + \beta_2 w(l)$, where β_1 stands for F_u and β_2 for F_w, then eq. (7.42) becomes a biquadratic algebraic equation in h. In that case the solution of the minimization problem reduces to computation of λ. This can be accomplished by substituting into the isoperimetric condition (7.34) the expression for h derived by solving (7.42).

7.6 Optimization of the Shape of Curved Elastic Beams

In all arguments presented in the preceding section, the geometry of the beam's axis was regarded as known, and only one design variable $[h(s)]$ was to be determined. In this section we regard as unknown both the distribution of thickness $h(s)$ and the shape of the beam's axis, that is, the pair of functions $\{x(s), z(s)\}$. We shall examine the problem of optimizing the functional (7.35) using the assumption of thinness (7.36). To compute the values of $u(l)$ and $w(l)$, we shall use (7.37). We shall also make the statement of the optimization problem more specific. The external load acting on the beam shall be a concentrated load P applied to the free end of the beam and acting in the direction parallel to the z axis. As our minimized functional, we take the absolute value of the vertical displacement of the unsupported end of the beam, that is, the magnitude of the displacement in the direction of the load.

We make use of these assumptions and, adopting x as the independent coordinate, we rewrite the basic relations of our variational problem (see [7.6]):

$$\frac{P}{A_\alpha}\int_0^{x_0} \frac{(x_0 - x)\sqrt{1 + z_x^2}}{h^\alpha}\,dx \to \min, \tag{7.43}$$

with

$$z(0) = 0, \qquad z(x_0) = z_0, \tag{7.44}$$

$$\int_0^{x_0} \sqrt{1 + z_x^2}\,dx = l, \tag{7.45}$$

$$b\int_0^{x_0} h(x)\sqrt{1 + z_x^2}\,dx = V, \tag{7.46}$$

where $z_x = dz/dx$ and $\alpha = 1, 2, 3$. The case $\alpha = 0$ is special, because for $\alpha = 0$ the rigidity of the beam is independent of h and optimization is carried out with respect to the shape of the beam's axis $z(x)$ only. In the case $\alpha = 0$, there is no need to consider condition (7.46).

The solution of the variational problem (7.43) to (7.46) is sought in the class of continuously differentiable functions $z(x)$ and continuous functions $h(x)$. Necessary conditions for an extremum (see [7.19] and [7.32]) for the functions $z(x)$ and $h(x)$ are transformed by elementary substitutions into

$$\{z_x/\sqrt{(1 + z_x^2)}\}[(x_0 - x)^{2/(1+\alpha)} + c_0] = c_1, \tag{7.47}$$

$$h = [\lambda_0(x_0 - x)^2]^{1/(1+\alpha)}, \tag{7.48}$$

where c_0, c_1, and λ_0, are arbitrary constants.

Equation (7.47) determines the optimal shape of the beam's axis $z = z(x)$, for the optimal choice of thickness $h(x)$. To compute the constants c_0, c_1, λ_0, and another arbitrary constant, which arises in integrating eq. (7.47), we can use the boundary conditions (7.44) and the two isoperimetric conditions (7.45) and (7.46).

It can be shown (see [7.6]) that in all cases, whether or not $z_0 = 0$, if, nevertheless, the length of the beam exceeds the magnitude of x_0 ($l \geq x_0$), the arbitrary constant arising in solving (7.47) is nonzero over the entire interval $0 < x \leq x_0$, while $z_x > 0$ on $0 \leq x \leq x_0$. It follows that the optimal solution $z(x)$ and its derivative z_x are positive ($z(x) > 0$ for $0 < x \leq x_0$ and $z_x(x) > 0$ for $0 \leq x \leq x_0$). In subsequent arguments we shall assume that $z_0 > 0$. Because z_x is nonnegative, the inequality $l < x_0 + z_0$ is a necessary condition for the existence of solutions to our problem.

After substituting (7.44) and (7.45), the integral of (7.47) may be represented in the form

$$\left.\begin{aligned} z &= c_1 \int_0^x \{1/\sqrt{[(x_0 - t)^{2/(\alpha+1)} + c_0]^2 - c_1^2}\}dt, \qquad c_1 > 0, \\ z_0 &= c_1 \int_0^{x_0} \{1/\sqrt{[(x_0 - t)^{2/(\alpha+1)} + c_0]^2 - c_1^2}\}dt, \\ l &= \int_0^{x_0} \{[(x_0 - t)^{2/(\alpha+1)} + c_0]/\sqrt{[(x_0 - t)^{2/(\alpha+1)} + c_0]^2 - c_1^2}\}dt, \end{aligned}\right\} \tag{7.49}$$

The last two equations in (7.49) may be regarded as two equations in the unknown variables c_0 and c_1, and may be solved to determine these constants. To find the value of λ_0 that appears in (7.48), we can substitute the expressions for h and z_x from (7.47) and (7.48) to derive the formula

$$\begin{aligned} \lambda_0 = (V/b)^{(1+\alpha)} \bigg\{ \int_0^{x_0} \{(x_0 - t)^{2/(\alpha+1)} \\ \times [(x_0 - t)^{2/(\alpha+1)} + c_0]/\sqrt{[(x_0 - t)^{2/(\alpha+1)} + c_0]^2 - c_1^2}\}dt\bigg\}^{-(1+\alpha)} \end{aligned} \tag{7.50}$$

If c_0 and c_1 are already known, this formula permits us to compute λ_0.

Equations (7.49) and (7.50) derived above depend on the values of the parameter α, which characterized the type of cross-section of the beam, but do not depend on the coefficient A_α. Therefore, the optimal shape of the axis will be the same for different types of beams that have the same exponent α in the relation $EI = A_\alpha h^\alpha$.

Now, let us briefly sum up some results of research into optimal shapes of curved beams for the case $\alpha = 1$. A more detailed exposition and also a discussion of several other cases is offered in [7.6].

For the sake of convenience, we shall introduce dimensionless variables and let $x' = x/x_0$, $z' = z/x_0$, $l' = l/x_0$, $\delta = z_0/x_0$, $c_0' = c_0/x_0$, and $c_1' = c_1/x_0$. Below, primes shall be omitted from these variables. We compute the integrals (7.49) and (7.50), changing to the new dimensionless variables and relations to determine the constants c_0 and c_1:

$$z = c_1 \ln\{[1 + c_0 + \sqrt{(1 + c_0)^2 - c_1^2}]/[1 + c_0 - x + \sqrt{(1 + c_0 - x)^2 - c_1^2}]\},$$
$$\delta = c_1 \ln\{[1 + c_0 + \sqrt{(1 + c_0)^2 - c_1^2}]/[c_0 + \sqrt{c_0^2 - c_1^2}]\},$$
$$l = \sqrt{(1 + c_0)^2 - c_1^2} - \sqrt{c_0^2 - c_1^2}. \tag{7.51}$$

The necessary conditions for the existence of solutions ($l < x_0 + z_0$) and the inequality $x_0^2 + z_0^2 = l^2$ can be written in the form $l - 1 < \delta \leq \sqrt{l^2 - 1}$.

After algebraic manipulation of this system of equations to solve for c_0 and c_1, we arrive at the formulas

$$\left.\begin{aligned} c_1 &= 1/(2l)\sqrt{l^2 - 1}[(2c_0 + 1 - l)(2c_0 + 1 + l)]^{1/2}, \\ \delta &= 1/(2l)\sqrt{l^2 - 1}[(2c_0 + 1 - l)(2c_0 + 1 + l)]^{1/2} \\ &\quad \times \ln\left[\left(1 + c_0 + \frac{1}{2l}|2c_0 + 1 + l^2|\right)\Big/\left(c_0 + \frac{1}{2l}|(2c_0 + 1 - l^2)|\right)\right]. \end{aligned}\right\} \tag{7.52}$$

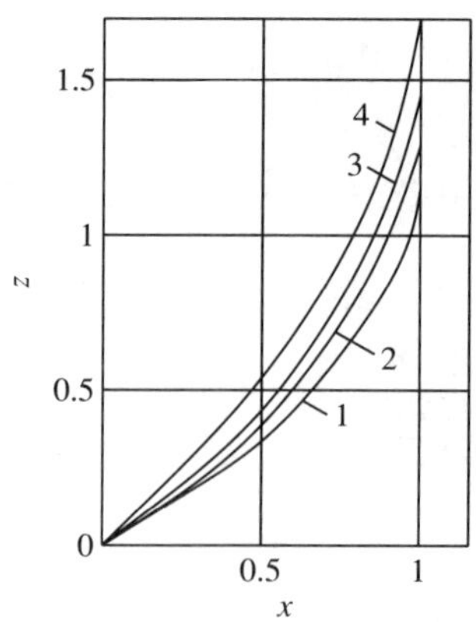

FIGURE 7.7. Optimal shape of the beam's axis.

The constant c_0 may be computed by solving the second equation in (7.52). Then we can use the first equality in (7.52) to compute c_1 directly. These computations of c_0 and c_1 were obtained numerically from a computer program.

Then, applying to a specific point the formula derived above allows us to establish the functional dependence $z = z(x)$.

The optimal shapes of the beam's axes derived from that program are illustrated in Figure 7.7 for $l = 2$. The curves labeled 1 to 4 correspond to the values 1.060, 1.308, 1.537, and 1.681 respectively of the parameter δ: A detailed analysis of the geometry of optimal shapes for the cases when the value of parameter δ is small compared to unity is carried out in [7.6] and [5].

7.7 Maximization of Rigidity for Piecewise Isotropic Bars Subjected to Torsion

The problem of maximizing the rigidity of elastic bars against torsion has been investigated in many research publications (see [2.2], [7.8] to [7.11], [7.16], [7.49] to [7.51], [7.59], [7.76], and [7.77]). We have already discussed the problem of designing homogeneous elastic bars to resist torsion in Section 4.4. Following the ideas of [7.8], we shall consider the problem of optimizing a cylindrical, piecewise-homogeneous elastic bar subjected to torsion. We shall denote by Ω the region of the bar's cross-section situated in the xy plane and by Γ the boundary of Ω (see Figure 7.8).

We assume that the bar is composed of two different materials filling regions Ω_0 and Ω_1, with moduli of torsional rigidity G_0 and G_1, respectively (where $\Omega = \Omega_0 + \Omega_1$). The interior region Ω_0 is convex and is separated from Ω_1 by a smooth boundary Γ_1. The contours Γ and Γ_1 are disjoint. We define in Ω a piecewise constant function $G(x, y)$, assuming that $G(x, y) = G_0$ if $(x, y) \in \Omega_0$, and $G(x, y) = G_1$ if $(x, y) \in \Omega_1$. To determine the stress function $\varphi(x, y)$, which is defined in $\Omega = \Omega_0 + \Omega_1$ and related to the stress components by the relations

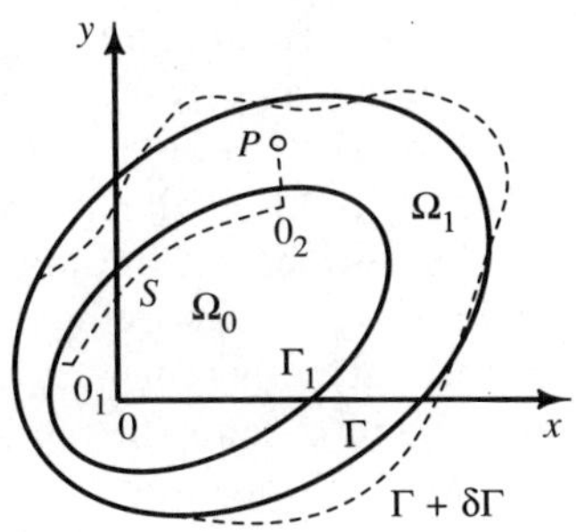

FIGURE 7.8. Effects of torsion on an elastic bar composed of two materials.

$$\tau_{xz} = \theta\varphi_y, \qquad \tau_{yz} = -\theta\varphi_x,$$

we need to solve the boundary-value problem

$$\left.\begin{aligned} &\Delta\varphi = -2G, && (\varphi)_\Gamma = 0, \\ &(\varphi)^+_{\Gamma_1} = (\varphi)^-_{\Gamma_1}, && \left[\frac{1}{G}\frac{\partial\varphi}{\partial n}\right]^+_{\Gamma_1} - \left[\frac{1}{G}\frac{\partial\varphi}{\partial n}\right]^-_{\Gamma_1} = 0, \end{aligned}\right\} \tag{7.53}$$

where Δ is the Laplace operator, $\partial/\partial n$ is the derivative in the direction normal to the boundary contour, and θ is the angle of torsion per unit length of the bar. The upper indices $(+)$ or $(-)$ in (7.53) denote the limits of the corresponding quantities as we approach Γ_1, respectively, from inside or outside the region bounded by it. We assume that Γ_1 is fixed. Now let S_0 stand for the area of the region Ω_0 bounded by Γ_1. We shall also assume that the area S_1 of Ω_1 remains fixed, that is,

$$\text{meas}(\Omega_1) = S_1.$$

Our optimization problem consists in finding the shape of the curve Γ satisfying the isoperimetric constraint $\text{meas}(\Omega_1) = S_1$ such that the torsional rigidity of the rod (our quality functional) attains its maximum

$$K = 2\int_\Omega \varphi(x, y)\mathrm{d}\Omega \to \max_\Gamma. \tag{7.54}$$

A peculiar property of this problem, distinguishing it from the problems discussed in Section 4.4 (see also [5], [2.2], and [7.50]) is the form of the equation that determines the stress function φ in Ω. The right-hand side of that equation is discontinuous. However, the necessary condition for the optimality of Γ remains exactly the same as before. It can be written in the form

$$(\nabla\varphi)^2 = \lambda^2. \tag{7.55}$$

Note that if the contour Γ is regarded as fixed and Γ_1 is regarded as unknown, and the functional (7.54) is maximized with respect to the choice of Γ_1, we again obtain the same problem of optimizing the distribution of elastic material for a beam inside some region Ω.

In the case when $G_0 = 0$, we can use (7.53) and (7.54) and by performing some elementary substitutions obtain a corresponding problem for a uniform cylindrical rod, having a fixed interior boundary contour, while the exterior contour is to be found. A solution to this problem is given in [2.2] and [7.50].

Using (7.55), we examine the problem of finding an optimal surface of a covering layer for a torsioned rod. This reinforcing layer is made from a more rigid material (i.e., from a material with much higher shear modulus). The region Ω_1 illustrated in Figure 7.8 represents the reinforcing layer. In addition to the usual Cartesian coordinates, we introduce also a curvilinear coordinate system $\{s, t\}$ related to the properties of the curve Γ_1. The coordinate s of a point $P \in \Omega$ is measured along Γ_1 from some fixed point $O_1 \in \Gamma_1$ to the point

O_2 which is the point of intersection of Γ_1 with a normal to Γ_1 passing through P. The coordinate t is equal to the length of the segment O_2P. Let $h = h(s)$ denote a function that defines in our curvilinear coordinate system the location of the unknown curve Γ, and let l be the total length of Γ. In solving this problem, we assume that

$$l^{-1} \max_s h(s) \sim \varepsilon, \qquad G_0/G_1 \sim \varepsilon, \qquad \min_s R(s) < l,$$

where ε is a small parameter and $R(s)$ denotes the radius of curvature of the curve Γ_1 at the point with coordinate s. As is usually done in reinforcement problems, we identify only the first-order terms of the infinite series in powers of the parameter ε (see [7.5]). We use the same designation for these unknown principal terms of these series as for the quantities represented by these series. In this approximation, a formula for the function φ in its domain Ω_1, where φ satisfies both eq. (7.53) and the boundary condition $\varphi = 0$ on Γ_1, is given in [7.5]:

$$\varphi(s, t) = \varphi(s, 0)[1 - t/h(s)].$$

Here $\varphi(s, 0)$ denotes the value of φ at points that lie on Γ_1. If we use this expression for φ in Ω_1, the computation of the solution to the torsion problem reduces to finding a function φ that provides a solution to the boundary-value problem in Ω_0:

$$\left.\begin{aligned} \Delta\varphi &= -2G_0, \qquad \forall(x, y) \in \Omega_0, \\ \varphi_t &= -G_0\varphi/(G_1 h), \qquad \forall(x, y) \in \Gamma_1. \end{aligned}\right\} \tag{7.56}$$

The symbol φ_t in (7.56) really denotes $(\varphi_t)^+$. The boundary condition in (7.56) is derived by substituting the formula for φ in Ω_1 in the relation $(G^{-1}\partial\varphi/\partial n)^+ = (G^{-1}\partial\varphi/\partial n)^-$, which is valid on the curve Γ_1. For given $h = h(s)$, the problem (7.56) is well posed.

Let us proceed with the optimization process. We transfer the optimality condition (7.55) from the contour Γ to the contour Γ_1. With accuracy up to terms of higher infinitesimal order, we obtain $(\varphi_t)_{\Gamma_1} = -\lambda$. Then it can be deduced from the formula for φ in Ω that

$$h(s) = \frac{1}{\lambda}(\varphi)_{\Gamma_1}. \tag{7.57}$$

Therefore, for an optimal distribution of the beam's reinforcing layer, the thickness $h(s)$ and the value of the function φ are proportional to each other along Γ_1. Formula (7.57) permits us to reduce a boundary-value problem of the third kind for the function $\varphi(x, y)$ to a boundary-value problem with the Neumann conditions

$$\left.\begin{aligned} \Delta\varphi &= -2G_0, \qquad \forall(x, y) \in \Omega; \\ \varphi_t &= -\lambda G_0/G_1, \qquad \forall(x, y) \in \Gamma_1. \end{aligned}\right\} \tag{7.58}$$

We denote the position radius vector for a point with coordinates (x, y) by $\mathbf{r}$

(where $\mathbf{r}^2$ denotes $\|\mathbf{r}\|^2 = (\mathbf{r}, \mathbf{r}) = x^2 + y^2$), and introduce an auxiliary function $\psi = \varphi/G_0 + \mathbf{r}^2/2$. Expressing (7.58) in terms of ψ instead of φ and using the identity $\dfrac{\partial(\mathbf{r}^2)}{\partial t} = (\nabla \mathbf{r}^2, \mathbf{n}) = 2(\mathbf{r}, \mathbf{n})$, we formulate a boundary-value problem for the function ψ:

$$\left.\begin{aligned} \Delta\psi &= 0, && \forall(x, y) \in \Omega_0, \\ \psi_t &= -\lambda/G_1 + (\mathbf{r}, \mathbf{n}), && \forall(x, y) \in \Gamma_1 \end{aligned}\right\} \tag{7.59}$$

(the Neumann problem for the Laplace equation).

In order to solve problem (7.59), it is necessary to satisfy the equality $\int \psi_t ds = 0$ with integration performed along the closed path Γ_1. This condition is satisfied if we choose the constant λ to be

$$\lambda = \frac{G_1}{l} \int_{\Gamma_1} (\mathbf{r}, \mathbf{n}) ds = \frac{2G_0 G_1}{l}. \tag{7.60}$$

Note that the constant λ is not determined, as is usually done, from an isoperimetric condition, but instead is found from the necessary condition for the existence of solutions to the boundary-value problem. To satisfy the isoperimetric criterion, we can use another arbitrary condition (i.e., another single arbitrary constant) that occurs in (7.59). This arbitrariness turns up in replacing the problem (7.56) by (7.58) and (7.59), and is caused by the fact that in applying condition (7.57) we managed to alter the type of boundary-value problem originally stated in (7.56) to a Neumann problem whose solutions can be determined only up to an arbitrary constant.

Denoting this arbitrary constant by γ, we represent ψ as a sum of two terms; $\psi = \Psi + \gamma$, $\Psi(x_0, y_0) = 1, \forall(x_0, y_0) \in \Omega_0$. The function Ψ satisfies, apart from the condition indicated above, the equation and boundary condition of (7.59) (with $\lambda = 2GS_0/l$). In our approximation the isoperimetric condition may be written as

$$\int_{\Gamma_1} h ds = S_1. \tag{7.61}$$

From the identity (7.61) and the relations (7.57) and (7.60), we can deduce a constraint on the behavior of the function φ:

$$\int_{\Gamma_1} \varphi ds = 2G_1 S_0 S_1/l. \tag{7.62}$$

Substituting (7.62) in the expression $\varphi = G(\Psi + \gamma - \mathbf{r}^2/2)$, we obtain a formula for the value of the constant γ:

$$\gamma = \frac{1}{l}\left[\frac{2G_1 S_0 S_1}{lG_0} - \int_{\Gamma_1} (\Psi - \mathbf{r}^2/2) ds\right].$$

After substituting these values of λ and γ, we arrive at the final version of the formulas determining the unknown quantities φ and h:

$$\left.\begin{aligned}
\varphi &= G_0\left[(\Psi - \mathbf{r}^2/2) - (1/l)\int_{\Gamma_1}(\Psi - \mathbf{r}^2/2)\mathrm{d}s\right] + 2S_0S_1G_1/l^2,\\
h &= \frac{l}{2G_1S_0}(\varphi)_{\Gamma_1},\\
K &= 2\int_{\Omega_0}\varphi\,\mathrm{d}x\mathrm{d}y\\
&= 2G_0\left[\int_{\Omega_0}(\Psi - \mathbf{r}^2/2)\mathrm{d}x\mathrm{d}y - (S_0/l)\int_{\Gamma_1}(\Psi - \mathbf{r}^2/2)\mathrm{d}s\right] + 4S_0^2S_1G_1/l^2.
\end{aligned}\right\} \tag{7.63}$$

In this fashion the process of finding the optimal solution has been reduced to finding a function Ψ satisfying the boundary-value problem (7.59) and (7.60) with side condition $\Psi(x_0, y_0) = 1$. The solution of this problem does not depend on the values of the physical constants G_0 and G_1 or the parameter S_1, and is only a function of the geometry of Ω_0.

In [7.8] and the monograph [5] the present author established that in this problem of optimal reinforcement the necessary optimality condition (7.55) is also a sufficient condition for a local optimum.

We shall introduce some solutions of very simple reinforcement problems. Let the region Ω_0 be a circular disk with radius R_0. In that case $(\mathbf{r}, \mathbf{n}) = R_0$ and $\lambda = G_1R_0$, and the boundary condition (7.59) assumes the form $\partial\psi/\partial t = 0$ for all $(x, y) \in \Gamma_1$. Consequently, ψ is independent of both x and y, and $\psi(x, y) \equiv 1$ (for all (x, y) in Ω_0). Using this fact and (7.63), we have

$$\left.\begin{aligned}
\varphi &= \tfrac{1}{2}G_0(R^2 - r^2) + \tfrac{1}{2}\pi S_1G_1,\\
h &= S_1/(2\pi R_0),\\
K &= (\pi/2)G_0R_0^4 + G_0S_1R_0^2.
\end{aligned}\right\}$$

Therefore, *for a circular region an optimal reinforcing layer has constant thickness*. This statement remains true even for thick layered coatings.

We shall now consider the problem of optimizing a reinforcing layer for a region Ω_0 that has the shape of an elliptical disk (see Figure 7.9). Let the center of the ellipse coincide with the origin of a Cartesian coordinate system xy, and let the function Ψ have assigned value at the origin $\Psi = 1$ (the origin x_0, y_0 coincides with the point 0). For the sake of convenience, we shall use elliptical coordinates $\{\alpha, \beta\}$, related to the Cartesian coordinates $\{x, y\}$ by the equalities $x = c\cosh\alpha\sin\beta$, $y = c\sinh\alpha\cos\beta$, where $2c$ is the focal distance. In elliptic coordinates the boundary Γ_1 of the elliptical region is given by $\alpha = \alpha_0$, $\beta \in [0, 2\pi)$, while the region Ω_0 is a rectangle: $0 \le \alpha \le \alpha_0$, $0 \le \beta \le 2\pi$ (with 0 and 2π identified with each other). Because of symmetry, we can seek the solution in the region $0 \le \alpha \le \alpha_0$, $0 \le \beta \le \pi$, representing one-half of the ellipse. We use the well-known formulas:

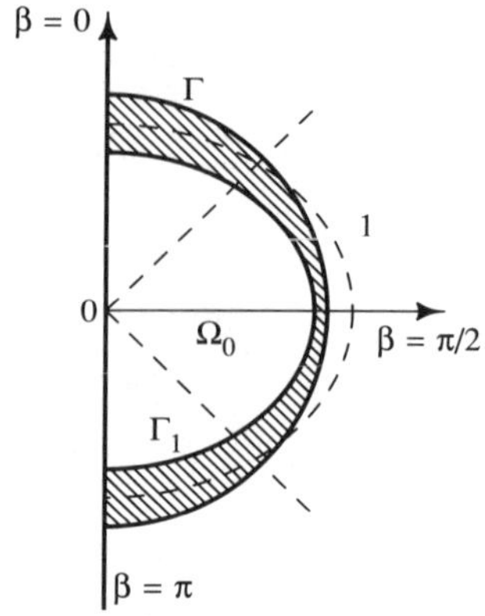

FIGURE 7.9. Reinforced elliptic disk.

$$\left.\begin{aligned} S_0 &= \frac{1}{2}\pi c^2 \sinh 2\alpha_0, \qquad \partial/\partial n = \frac{1}{g}\frac{\partial}{\partial\alpha}, \qquad \partial/\partial s = -\frac{1}{g}\frac{\partial}{\partial\beta}, \\ (\mathbf{r}, \mathbf{n})_{\alpha=\alpha_0} &= (c^2/2g)\sinh 2\alpha_0, \qquad g = c\sqrt{\tfrac{1}{2}(\cosh 2\alpha + \cos 2\beta)}, \end{aligned}\right\} \tag{7.64}$$

and we write in elliptical coordinates the equation and boundary conditions that define the function Ψ:

$$\Psi_{\alpha\alpha} + \Psi_{\beta\beta} = 0, \qquad 0 \le \alpha \le \alpha_0, \qquad 0 \le \beta \le \pi; \tag{7.65}$$

$$\Psi = 1 \quad \text{when } \alpha = 0 \quad \text{and} \quad \beta = 0; \tag{7.66}$$

$$\Psi_\alpha = \tfrac{1}{2}c^2 \sinh 2\alpha_0(1 - 2\pi g/l), \quad \text{when } \alpha = \alpha_0, \qquad 0 \le \beta \le \pi; \tag{7.67}$$

$$\Psi_\alpha = 0 \quad \text{when } \alpha = 0 \quad \text{and} \quad 0 \le \beta \le \pi; \tag{7.68}$$

$$\Psi_\beta = 0 \quad \text{when } 0 \le \alpha \le \alpha_0, \quad \beta = 0; \quad \text{or when } 0 \le \alpha \le \alpha_0, \quad \beta = \pi; \tag{7.69}$$

We represent the function Ψ in a Fourier series

$$\Psi = \sum_{n=0}^{\infty} a_n \cos n\beta \cosh n\alpha. \tag{7.70}$$

This infinite series must satisfy conditions (7.68) and (7.69).

The coefficients of the terms in the Fourier series are determined from conditions (7.66) and (7.67). We derive

$$a_0 = 1 - \sum_{n=1}^{\infty} a_n,$$

$$a_n = -\frac{2c^2 \sinh 2\alpha}{nl \sinh n\alpha_0} \int_0^\pi g \cos n\beta \mathrm{d}\beta \tag{7.71}$$

Substituting (7.70) and (7.71) in (7.63) and recalling the basic formulas $\mathrm{d}x\mathrm{d}y = g^2\mathrm{d}\alpha\mathrm{d}\beta$ and $\mathrm{d}s^2 = g\mathrm{d}\beta$ on Γ_1, and (7.64), we obtain

$$\left.\begin{aligned}
\varphi &= G_0\left[\frac{1}{2}a_2 + \frac{c^2}{4}(\cosh 2\alpha_0 - \cosh 2\alpha + \cos 2\beta) + \sum_{n=1}^{\infty}(\cos n\beta \cosh n\alpha)\right.\\
&\quad \left. + (na_n \sinh 2n\alpha_0)/(2c^2 \sinh 2\alpha_0)\right] + 2G_1 S_0 S_1/l^2.\\
h &= (G_0 l/(2G_1 S_0))\left\{a_2/2 + \frac{1}{4}c^2 \cos 2\beta + \sum_{n=1}^{\infty} a_n[(\cos n\beta \cosh n\alpha_0\right.\\
&\quad \left. + (na_n \sinh 2n\alpha_0)/(2c^2 \sinh 2\alpha_0)]\right\} + S_1/l,\\
K &= \pi c^2 G_0 \sinh 2\alpha_0\left[a_2 + (c^2/8)\cosh 2\alpha_0\right.\\
&\quad \left. + \sum_{n=1}^{\infty}(na_n^2 \sinh 2n\alpha_0)/(2c^2 \sinh 2\alpha_0)\right] + 4G_1 S_1 S_0^2/l^2.
\end{aligned}\right\} \tag{7.72}$$

The expressions in (7.72) for the unknown quantities φ, h, and K may be simplified if we use the above assumption that min $R_s \sim l$, which permits us to regard the quantity $1/\cosh 2\alpha_0$ as "small" as compared to unity. Expanding the integrand in (7.71) in infinite series of powers of the parameter $1/\cosh 2\alpha_0$, it is possible to show that within the accuracy of $(1/\cosh 2\alpha_0)^2$,

$$\left.\begin{aligned}
&a_0 = 1 - a_2, \qquad a_2 = -\pi c^3/4\sqrt{2l}\sqrt{\cosh 2\alpha_0} = -(c^3/8)\cosh 2\alpha_0,\\
&a_1 = a_3 = \cdots = 0.
\end{aligned}\right\} \tag{7.73}$$

After making use of (7.73) and performing some elementary substitutions, the expressions for h and K assume the form

$$\left.\begin{aligned}
h &= S_1/l + [(cG_0\sqrt{\cosh 2\alpha_0})/(4\sqrt{2}G_1 \sinh 2\alpha_0)]\cos 2\beta,\\
K &= \tfrac{1}{16}\pi c^4 G_0 \sinh 4\alpha_0 + 4G_1 S_1 S_0^2 l^{-2}.
\end{aligned}\right\} \tag{7.74}$$

Here $l \approx c\pi\sqrt{2\cosh 2\alpha_0}$. From (7.74), we can deduce the shape of h. Figure 7.9 illustrates the best distribution of the reinforcing layer Ω_1. We see that the thickness h decreases with an increase in the radius of curvature as we move along Γ_1. Figure 7.9 indicates the region occupied by the reinforcing layer by cross-hatching, while the broken line indicates the curve Γ that is the boundary of Ω for constant thickness $h = S_1/l$. This constant distribution is optimal if the region Ω_0 is a circular disk.

7.8 Shells with Maximal Rigidity

The equilibrium equations for a shell of revolution supported along an edge (say, $z = 0$) and loaded by torsion producing forces q are

$$d/d\theta[R^2(S + 2H/R_2)] = -qR_1R^2,$$

$$H = [(Gh^3R)/(6R_1R_2)]d/d\theta(u/R), \qquad S = [(GhR)/R_2]d/d\theta(u/R),$$

where u is the displacement in the circumferential direction (along a parallel surface of the shell); G is the shear modulus; h is the shell's thickness; R, R_1, and R_2 denote, respectively, the local radius of the shell, the radius of curvature of the meridian, and the distance, measured in the normal direction, between the surface and the axis of rotation; and θ is the angle between the normal to the surface and the axis of rotation of the shell z. We substitute the formulas for H and S in the equation of equilibrium and discard from the resulting expression the term of order $(h/R_2)^2$, which we regard as a quantity of higher infinitesimal order than the errors permitted in the derivation of equations of thin shell theory. We have

$$\frac{\mathrm{d}}{\mathrm{d}\theta}\left[\frac{Gh}{R_1}\frac{\mathrm{d}}{\mathrm{d}\theta}\left(\frac{u}{R}\right)\right] = -qR_1R^2.$$

We recall that

$$\mathrm{d}R/\mathrm{d}\theta = R_1 \cos\theta, \qquad \mathrm{d}z/\mathrm{d}R = \tan\theta,$$

and replacing derivatives with respect to θ by derivatives with respect to z, derive an equality expressing the dependence of the displacement u on the coordinate z:

$$\left.\begin{aligned} &\frac{\mathrm{d}}{\mathrm{d}z}\left[\frac{GhR^3}{\sqrt{1+(\mathrm{d}R/\mathrm{d}z)^2}}\frac{\mathrm{d}}{\mathrm{d}z}\left(\frac{u}{R}\right)\right] = m,\\ &\text{where } m = -qR^2\sqrt{1+(\mathrm{d}R/\mathrm{d}z)^2}. \end{aligned}\right\} \tag{7.75}$$

The boundary conditions for $z = 0$ and at the free end assume the form

$$u(0) = 0, \qquad \left[\frac{GhR^3}{\sqrt{1+(\mathrm{d}R/\mathrm{d}z)^2}}\frac{\mathrm{d}}{\mathrm{d}z}\left(\frac{u}{R}\right)\right]_{z=l} = 0. \tag{7.76}$$

Integrating eq. (7.75) and choosing constants of integration to satisfy conditions (7.76), we derive a formula for the displacement function $u = R\int_0^l N(z,\zeta)m(\zeta)\mathrm{d}\zeta$,

$$N(z,\zeta) = \begin{cases} \displaystyle\int_0^z (1/\psi(\eta))\mathrm{d}\eta, & 0 \le z \le \zeta \le l,\\ \displaystyle\int_0^\zeta (1/\psi(\eta))\mathrm{d}\eta, & 0 \le \zeta \le z \le l, \end{cases}$$

$$\psi(\eta) = \left[\frac{G(\eta)h(\eta)R^3(\eta)}{\sqrt{1+(\mathrm{d}R/\mathrm{d}\eta)^2}}\right]. \tag{7.77}$$

To be more specific , we shall consider cases when the load is applied to the free end of the shell. Parallel displacement of the free edge caused by the applied torque $M = q_0(R^3\sqrt{1+(\mathrm{d}R/\mathrm{d}z)^2})_{z=l}$ is given by the formula

$$u(l) = M\int_0^l [(\sqrt{1+(\mathrm{d}R/\mathrm{d}z)^2})/(GhR^3)]\mathrm{d}z \tag{7.78}$$

The volume of the shell's material is regarded as fixed and equal to a constant V,

$$2\pi \int_0^l hR(\sqrt{1 + (\mathrm{d}R/\mathrm{d}z)^2})\mathrm{d}z = V. \tag{7.79}$$

We postulate that the shell has maximal rigidity if the displacement $u(l)$ is the least possible. Therefore, the problem of optimizing the rigidity of a shell (see [6.7] and [7.31]) reduces to finding the minimum of the functional (7.78) with constraint (7.79).

We define the optimization project for our shell by regarding the thickness function $h(x)$ as our control, while the function $R = R(z)$ is regarded as known. It is possible to show that a necessary and sufficient condition for a minimum of the functional (7.78) with constraint (7.79) is the equality

$$hR^2 = \sqrt{v}, \tag{7.80}$$

where v is a positive constant.

Making use of the optimality condition (7.80) and the isoperimetric condition (7.79), we find the optimal distribution of thickness

$$h = \frac{V}{2\pi R^2} \int_0^l \left(\frac{\sqrt{1 + (\mathrm{d}R/\mathrm{d}z)^2}}{R} \mathrm{d}z \right)^{-1}. \tag{7.81}$$

To estimate the effectiveness of the optimization, we compare our optimal shell with a shell whose meridian has the same shape, but has a constant thickness. The gain secured by optimizing is given by the formula

$$\eta = 1 - u(l)/u_0(l), \tag{7.82}$$

where $u_0(l)$ and $u(l)$ denote, respectively, the circumferential displacements of the edge of shell with constant thickness and the optimal shell. The volumes are assumed to be equal. After some computations, we derive an expression for η defined by (7.82), expressing η as a function of $R(z)$:

$$\eta = 1 - \frac{\int_0^l (R\sqrt{1 + (\mathrm{d}R/\mathrm{d}z)^2}\,\mathrm{d}z) \int_0^l (\sqrt{1 + (\mathrm{d}R/\mathrm{d}z)^2}/R^3 \mathrm{d}z)}{\int_0^l (\sqrt{1 + (\mathrm{d}R/\mathrm{d}z)^2})/R)\mathrm{d}z}.$$

We shall now examine the problem of finding an optimal shape for the meridian. The distribution of thickness $h = h(z)$ is regarded as known. Using relations (7.78) and (7.79) and Euler's equations, it is not hard to derive necessary optimality conditions (in a differential form) such that the first integral of this condition is of the form

$$\{1/(\sqrt{1 + (\mathrm{d}R/\mathrm{d}z)^2}\}\{v/hR^3) + Rh\} = C. \tag{7.83}$$

The value of the constant C can be determined from the boundary condition $R(l) = R_l$, while v can be found from the isoperimetric condition. If $v = 0$, then the optimal shell also constitutes a surface of minimal area. The boundary-value problem is solved numerically. As a result of such computations, we conclude that in some region of values for these parameters there exist two

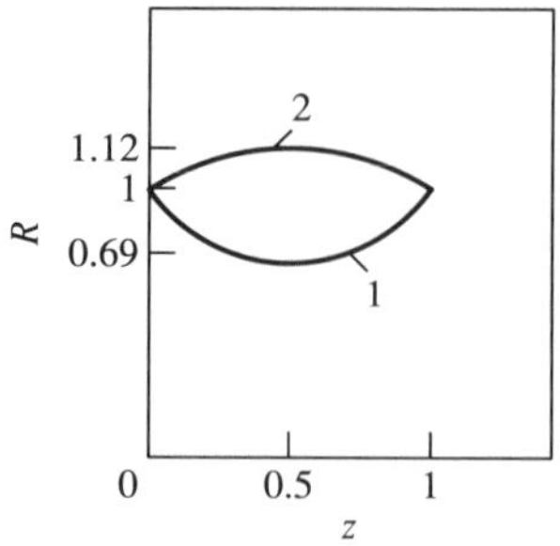

FIGURE 7.10. Optimal shapes of a shell for values $\nu = 0$ and $\nu = \infty$.

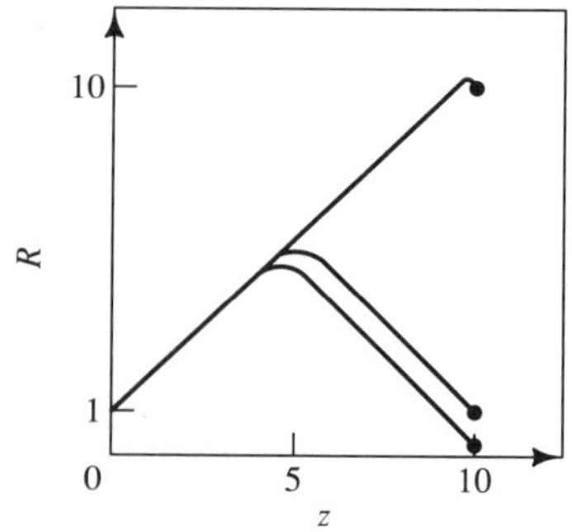

FIGURE 7.11. Effects of varying h_l.

extremal solutions. One of these solutions (with $C > 0$) corresponds to a minimum of the functional (7.78), while the other one (with $C < 0$) corresponds to its maximum. Figure 7.10 illustrates the optimal outline of a shell for different values of the parameter ν. Curve 1 corresponds to $\nu = 0$, while curve 2 to $\nu = \infty$.

We examine a case when the "controlled quantities" are both the thickness and the radius of the shell. Listing the separate optimality conditions for h and R, we obtain for this problem two conditions (7.81) and (7.83) that can be used to find the unknown design variables. It is easy to see that for a distribution of thickness $h(z)$ that satisfies condition (7.81) this equation can be integrated in terms of elementary functions,

$$R = \sqrt{-z^2 + R_0^2(1 - z/l) + (R_l^2 + l^2)z/l}. \tag{7.84}$$

Looking at (7.84), it is obvious that the shape of an optimal shell is a function only of the quantities R_0, R_l, and l. The constant ν can be determined from the isoperimetric condition. Figure 7.11 shows curves 1 to 3 representing a family of optimal shapes ($R = R(z)$) of elastic shells for values of $R_0 = 1, l = 1$ and for $R_l = 0.1$, $R_l = 1$, $R_l = 10$.

7.9 Design of Plates Having Variable Thickness and Subjected to Tension in the Plane of the Plate

Let the median surface of a plate with variable thickness be located in the xy plane and occupy a region Ω. The boundary of Ω consists of two segments: Γ_u and Γ_q ($\Gamma = \Gamma_u + \Gamma_q$). Displacements of the plate are assigned to the segment Γ_u, while external loads $q^s = (q_x^s, q_y^s)$ acting on the plate in the xy plane are assigned on Γ_q. Also, other forces $q^v = (q_x^v, q_y^v)$ are applied to the plate in the xy plane. We assume that the plate is in a two-dimensional stressed state and that during the process of deforming the plate no significant buckling out of the xy plane will take place.

We denote by $u = u(x, y)$ and $v = v(x, y)$ the displacements of the median surface of the plate in the direction of the x and y coordinates, respectively, and by $h = h(x, y)$ the thickness of the plate, which is assumed to be symmetric with respect to the median surface. The functions $u(x, y)$ and $v(x, y)$ satisfy the system of differential equations of equilibrium and boundary conditions

$$\left.\begin{aligned}
&L_1(u, v, h) + q_x^v = 0,\\
&L_2(u, v, h) + q_y^v = 0,\\
&L_1(u, v, h) \equiv \frac{E}{1-\nu^2}\left([h(u_x + \nu v_y)]_x + \frac{1-\nu}{2}[h(u_y + v_x)]_y\right),\\
&L_2(u, v, h) \equiv \frac{E}{1-\nu^2}\left([h(\nu u_x + v_y)]_y + \frac{1-\nu}{2}[h(u_y + v_x)]_x\right),\\
&u = v = 0 \quad \text{for all } (x, y) \in \Gamma_u.\\
&N_1(u, v, h) - q_x^v = 0, \qquad N_2(u, v, h) - q_y^v = 0 \quad \text{for all } (x, y) \in \Gamma_q,\\
&N_1(u, v, h) \equiv \frac{E}{1-\nu^2}\left([(u_x + \nu v_y)]\cos(n, x)\right.\\
&\qquad\qquad \left. + \frac{1-\nu}{2}[(u_y + v_x)\cos(n, y)]\right),\\
&N_2(u, v, h) \equiv \frac{E}{1-\nu^2}\left([(\nu u_x + v_y)]\cos(n, y)\right.\\
&\qquad\qquad \left. + \frac{1-\nu}{2}[(u_y + v_x)\cos(n, x)]\right).
\end{aligned}\right\} \tag{7.85}$$

Here E is Young's modulus, ν is Poisson's ratio, n is the direction normal to the boundary of Ω, and the subscripted variables x and y in the expressions for the operators L_1, L_2, N_1, and N_2 denote (as usual) partial differentiation.

The optimization problem discussed here (see [7.12]) consists in finding a distribution of thickness for $h(x, y) > h_{\min}$ (where $h_{\min}$ is a given constant) that assigns a minimum to the volume of the plate's material,

$$J = \int_\Omega h(x, y)\mathrm{d}\Omega \to \min_h, \tag{7.86}$$

with a constraint on the gradient of the plate's thickness,

$$J_1 = \max_{x,y}(h_x^2 + h_y^2) = \alpha, \tag{7.87}$$

and in satisfying a bound on its rigidity,

$$J_2 = \max_{x,y} (u_x^2 + v_y^2) = \beta, \tag{7.88}$$

where $\alpha > 0$, $\beta > 0$ are some given constants, and $(x, y) \in \Omega$. The constraint (7.88) implies that the displacement must not exceed in magnitude the value of $\beta^{1/2}$, while (7.87) implies a bound on the rate of change of the function $h(x, y)$.

For the sake of convenience in solving this problem, these constraints are approximated by the constraints

$$I_1 = \left[\frac{1}{S}\int_\Omega (h_x^2 + h_y^2)^t d\Omega\right]^{1/t} = \alpha, \tag{7.89}$$

$$I_2 = \left[\frac{1}{S}\int_\Omega (u^2 + v^2)^p d\Omega\right]^{1/p} = \beta. \tag{7.90}$$

where S is the area of Ω, and p and t are positive integers.

Questions concerning the application of this technique to optimization problems were discussed in Section 2.3 (also see [5] and [2.3]). Now we derive optimality conditions for the modified problem (7.85), (7.86), (7.89), and (7.90). The variation of the cost functional is given by

$$\delta J = \int_\Omega \delta h d\Omega. \tag{7.91}$$

The formula for the variation of the functional (7.89) may be restated as

$$\begin{aligned} \delta I_1 &= -\int_\Omega Q\delta h d\Omega + (2I_1^{1-t}/S)\int_\Gamma \{(h_x^2 + h_y^2)^{t-1}[h_x \cos(n, x) \\ &\quad + h_y \cos(n, y)]\} dh d\Gamma, \\ Q &\equiv \{[h_x^2 + h_y^2)^{t-1}h_x]_x + [(h_x^2 + h_y^2)^{t-1}h_y]_y\}(2I_1^{1-t}/S). \end{aligned} \tag{7.92}$$

The function h is assumed to be known on Γ. In that case the contour integral introduced in (7.92) for δI_1 is equal to 0, because $(\delta h)_\Gamma = 0$. Therefore,

$$\delta I_1 = -\int_\Gamma Q\delta h d\Omega. \tag{7.93}$$

To compute the first variation of the functional (7.90), we make use of the substitution

$$\left.\begin{aligned} \delta I_2 &= \int_\Omega \Phi(u\delta u + v\delta v) d\Omega, \\ \Phi &\equiv (2I_2^{1-p}/S)(u^2 + v^2)^{p-1}. \end{aligned}\right\} \tag{7.94}$$

The variations of the displacement variables δu and δv that appear in the formula for I_2 can be expressed in terms of the variation δh of the control function. To do so, we introduce adjoint variables $\varphi = \varphi(x, y)$ and $\psi = \psi(x, y)$ and vary both the equations and boundary conditions in (7.85):

$$\left.\begin{aligned}\delta L_1 &= L_1(u, v, \delta h) + L_1(\delta u, \delta v, h) = 0,\\ \delta L_2 &= L_2(u, v, \delta h) + L_2(\delta u, \delta v, h) = 0,\\ \delta u &= \delta v = 0 \quad \text{for all } (x, y) \in \Gamma_u,\\ \delta N_1 &= N_1(u, v, \delta h) + N_1(\delta u, \delta v, h) = 0,\\ \delta N_2 &= N_2(u, v, \delta h) + N_2(\delta u, \delta v, h) = 0.\end{aligned}\right\} \tag{7.95}$$

The expressions for δL_1 and δL_2 are multiplied, respectively, by φ and ψ and the sum $\varphi\delta L_1 + \psi\delta L_2$ is introduced. This sum is integrated over the region Ω, and after integration by parts, using the boundary conditions assigned to u and v, the following expression is derived:

$$\left.\begin{aligned}0 = &\int_\Omega M\delta h d\Omega - \int_\Omega [L_1(\varphi, \psi, h)\delta u + L_2(\varphi, \psi, h)\delta v] d\Omega\\ &+ \int_{\Gamma_u} [\varphi\delta N_1(u, v, h) + \psi\delta N_2(u, v, h)] d\Gamma\\ &+ \int_{\Gamma_q} [N_1(\varphi, \psi, h)\delta u + N_2(\varphi, \psi, h)\delta v] d\Gamma,\end{aligned}\right\} \tag{7.96}$$

where

$$M = \frac{E}{1 - \nu^2}\left[u_x\varphi_x + v_y\psi_y + \nu(u_x\psi_y + v_y\varphi_x) + \frac{1 - \nu}{2}(u_y + v_x)(\varphi_y + \psi_x)\right].$$

This formula indicates that M is a quadratic form in the variables u_x, u_y, v_x, v_y, φ_x, φ_y, ψ_x, and ψ_y. The Relation (7.96) allows us to represent the formula for δI_2 in the form

$$\left.\begin{aligned}\delta I_2 = &\int_\Omega M\delta h d\Omega\\ &+ \int_\Omega \{[u\Phi - L_1(\varphi, \psi, h)]\delta u + [v\Phi - L_2(\varphi, \psi, h)\delta v]\} d\Omega\\ &+ \int_{\Gamma_u} [\varphi\delta N_1(u, v, h) + \psi\delta N_2(u, v, h)] d\Gamma\\ &+ \int_{\Gamma_q} [N_1(\varphi, \psi, h)\delta u + N_2(\varphi, \psi, h)\delta v] d\Gamma.\end{aligned}\right\}$$

We specify the following equations and boundary conditions to be obeyed by the adjoint variables:

$$\left.\begin{aligned}L_1(\varphi, \psi, h) &= u\Phi, & L_2(\varphi, \psi, h) &= v\Phi,\\ \varphi &= 0, \quad \psi = 0 & \text{for all } &(x, y) \in \Gamma_u,\\ N_1(\varphi, \psi, h) &= 0, & N_2(\varphi, \psi, h) &= 0,\end{aligned}\right\} \tag{7.97}$$
$$\text{(for all } (x, y) \in \Gamma_q).$$

Note that unlike eqs. (7.85), which are linear, eqs. (7.97) are nonlinear, because of the nonlinear dependence of the right-hand sides on the variables u and v. For functions φ and ψ obeying (7.97), all integrals and expressions for I_2, with the exception of the first term, are equal to 0. Consequently, we derive a formula for the first variation of the functional I_2 in terms of the first variation of the control function

$$\delta I_2 = \int_\Omega M\delta h \mathrm{d}\Omega. \tag{7.98}$$

We construct a generalized Lagrangian functional and set its first variation equal to 0: $\delta J + \lambda_1 \delta J_1 - \lambda_2 \delta J_2 = 0$.

As a result we obtain a necessary optimality condition

$$(\lambda_1 Q + \lambda_2 M - 1)\chi = 0. \tag{7.99}$$

Here λ_1 and λ_2 are Lagrangian multipliers, while χ is an auxiliary design variable related to h by the formula $h = \frac{1}{2}\chi^2 + h_{\min}$, thus "automatically" assuring the validity of the constraint $h \geq h_{\min}$. Other techniques for introducing such auxiliary variables were discussed in Section 2.1. For given values of p and t, the optimization problem is reduced to the solution of the coupled equations (7.85) and (7.97) for the variables u and v and the adjoint variales φ and ψ. The design variable h may be found from the optimality condition (7.99), while the values of the unknown constants λ_1 and λ_2 may be determined from the isoperimetric conditions (7.89) and (7.90). Thus, an optimal distribution of the plate's thickness can be found from a solution of a well-posed problem, which happens to be nonlinear due to the nonlinearity of (7.97) and (7.99).

The formula for an improvement producing a variation of the design variable can be derived by making use of the conditions

$$\delta J < 0, \qquad \delta I_1 = 0, \qquad \delta I_2 = 0, \tag{7.100}$$

and within a specified accuracy assumes the form (4.67), (4.69).

It should be noted that in carrying out computations based on certain algorithms of succesive optimization it is impossible to avoid computational errors arising from inaccuracies in eqs. (7.89) and (7.90), that is, from the fact that $\alpha - I_1 = \Delta_1 \neq 0$ and $\beta - I_2 = \Delta_2 \neq 0$. In this regard, the following corrective steps were introduced in [7.12]. The second and third relation in (7.100) can be replaced by $\delta I_1 = \Delta_1$ and $\delta I_2 = \Delta_2$, thus permitting us to compute the values of the multipliers λ_1 and λ_2 that occur in the formula (4.67) for the variation of the thickness δh.

Finding the displacement field $u(x, y)$, $v(x, y)$ for the plate for current values of $h(x, y)$ is carried out using a direct variational–finite-difference technique for minimization of the potential energy functional for the system. This variational problem may be approximated by a strictly convex, finite-difference quadratic scheme, with second-order accuracy. In particular, this scheme guarantees the uniqueness of solutions for the direct problem of finding the values u_{ij} and v_{ij} of displacements at the vertices of a grid. Computations used the upper relaxation technique (see [7.60]). We can apply an effective technique in which the quantities u_{ij} and v_{ij} are pairwise simultaneously varied:

$$u_{ij}^{k+1} = u_{ij}^{k} + \gamma(r_u^k)_{ij}, \qquad v_{ij}^{k+1} = v_{ij}^{k} + \gamma(r_v^k)_{ij},$$

where γ is a relaxation parameter and $(r_u)_{ij}$ and $(r_v)_{ij}$ are quantities defined from the conditions for maximizing the decrease in the value of the finite-difference approximating functional.

The solution of the adjoint problem of finding the functions φ and ψ in (7.97) is also accomplished by finite difference techniques. For the sake of convenience and brevity, identical computational techniques were used for computing the values of the pairs (u_{ij}, v_{ij}) and $(\varphi_{ij}, \psi_{ij})$ in [7.12].

We introduce some data related to the specific problem of optimizing the distribution of thickness $h(x, y)$ with $0 < x < 1$, $0 < y < 1$, for a square plate $(0 \leq x \leq 1, 0 \leq y \leq 1)$ rigidly built-in along its boundary and subjected to two concentrated loads of unit strength positioned at the points (0.7, 0.5) and (0.3, 0.5), and acting, respectively, in the direction of the positive x axis and in the opposite direction. Along the boundary Γ the thickness $h(x, y)$ is given by

$$h(0, y) = h(1, y) = 3, \qquad h(x, 1) = h(x, 0) = \cos(2\pi x) + 2.$$

This thickness does not change during the solution process. The parameters α and β are equal, respectively, to 2π and 0.07. During this computation we assume that $t = p = 40$. Use of the methodology of [7.12] turns out to be very economical. Only a few iterations were necessary to obtain solutions of both the direct and the adjoint problem. With the choice of the best relaxation parameter γ the number of iterations did not exceed 30. As a result of the iterative process, the derived thickness distribution indicates that the load-bearing material of the plate is concentrated at the center and near the locations of the loads. In other regions the utilization of the material is less effective. Regions having the greatest values of the gradient of thickness are close to the edges of the plate. It is important to note that the presence of the constraint (7.87) leads to smaller jumps in the values of the thickness. A

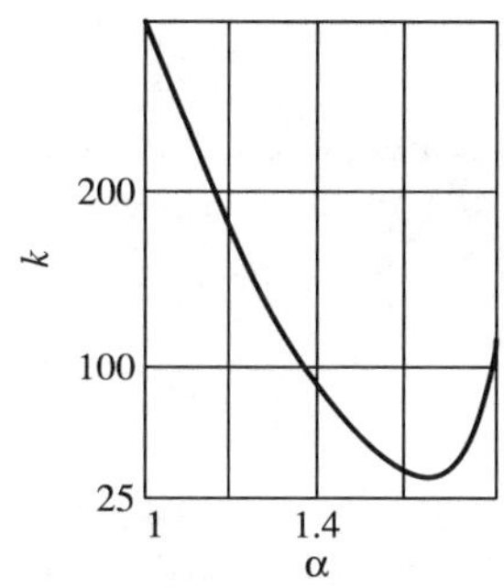

FIGURE 7.12. Optimization of a square plate: number of iterations (k) versus the relaxation parameter (α).

constraint on the rate of variation of the thickness was also adopted in [7.66] in investigating the one-dimensional case of a round plate subjected to bending loads. We should also remark that the introduction of the constraint (7.87) comes naturally. Allowing large (unrestricted) values of thickness gradients for the purpose of improving the effects of optimization (reduction of weight) generally requires corrections to the mathematical model itself. In particular, to obtain reliable results one needs to introduce additional hypotheses to describe the response of plates that exhibit sharp changes in their rigidity properties.

7.10 Some Problems Concerning Plates of Variable Thickness Subjected to Bending

Let a plate with variable thickness $h(x, y)$ be supported along its boundary curve Γ situated in the xy plane and be subjected to static bending action caused by a transverse load $q(x, y)$. The curve Γ consists of two segments, Γ_1 and Γ_2 ($\Gamma = \Gamma_1 + \Gamma_2$), and forms the boundary of a region Ω. On the segment Γ_1, the plate is built-in, while on the segment Γ_2, it is freely supported. The plate's displacement function is denoted by $w(x, y)$. The quality of the structure and its response is characterized by the functionals $J_0(w(h), h)$, $J_1(w(h), h), \ldots,$ $J_l(w(h), h)$. We assume that the distribution of thickness satisfies the geometric constraints

$$h_{\min} \le h(x, y) \le h_{\max}, \qquad \forall (x, y) \in \Omega, \tag{7.101}$$

$$\left(\frac{1}{\operatorname{meas} \Omega} \int |\nabla h|^p d\Omega\right)^{1/p} \le \alpha, \qquad p > 2, \tag{7.102}$$

where $h_{\min}$, $h_{\max}$, and α are given positive constants, and ∇ is a gradient operator represented in the xy coordinate system. For sufficiently large values of the parameter p, the expression on the left-hand side of (7.102) approximates the magnitude of $\max_{x,y} |\nabla h|$.

The response of the plate is constrained by the system of inequalities

$$J_k(w(h), h) \le 0, \qquad k = 1, 2, \ldots, l. \tag{7.103}$$

The function $h(x, y)$ is an element of the Sobolev space $W_p^1(\Omega)$ and satisfies the system of inequalities (7.101) to (7.103), defining the set of admissible thickness functions H_∂. For a thickness function $h \in H_\partial$, the state function (displacement function) for the plate $w(x, y)$ can be determined from conditions given in [1.27] and [7.54]

$$w \in V, \qquad a_h(w, v) = \int_\Omega qv d\Omega \quad \text{for all } v \in V, \tag{7.104}$$

where V is a (closed) subspace of the Sobolev space $W_2^2(\Omega)$ determined by the boundary support conditions and a_h is a symmetric bilinear form on $V \times V$

which originated with the total strain energy of the plate

$$\left.\begin{aligned} a_h(u, v) &= \int_\Omega \frac{Eh^3}{12(1-\nu^2)} A(u, v)\mathrm{d}\Omega, \\ A(u, v) &\equiv (u_{xx} + u_{yy})(v_{xx} + v_{yy}) \\ &\quad - (1-\nu)(u_{xx}v_{yy} + u_{yy}v_{xx} - 2u_{xy}v_{xy}). \end{aligned}\right\} \tag{7.105}$$

As usual, the subscripts x and y in (7.105) denote partial differentiation with respect to the indicated variables.

Therefore, for each choice of $h \in H_\partial$ there corresponds, according to (7.104), a unique displacement function w. The quality functional can be written as

$$J(h) = J_0(w(h), h). \tag{7.106}$$

The optimization problem considered next (see [7.13]) consists in finding a function $h_0 \in H_\partial$ that optimizes the quality functional

$$h_0 \in H_\partial, \qquad J(h_0) = \inf_{h \in H_\partial} J_0(h). \tag{7.107}$$

Following [7.13] we shall introduce basic concepts relevant to the optimization problem (7.101) to (7.107). First we shall show that an optimal solution exists. Note that the conditions $w \in V$ and $a_h(w, w) = 0$ imply that $w \equiv 0$. This means that for a plate supported on its boundary, if the strain energy is equal to 0, then no deformations exist. This statement and certain results of [1.27] and [7.54] imply both coercivity and uniform continuity of the bilinear form a_h (defined on $V \times V$), that is, the existence of positive constants C_1, C_2, and C_3, such that,

$$|a_h(u, v)| \leq C_1 \|u\|_V \|v\|_V, \qquad C_2 \|u\|_V^2 \leq a_h(u, u) \leq C_3 \|u\|_V^2, \tag{7.108}$$

for arbitrary $u, v \in V$, and for any h satisfying inequalities (7.101) and (7.102).

Let us denote by $Y = \{h; h \in C(\Omega), h_{\min} \leq h(x, y) \leq h_{\max}\}$ the set of functions endowed with the topology $C(\bar{\Omega})$. For any $h \in Y$ and for a realizable load $q \in L_2(\Omega)$, eq. (7.108) implies the existence of a unique solution to the "direct" problem given by (7.104) and (7.105). We assume that the functionals $J_k(w(h), h)$, $k = 1, \ldots, l$, represent continuous maps from $Y \times V$ into $\mathbf{R}$. Then the following theorem is true.

THEOREM. *Let the bilinear form a_h, the quality functional $J(h)$ and the admissible (nonempty) set of functions H_∂ satisfy* (7.101) *to* (7.103), (7.105), *and* (7.106) *and let the functionals $J_k(w, h)$ be continuous maps from $Y \times V$ into* $\mathbf{R}$ *for* k $= 1, \ldots, l$. *Then there exists a solution of the problem* (7.107).

PROOF. The set H_∂ contains a minimizing sequence $\{h^n\}$, $n = 1, 2, \ldots$ such that $\{h^n\} \in H_\partial$ and

$$\lim_{n \to \infty} J(h^n) = \inf_{h \in H_\partial} J(h). \tag{7.109}$$

Inequalities (7.101) and (7.102) imply that the sequence $\{h^n\}$ is bounded in $W_p^1(\Omega)$. Therefore, it is possible to find a subsequence $\{h^m\}$ such that $h^m \to h^*$ weakly in W_p^1, and by the Sobolev imbedding theorem $h^m \to h$ converges strongly in $C(\bar{\Omega})$. Making use of this statement, and the fact that $\{h^m\} \in H_\partial$, we conclude that h^* also satisfies inequalities (7.101) and (7.102). Because of the strong convergence of $\{h^m\}$ to h^* in $C(\bar{\Omega})$, by [7.56] we have the convergence $w(h^m) \to w(h^*)$ in V as $m \to \infty$. Here h^m and h^* denote, respectively, the solutions of problem (7.104) and (7.105) for $h = h^m$ and $h = h^*$. From the convergence of $w(h^m)$ to $w(h^*)$ and the strong convergence of h^m to h^* in $C(\bar{\Omega})$ and the continuity of the maps $Y \times V$ into $\mathbf{R}$ for the corresponding to the functionals $J_k(u, h)$ $(u \in V)$, it follows that h^* satisfies inequalities (7.103). An identical argument indicates that $h^* \in H_\partial$. Now, using (7.106) and (7.109), the continuity of the maps represented by the functionals $J_k(u, h)$, the strong convergence of h^m to h^* in $C(\bar{\Omega})$, and the convergence of $w(h^m)$ to $w(h^*)$ in V, we obtain

$$J(h^*) = \lim_{m \to \infty} J(h^m) = \inf_{h \in H_\partial} J(h).$$

Thus, the function h^* solves the optimization problem (7.107).

We shall now examine the optimization problem for a square plate (Ω: $0 < x < d, 0 < y < d$) freely supported along its entire boundary contour Γ. As our optimized functional J_0, we choose the total volume of the plate, while inequalities (7.103) are replaced by a bound on the maximum deflection $\max_{\bar{\Omega}} |w| \le \beta$, where β is a given positive number. The load $q(x, y)$ applied to the plate is fixed. In dimensionless variables the basic relations of this optimization problem (i.e., the differential equations of bending, the boundary conditions, constraints on admissible values of the thickness function and on its gradients, bounds on the maximum deflections, and the specific form of the minimized volume functional) may be written as

$$L(h)w = [h^3(w_{xx} + \nu w_{yy})]_{xx} + [h^3(w_{yy} + \nu w_{xx})]_{yy} + 2(1 - \nu)(h^3 w_{xy})_{xy} = q, \tag{7.110}$$

$$\begin{aligned} w(0, y) &= w(1, y) = w(x, 0) = w(x, 1) = w_{xx}(0, y) = w_{xx}(1, y) \\ &= w_{yy}(x, 0) = w_{yy}(x, 1) = 0, \end{aligned} \tag{7.111}$$

$$h_{\min} \le h(x, y) \le h_{\max}, \qquad (x, y) \in \Omega, \tag{7.112}$$

$$h(x, y) = g(x, y), \qquad (x, y) \in \Gamma, \tag{7.113}$$

$$J_2 = \left(\int_\Omega |\nabla h|^p d\Omega \right)^{1/p} \le 1, \qquad (p \ge 2), \tag{7.114}$$

$$J_1 = \left(\int_\Omega |w|^p d\Omega \right)^{1/p} \le 1, \qquad (p \ge 2), \tag{7.115}$$

$$J_0 = \int_\Omega h d\Omega. \tag{7.116}$$

Here $g(x, y)$ is an a priori given function determining the thickness along the curve Γ that forms the boundary of the region Ω: $0 < x < 1, 0 < y < 1$.

To solve problem (7.110) to (7.116), we employ the technique of successive optimization based on variation of the gradients with respect to the control function, and subsequent computations of the displacement field corresponding to the current distribution of thickness. To apply this algorithm, we need formulas for the variation of relevant functionals and equations for the adjoint variable $v(x, y)$. The constraint (7.112) is taken care of by introduction of an auxiliary design function $\varphi(x, y)$ related to $h(x, y)$ by the formula $h = \frac{1}{2}[(h_{\max} + h_{\min}) - (h_{\max} - h_{\min}) \sin \varphi]$. Noting the boundary conditions, and setting the variations of our functionals equal to 0, we derive

$$\delta J_0 = \int \delta h d\Omega, \qquad \delta J_2 = -\int_\Omega \operatorname{div}\left[\left(\frac{b}{\|b\|_{L_p}}\right)^{p-1} \frac{\nabla h}{b}\right] \delta h d\Omega, \quad (7.117)$$

$$\delta J_1 = \int_\Omega \frac{3Eh^2}{12(1 - \nu^2)} A(w, v) \delta h d\Omega, \quad (7.118)$$

$$L(h)v = -\left(\frac{|w|}{\|w\|_{L_p}}\right)^{p-1} \operatorname{sign} w, \quad (7.119)$$

$$v(0, y) = v(1, y) = v(x, 0) = v(x, 1) = v_{xx}(0, y) = v_{xx}(1, y)$$
$$= v_{yy}(x, 0) = v_{yy}(x, 1) = 0. \quad (7.120)$$

Here $\delta h = \frac{1}{2}(h_{\max} - h_{\min}) \cos \varphi \delta \varphi$ and $b = |\nabla h|$.

In deriving (7.117) and (7.118), we used integration by parts and regarded the adjoint variable v as a solution of the boundary-value problem (7.119) and (7.120). Formula (7.118) relating δJ_1 to δh represents a sensitivity analysis formulation of the constraint functional (7.115) subject to (7.110) and (7.112). We observe that in the boundary-value problem (7.119) and (7.120) for the adjoint variable v, only the right-hand side of eq. (7.119) differs from the corresponding boundary-value problem for the bending of a plate. This property can be exploited in formulating numerical variational techniques for finding w and v.

To compute an optimal distribution of thickness using an algorithm of successive optimization we apply formulas (7.117) and (7.118) and relations (7.110), (7.111), (7.119), and (7.120). A detailed description of this algorithm was given in Section 4.7. We used as our initial approximation for the thickness function $h_0 = (1 + \varepsilon \cos 2\pi x \cos 2\pi y)/2\pi$. During the process of numerical solution by iteration, guided by the sensitivity analysis formulas, we realized a monotone decrease in the value of the functional J_0 (i.e., in the plate's volume). To improve the distribution of thickness, which was derived from computations using the values of the parameters $\nu = 1/3$ and $\varepsilon = .2$ $(0 < \varepsilon < 1)$, we note that the constraints $h_{\min} = 1 - \varepsilon$, $h_{\max} = 1 + \varepsilon$, and $g = h^0(x, y)|_{x,y \in \Gamma}$ are active, and so are the constraints consisting of bounds on the admissible values of the thickness and its gradients. The distribution of thickness $h(x, y)$

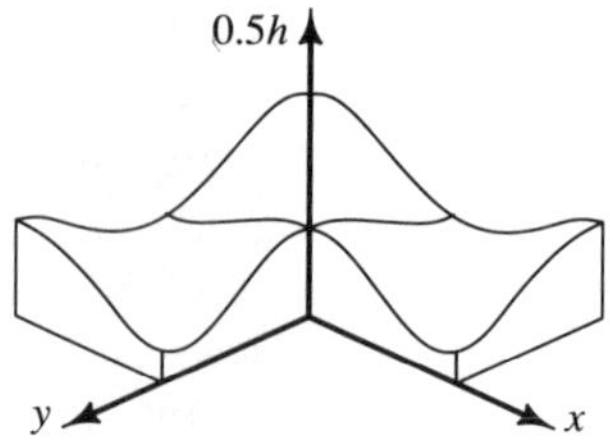

FIGURE 7.13. Thickness of optimal plate.

derived in this manner is illustrated in Figure 7.13. Numerical results for other values of parameters indicate that, qualitatively, the general shape of the optimal plate remains unchanged.

7.11 Initially Deformed Plates with Maximal Rigidity

Following [7.17] we shall investigate the bending of an initially bent plate supported along its boundary curve Γ, which lies in the xy plane and is loaded by transverse forces $q(x, y)$ and also by some loads acting in the xy plane applied to its edge (see Figure 7.14). On the boundary segment of the plate Γ_1, the plate is rigidly built-in, while on the segment Γ_2 it is freely supported ($\Gamma = \Gamma_1 + \Gamma_2$). The curve Γ is the boundary of a region Ω located in the xy plane. Let $f(x, y)$ denote the shape of the middle surface of this plate, which is initially deformed in the absence of external loads. Let $w(x, y)$ be a function describing deflections in the direction of the z axis of the middle surface of the plate caused by external loads. The magnitude of $w(x, y)$ is measured from the middle surface of the initially deformed plate. Therefore, the full deflection of the median surface is given by the function $w(x, y) + f(x, y)$. Let us assume that the deflections are "small," that is, the typical deflection is smaller than

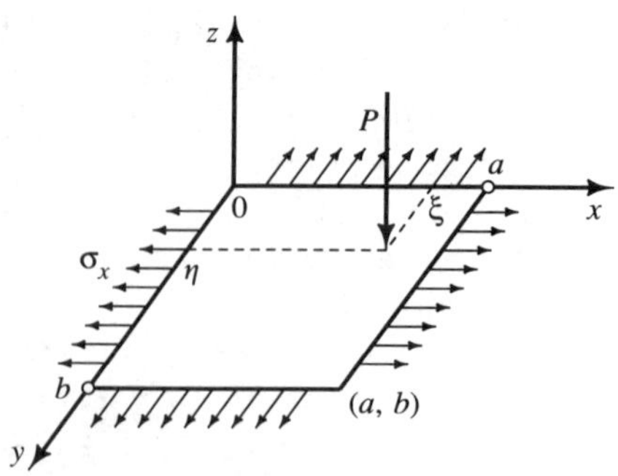

FIGURE 7.14. A system of loads applied to a plate.

the thickness of the plate. Then the equilibrium equation and boundary conditions are of the form

$$Lw - K(\varphi)w = q + K(\varphi)f, \tag{7.121}$$

$$(w)_\Gamma = 0, \qquad \left(\frac{\partial w}{\partial n}\right)_{\Gamma_1} = 0, \qquad D\left[\Delta w - \frac{1-\nu}{R}\frac{\partial w}{\partial n}\right]_{\Gamma_2} = 0, \tag{7.122}$$

where $\partial w/\partial n$, R, Δ, and φ are, respectively, the derivative of $w(x, y)$ in the direction normal to the boundary curve, the radius of curvature of the boundary, the Laplace operator, and the stress function for stresses acting in the plane of the plate. The cylindrical rigidity of the plate $D = D(x, y)$ is a given function of the xy coordinates; and K and L are linear differential operators

$$\left.\begin{aligned} L(h)w &= (Dw_{xx})_{xx} + (Dw_{yy})_{yy} + \nu(Dw_{yy})_{xx} + \nu(Dw_{xx})_{yy} \\ &\quad + 2(1-\nu)(Dw_{xy})_{xy'} \\ K(\varphi)w &= h(\varphi_{yy}w_{xx} + \varphi_{xx}w_{yy} - 2\varphi_{xy}w_{xy}). \end{aligned}\right\} \tag{7.123}$$

As usual, the subscripts x and y denote partial differentiation with respect to the indicated variables. The stress function φ is determined by solving the boundary-value problem for the partial differential equation

$$\Delta\left(\frac{1}{Gh}\Delta\varphi\right) - \frac{1}{2}\left[\left(\frac{1}{Gh}\varphi_{yy}\right)_{xx} + \left(\frac{1}{Gh}\varphi_{xx}\right)_{yy} - 2\left(\frac{1}{Gh}\varphi_{xy}\right)_{xy}\right] = 0, \tag{7.124}$$

with appropriate boundary conditions on Γ (replicating classical problems of the linear two-dimensional theory of elasticity). The function φ does not depend on either w or f. Therefore, this function can be determined early in the solution process. In subsequent arguments leading to the determination of the displacement function $w(x, y)$ and of the best shape of the median surface $f(x, y)$, we shall assume that the function φ is known.

The magnitude of the area of the median surface is given and is equal to S. This assumption is equivalent to the isoperimetric condition assigned to the function $f(x, y)$

$$\int_\Omega \sqrt{1 + f_x^2 + f_y^2}\, d\Omega = S. \tag{7.125}$$

Using the assumption that the function $f(x, y)$ is "small," we transform (7.125) into the form

$$\int_\Omega (f_x^2 + f_y^2)\, d\Omega = 2(S - S_0), \tag{7.126}$$

where S_0 is the area of Ω. The function f in (7.126) describing the initial deflection of the plate must satisfy the boundary condition

$$(f)_\Gamma = 0. \tag{7.127}$$

A continuous function $f(x, y)$ satisfying the isoperimetric condition (7.126) with boundary condition (7.127) can be regarded as a candidate for the unknown design variable.

We formulate the following optimization problem. We need to find a function $f(x, y)$ that satisfies the isoperimetric condition (7.126) and boundary condition (7.127) such that the displacement function $w(x, y)$ derived from the boundary-value problem (7.121) and (7.122) assigns a minimum to the functional

$$J = \int_{\Omega} q(x, y)w(x, y)\mathrm{d}\Omega \to \min_{f}. \tag{7.128}$$

To derive the necessary optimality condition and the corresponding equations for the adjoint variable, we write the variational equation directly derived from (7.121):

$$L\delta w - K(\varphi)\delta w - K(\varphi)\delta f = 0.$$

Multiplying the left-hand side of this equation by the adjoint function $v(x, y)$ and integrating over the region Ω, we obtain

$$\int_{\Omega} v[L\delta w - K(\varphi)\delta w - K(\varphi)\delta f]\mathrm{d}\Omega = 0. \tag{7.129}$$

Let us assume that the function $v(x, y)$ satisfies the boundary conditions (7.122). For functions satisfying these conditions, the operator L is self-adjoint. Therefore, $(v, L\delta w) = (\delta w, Lv)$, where, as usual, the brackets denote the inner product. It can be shown (see [7.17]) that the operator K is also self-adjoint for the class of functions under discussion (i.e., $(v, K(\varphi)\delta w) - (K(\varphi)v, \delta w) = 0$). Noting these properties of L and K, we rewrite eq. (7.129):

$$\int_{\Omega} [Lv - K(\varphi)v]\delta w\mathrm{d}\Omega - \int_{\Omega} [\delta f K(\varphi)v]\mathrm{d}\Omega = 0. \tag{7.130}$$

Equation (7.130) can be used to represent the variation of the minimized compliance functional in the form

$$\delta J = \int_{\Omega} [Lv - K(\varphi)v + q]\delta w\mathrm{d}\Omega - \int_{\Omega} [\delta f K(\varphi)v]\mathrm{d}\Omega = 0. \tag{7.131}$$

We derive the adjoint function $v(x, y)$ by solving the boundary-value problem

$$Lv - K(\varphi)v = -q, \tag{7.132}$$

$$(v)_{\Gamma} = 0, \qquad \left(\frac{\partial v}{\partial n}\right)_{\Gamma_1} = 0, \qquad D\left[\Delta v - \frac{1 - \nu}{R}\frac{\partial v}{\partial n}\right]_{\Gamma_2} = 0. \tag{7.133}$$

For a function $v(x, y)$ determined in this fashion, the first integral on the right-hand side of (7.131) is equal to 0. As a result, we arrive at the required formula which expresses the first variation of the functional J in terms of the variation of the design variable:

$$\delta J = -\int_{\Omega} K(\varphi) v \delta f \, d\Omega. \tag{7.134}$$

It follows from condition (7.126) that for an arbitrary variation of the design variable δf substituted in the integrand of (7.134), the following equality must be true:

$$\int_{\Omega} (f_x \delta f_x + f_y \delta f_y) d\Omega = 0.$$

Making use of boundary condition (7.127) assigned to the function $f(x, y)$, we transform this equality into the form

$$\int_{\Omega} \Delta f \delta f \, d\Omega = 0. \tag{7.135}$$

A necessary condition for a minimum of the functional J is $\delta J = 0$. Therefore, recalling conditions (7.134) and (7.135) we obtain the optimality criterion

$$\Delta f = \lambda v, \tag{7.136}$$

where the constant λ is a Lagrangian multiplier.

In this manner the derivation of an optimal distribution of thickness for a plate has been reduced to the solution of three simultaneous boundary-value problems. First we solve a boundary-value problem of the two-dimensional theory of elasticity and obtain a stress function. Then we solve eq. (7.132) with boundary conditions (7.133) and derive the adjoint function $v(x, y)$.

Afterward we substitute this expression for $v(x, y)$ into the right-hand side of eq. (7.136) to find the optimal shape $f(x, y)$. This (last) problem has been reduced to the classical formulation of Dirichlet's problem for the Poisson equation (see (7.127) and (7.136)).

Note that (as is well known) the boundary-value problem (7.132) and (7.133) for the adjoint variable $v(x, y)$ determines the deflections of an initially straight plate subjected to both the loads $q(x, y)$ and loads acting in the xy plane. Therefore, to determine $v(x, y)$ for a given type of load $q(x, y)$ one may use well-known analytical and numerical techniques for solving problems in the simultaneous bending and tension of plates. For a majority of practically important cases of applied loads, these solutions may be found in popular books or monographs. Solving the Dirichlet problem for the Poisson equation presents no difficulties, and for some types of regions Ω may be accomplished by purely analytical techniques.

We shall now consider some examples of optimal designs of elastic plates whose shapes projected into the xy plane form a rectangle: $0 \leq x \leq a, 0 \leq y \leq b$. Let the plate be supported along a boundary curve Γ and loaded at a point (ξ, η) (with $0 < \xi < a$, $0 < \eta < b$) by a concentrated force q (i.e., $q = P\delta(x - \xi, y - \eta)$). Assuming that the thickness of the plate is constant, we can write the adjoint equation (7.132) and the boundary conditions (7.133) for the function $v(x, y)$ (see Figure 7.14)

$$D\Delta^2 v = K(\varphi)v - P\delta(x-\xi, y-\eta), \tag{7.137}$$

$$\left.\begin{aligned}(v)_{x=0} = (v)_{x=a} = (v)_{y=0} = (v)_{y=b} = 0,\\ (v_{xx})_{x=0} = (v_{xx})_{x=a} = (v_{yy})_{y=0} = (v_{yy})_{y=b} = 0.\end{aligned}\right\} \tag{7.138}$$

We assume that uniformly distributed tensile loads are applied to the edges of the plate, $\sigma_x = \varphi_{yy} = \sigma_1$ for $x = 0$ and $x = a$, and $\sigma_y = \varphi_{xx} = \sigma_2$ for $y = 0$ and $y = b$, where σ_1 and σ_2 are given constants. In this case the expression for $K(\varphi)v$ is given by $h\sigma_1 v_{xx} + h\sigma_2 v_{yy}$.

The solution of this boundary-value problem was given in [6.94]:

$$\left.\begin{aligned}v &= \sum_{m,n=1}^{\infty} v_{mn} \sin\frac{m\pi x}{a}\sin\frac{n\pi y}{b},\\ v_{mn} &= -\frac{4P}{ab}\gamma_{mn}\sin\frac{m\pi x}{a}\sin\frac{n\pi y}{b},\\ \gamma_{mn} &= \left[D\left(\frac{m^2\pi^2}{a^2}+\frac{n^2\pi^2}{b^2}\right)^2 + \left[h\left(\sigma_1\frac{m^2\pi^2}{a\pi^2}+\sigma_2\frac{n^2\pi^2}{b\pi^2}\right)\right]\right]^{-1}.\end{aligned}\right\} \tag{7.139}$$

The solution of the Dirichlet problem (7.127) and (7.136) for the case of a rectangular region Ω is given by the formulas

$$\left.\begin{aligned}f &= -\lambda\int_{\Omega}(G(x,y,t,s)K(\varphi)v)\,d\Omega,\\ G(x,y,t,s) &= \frac{1}{\pi^2 ab}\sum_{m,n=1}^{\infty}\beta_{mn}\sin\frac{m\pi x}{a}\sin\frac{n\pi y}{b}\sin\frac{m\pi t}{a}\sin\frac{n\pi s}{b},\\ \beta_{mn} &= \left(\frac{m^2}{a^2}+\frac{n^2}{b^2}\right)^{-1},\end{aligned}\right\} \tag{7.140}$$

where $G(x, y, t, s)$ is Green's function. We substitute (7.139) in the formula (7.140) for v and perform elementary algebra. The constant λ is determined from the isoperimetric condition (7.126). Thus, we obtain

$$f = \sum_{m,n=1}^{\infty} f_{mn}\sin\frac{m\pi x}{a}\sin\frac{n\pi y}{b},$$

$$\left.\begin{aligned}f_{mn} &= \frac{2}{\pi}\zeta_{mn}\beta_{mn}\sin\frac{m\pi x}{a}\sin\frac{n\pi y}{b}\cdot\left\{\frac{ab}{2(S-S_0)}\right.\\ &\quad\left.\cdot\sum_{k,l=1}^{\infty}\zeta_{kl}^2\beta_{kl}\sin^2\frac{m\pi\xi}{a}\sin^2\frac{n\pi\eta}{b}\right\}^{-1/2},\\ \zeta_{mn} &= \gamma_{mn}\left[-h\left(\sigma_1\frac{m^2\pi^2}{a^2}+\sigma_2\frac{n^2\pi^2}{b^2}\right)\right].\end{aligned}\right\}$$

In trying to achieve an optimal design for a plate, we first decompose the deflection function $w(x, y)$ and the magitude of the optimized functional in a formal trigonometric series:

$$w = \sum_{m,n=1}^{\infty} w_{mn} \sin \frac{m\pi x}{a} \sin \frac{n\pi y}{b},$$

$$w_{mn} = \gamma_{mn}\left[\frac{4P}{ab} \sin \frac{m\pi\xi}{a} \sin \frac{n\pi\eta}{b} - h\left(\sigma_1 \frac{m^2\pi^2}{a^2} + \sigma_2 \frac{n^2\pi^2}{b^2}\right)\right],$$

$$J = P \sum_{m,n=1}^{\infty} w_{mn} \sin \frac{m\pi\xi}{a} \sin \frac{n\pi\eta}{b}.$$

We shall consider the bending of an initially distorted plate that lies on a Winkler foundation with elastic compliance coefficient c. The plate is loaded by transverse forces $q(x, y)$ and is not subjected to any loads acting in the xy plane. Using our assumptions and notation similar to one used when the effects of elastic foundation were absent, we obtain an expression for the reaction of the foundation and for the state equation in bending, respectively:

$$c[w(x, y) + f(x, y)];$$

$$Lw = q(x, y) + c[w(x, y) + f(x, y)]. \tag{7.141}$$

As in the preceding arguments, we choose as our design variable the initial deformation function $f(x, y)$ for the plate, and as our quality criterion the compliance integral, which is to be minimized under the constraints (7.126) and (7.127). To solve this optimization problem, we follow the same approach as before. Performing the same substitutions, we derive an equation for the adjoint variable $v(x, y)$,

$$Lv + cv = -q(x, y) \tag{7.142}$$

with boundary condition (7.133), and necessary optimality condition

$$\Delta f = \lambda v(x, y). \tag{7.143}$$

The optimal shape of the initially deformed plate is determined by solving a Dirichlet problem for eq. (7.143) with boundary condition (7.127). We arrive at an analogous result. The optimal design of an initially bent plate supported on an elastic foundation is reduced to finding the values of the adjoint variable by solving problem (7.133) and (7.142), and the subsequent determination of the shape function $f(x, y)$ from (7.127) and (7.143).

Now, let us suppose that a rectangular, flat plate ($0 \leq x \leq a$, $0 \leq y \leq b$) is subjected to a concentrated load applied at the point (ξ, η), that is, $q = P\delta(x - \xi, y - \eta)$. This plate has a constant thickness with cylindrical modulus of rigidity D and is supported freely along the boundary curve Γ. The solution of the boundary-value problem (7.133) and (7.142) is of the form

$$v = \sum_{m,n=1}^{\infty} v_{mn} \sin \frac{m\pi x}{a} \sin \frac{n\pi y}{b}, \tag{7.144}$$

$$\left.\begin{aligned} v_{mn} &= -\frac{4P}{ab}\alpha_{mn} \sin \frac{m\pi\xi}{a} \sin \frac{n\pi\eta}{b}, \\ \alpha_{mn} &= \left[D\left(\frac{m^2\pi^2}{a^2} + \frac{n^2\pi^2}{b^2}\right)^2 + c\right]^{-1}. \end{aligned}\right\} \tag{7.145}$$

Using the expression for Green's function in solving the Dirichlet problem (7.133) and (7.142), we derive formulas defining the shape function

$$f = \sum_{m,n=1}^{\infty} f_{mn} \sin\frac{m\pi x}{a} \sin\frac{n\pi y}{b}, \tag{7.146}$$

where

$$\left.\begin{aligned} f_{mn} &= \frac{2}{\pi}\alpha_{mn}\beta_{mn} \sin\frac{m\pi\xi}{a} \sin\frac{n\pi\eta}{b} \cdot \left\{\frac{ab}{2(S-S_0)}\right. \\ &\quad \cdot \left.\sum_{k,l=1}^{\infty} \alpha_{kl}^2 \beta_{kl} \sin^2\left(\frac{k\pi\xi}{a}\right) \sin^2\left(\frac{l\pi\eta}{b}\right)\right\}^{-1/2}, \\ \beta_{mn} &= \left(\frac{m^2}{a^2} + \frac{n^2}{b^2}\right)^{-1}. \end{aligned}\right\}$$

Then the formula for the magnitude of the optimized functional can be written in the form

$$J = P \sum_{m,n=1}^{\infty} w_{mn} \sin\frac{m\pi x}{a} \sin\frac{n\pi y}{b},$$

$$w_{mn} = \alpha_{mn}\left(\frac{4P}{ab} \sin\frac{m\pi\xi}{a} \sin\frac{n\pi\eta}{b} - cf_{mn}\right).$$

Using identical techniques we can solve the optimal design problem for a flat rectangular plate subjected to a distributed load $q(x, y) = q_0$. In that case the optimal solution is given by

$$f = \sum_{m,n=1,3,5,\ldots}^{\infty} f_{mn} \sin\frac{m\pi x}{a} \sin\frac{n\pi y}{b},$$

$$f_{mn} = \frac{2}{\pi mn}\alpha_{mn}\beta_{mn} \sin\frac{m\pi\xi}{a} \sin\frac{n\pi\eta}{b} \cdot \left[\frac{ab}{2(S-S_0)} \sum_{k,l=1,3,5,\ldots}^{\infty} \alpha_{kl}^2\beta_{kl}/(k^2l^2)\right]^{-1/2},$$

$$J = \frac{4}{\pi^2} abq_0 \sum_{m,n=1,3,5,\ldots}^{\infty} w_{mn}/(mn),$$

$$w_{mn} = \alpha_{mn}(16q_0/(\pi^2 mn) - cf_{mn}).$$

The coefficients α_{mn} and β_{mn} can be computed from (7.145) and (7.146).

7.12 Dynamic Problems in Optimization of Beams and Shells with Constraints on Deflections

A broad class of structural problems related to weight minimization with constraints on rigidity has been computed for dynamic loading conditions, and specifically for impact loads. Independently, problems of optimizing structural elements performing forced harmonic vibrations have been studied in great detail. Topics of research in that optimal design of dynamic systems conducted prior to 1972 may be found in the expository article [35]. In recent years the volume of research into the optimization of dynamic systems has increased dramaticaly. The main topics of this research can be arbitrarily grouped together as follows: design of structures subjected to variable (non-static) loads; design of structures undergoing harmonic vibrations; design of structures undergoing free vibrations; and design of nonconservative systems with constraints on stability. Note that some research on the design of structures with dynamic loads have a direct bearing on the results presented in this chapter.

In the earliest publications dealing with the optimization of structures subjected to nonstatic or dynamic interactions one should include the articles [7.20] and [7.21], which considered the minimization of the maximal deflection for a beam with a variable cross-section. The theory of optimization of dynamic elastic systems with distributed parameters has progressed with the application of sensitivity analysis (see [24], [21], [7.4], and [7.33] to [7.37]). The Optimal design of one-dimensional structural elements was discussed in [7.29]. In [7.23], [7.26], [7.43], [7.72], and [7.98], problems of optimization of discrete dynamic systems and of sequential improvements in the value of the quality functional were considered using the technique of parametric optimization. In [1.6] and [1.7], the authors developed a methodology for solving problems of optimal design for dynamic systems based on optimization theory for systems with distributed parameters. Analytic solutions for some one-dimensional systems composed of structural elastic elements performing forced harmonic oscillations were derived in [1.14], [1.28], and [7.28]. To solve the problem of minimizing the maximal deflection for a beam lying on a Winkler–Pasternak foundation and performing forced harmonic oscillations, techniques of mathematical programming [1.1] have also been applied. A solution for a two-dimensional problem of optimization for elastic plates subjected to harmonic loading is given in [1.15].

Various optimal design problems related to weight minimization for dynamic structural systems with constraints assigned to the natural frequencies of the vibrations were examined in numerous research publications. (We mention only the following: [7.1], [7.3], [7.18], [7.22], [7.30], [7.42], [7.44], [7.45], [7.48], [7.55], [7.62]–[7.73], [7.86], [7.87], [7.89] to [7.91], [7.95], [7.96], and [7.102], and also the monographs and expository studies [11], [19], [23], [32] to [34], and [48].)

In designing structures subjected to dynamic interactions, it is of interest to examine the influence of the weight, rigidity, and geometric properties of a structure on its dynamic response. As a measure of the dynamic response, we can use the maximum deflection expected during the entire service of a structure, identifying structural rigidity with the magnitude of this functional. The magnitude of the maximum deflection is a "natural" measure of structural rigidity. Indeed, if this functional is set equal to 0, we see that no deformations or strains exist; the absence of deformations or strains implies that the total strain energy of the structure is also equal to 0. We also remark that the time instant and location of the point of maximum deflection are not initially known. This maximum is not attained at some unique time and does not occur at the same location. Instead, it generally occurs at various time instances and at different locations within the structure. This aspect complicates computations related to optimization of weight and rigidity properties.

In general, for nonstatic loads estimating the influence of design parameters on the magnitude of a functional crucial to the optimization process is tied to the evaluation of the response of the structure on the entire time interval $[0, T]$ of its functioning. Because an explicit dependence of the dynamic response on the design parameters is absent, a "direct" sensitivity analysis leads to very complicated computations involving a large number of feasible variations. The technique discussed in this section is far more effective. This technique is essentially based on the introduction of adjoint variables and of corresponding differential equations used for determining the values of these variables.

Before we proceed to describe solutions of specific problems, we comment that the magnitude of maximal deflection was regarded as the optimized criterion in several articles (see [5] and [24]), while in [7.20] and [7.33] it served as a constraint in optimizing the weight of the structure.

To solve such optimization problems, we shall use results introduced in Section 3.6. Let us consider the problem of optimizing the design of a beam subjected to a nonstatic load $q(x, y)$. The equations of motion and the boundary and initial conditions for a freely supported beam are

$$\left.\begin{aligned}&(\alpha E h^2 u_{xx})_{xx} + \rho h u_{tt} = q(x, t),\\ &u(0, t) = u(l, t) = u_{xx}(0, t) = u_{xx}(l, t) = 0,\\ &u(x, 0) = u_t(x, 0) = 0,\end{aligned}\right\} \tag{7.147}$$

where $u(x, t)$ is the displacement function; $h(x)$ the area of the tranverse cross-section; and ρ and E are, respectively, the material density and Young's modulus for the beam. As usual, the subscripted indices t and x stand for partial differentiation with respect to time and the coordinate measuring the distance along the axis of the undeformed beam ($t \in [0, T]$ and $x \in [0, l]$, where T and l are known constants).

The optimization problem consists in finding a function $h(x)$ such that the total mass of the beam is minimized:

$$J = \int_0^l \rho h \mathrm{d}x = \frac{1}{T} \int_0^T \int_0^l \rho h \mathrm{d}x \mathrm{d}t \to \min_h, \tag{7.148}$$

with a constraint assigned to the maximum deflection

$$J_1 = \max_{x,t} |u(x, t)| = c_1, \tag{7.149}$$

and bounds assigned to the minimum and maximum values of the control variable

$$h_{\min} \leq h(x) \leq h_{\max}. \tag{7.150}$$

In studying the optimization problem (7.147) to (7.150), we replace the local constraint (7.149) by the (global) integral constraint

$$J_p = \left(\frac{1}{lT}\int_0^T\int_0^l |u(x, t)|^p \mathrm{d}x\mathrm{d}t\right)^{1/p} = c_1. \tag{7.151}$$

The adjoint variable $v(x, t)$ is defined as a solution to the boundary-value problem

$$\left.\begin{aligned}(\alpha E h^2 v_{xx})_{xx} + \rho h v_{tt} &= -\frac{1}{lT}\left(\frac{u}{J_p}\right)^{p-1},\\ v(0, t) = v(l, t) = v_{xx}(0, t) &= v_{xx}(l, t) = 0,\\ v(x, T) = v_t(x, T) &= 0.\end{aligned}\right\} \tag{7.152}$$

The first variations of the functionals δJ and δJ_p are given as

$$\left.\begin{aligned}\delta J &= \rho\int_0^l \delta h \mathrm{d}x = \langle e, \delta h\rangle,\\ \delta J_p &= \int_0^l\left[\int_0^T (2\alpha E h u_{xx} v_{xx} - \rho u_t v_t)\mathrm{d}t\right]\delta h \mathrm{d}x = \langle e_1, \delta h\rangle.\end{aligned}\right\} \tag{7.153}$$

The bounds (7.150) are satisfied if we introduce an auxiliary function $\varphi(x)$ (called "Valentine's substitution") such that $h = \frac{1}{2}(h_{\max} + h_{\min}) + \frac{1}{2}(h_{\max} - h_{\min}) \cos \varphi(x)$. In accordance with this transformation, the variation of the design variable δh in (7.153) is computed from the formula

$$\delta h = -\tfrac{1}{2}(h_{\max} - h_{\min}) \sin \varphi \delta\varphi.$$

The solution of the optimization problem (7.147) to (7.151) is constructed numerically using an iterative algorithm for dynamically loaded structures that was introduced in [1.6]. It is a generalization of corresponding algorithms used for solving static optimization problems (see [5]). Without getting into the details of this algorithm, let us discuss some of its properties. It is an iterative algorithm that reduces to sequential solutions of the direct and adjoint boundary-value problems for a given function h, and then makes improvements δh in the control function h. To solve both the direct and adjoint problem for given values of the control function h, we subdivide the time interval $[0, T]$ into subintervals with endpoints $t = t_s$, with $s = 0, 1, 2, \ldots$. Now, making use of the variational formulation of our problem, we determine, after computation of quadratures and the minimum of their values, a solution to the direct problem at time t_s, after determining this solution at time t_{s-1}. The solution of the adjoint problem is determined by going back-

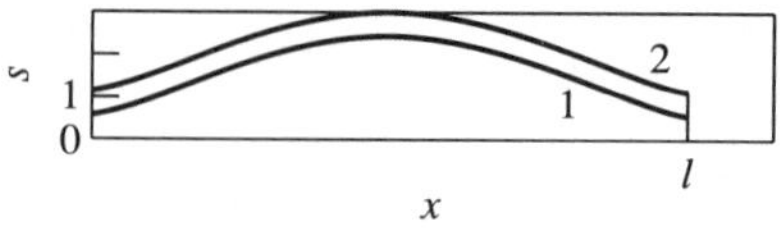

FIGURE 7.15. Optimal beam shapes for different loads.

wards in time. In solving the variational problems, we use convex finite-difference approximations and a gradient technique to accomplish minimization. This improved the economy of computational scheme and assured sufficient accuracy of the computation.

The computation of optimal beams was carried out for different loads and for the following choices of the parameters: $\alpha = 0.3$, $E = 7 \times 10^5$ g/cm^2, $T = 1$ s, $q_0 = 10$ g/cm, $h_{\min} = 0.5$ cm^2, $h_{\max} = 3$ cm^2, $\rho = 1.2$ g/cm^2, $c_1 = 0.23$ cm, and $l = 40$ cm.

The numerical computations for a harmonic load

$$q = q_0 \sin 2\pi t, \qquad 0 \leq t \leq T, \tag{7.154}$$

resulted in values of $h(x)$ illustrated in Figure 7.15, curve 1. The curve marked 2 in Figure 7.15 corresponds to an optimal distribution of area for the applied load:

$$q = \begin{cases} 2q_0 t, & \text{if } 0 \leq t \leq T/4, \\ q_0 T/2, & \text{if } T/4 \leq t \leq T. \end{cases} \tag{7.155}$$

The characteristic property of an optimal solution to loads such as (7.154) and (7.155) is the concentration of load-bearing material in the middle of the beam. The optimal mass is compared with the mass of a beam with constant thickness and having the same rigidity. The relative gain in the total mass obtained by optimizing amounts to 18% and 19.2%, respectively.

We shall consider an axially symmetric shell of variable thickness, having length l and radius R. The x axis coincides with the axis of the shell. A time-dependent radial load $q(x, t)$ is applied to the shell. Let $w(x, t)$ denote the tangential displacement. We write all the basic relations using dimensionless coordinates:

$$x' = x/R, \qquad h' = h/R, \qquad v' = v/R, \qquad w' = w/R, \qquad t' = t/T,$$

$$\rho' = \rho((1 - \nu^2)R^2/ET^2), \qquad q' = q/E, \qquad l' = l/R, \qquad T' = 1.$$

Here ρ, E, and ν are the density, Young's modulus, and Poisson's ratio of the shell's material, respectively, and $[0, T]$ is a given time interval. Equations describing axisymmetric motion of a cylindrical shell may be written as a system of two partial differential equations (with primes omitted);

$$\left.\begin{aligned} [h(v_x + \nu w)]_x - \rho h v_{tt} &= 0, \\ h(w + \nu v_x) + 1/12\,(h^3 w_{xx})_{xx} + \rho h w_{tt} &= q. \end{aligned}\right\} \tag{7.156}$$

The shell is freely supported at both ends. Therefore, the boundary and initial conditions are

$$\left.\begin{aligned} &w(0, t) = w(l, t) = w_{xx}(0, t) = w_{xx}(l, t) = 0, \\ &v(0, l) = v(l, t) = 0, \\ &v(x, 0) = v_t(x, 0) = 0, \qquad w(x, 0) = w_t(x, 0) = 0. \end{aligned}\right\} \tag{7.157}$$

The optimization problem consists in minimizing the volume of the shell's material,

$$J = \int_0^l h(x)dx \to \min_h, \tag{7.158}$$

with the two-sided constraint (7.150) on the unknown design variable and a local constraint on the magnitude of the maximal deflection:

$$\max_{x,t} |w(x, t)| \leq d. \tag{7.159}$$

In solving problem (7.156) to (7.159), it is convenient to use a device described in Section 2.2 and replace the local condition (7.150) by the integral equality

$$J_1 = \int_0^T \int_0^l \tfrac{1}{2}\{|w(x, t) - d| + (w(x, t) - d)\}dxdt = 0. \tag{7.160}$$

To derive effective formulas of sensitivity analysis, which relate variations of relevant functionals to variations of the design variable, we introduce two adjoint variables $r = r(x, t)$ and $s = s(x, t)$. We define these functions as solutions of the system of partial differential equations

$$\left.\begin{aligned} &[h(r_x - vs)]_x - \rho h r_{tt} = 0, \\ &h(s - vr_x) + 1/12(h^3 s_{xx})_{xx} + \rho h s_{tt} = Q, \\ &Q = \tfrac{1}{2}[1 + \operatorname{sign}(|w(x, t)| - d) \operatorname{sign} w(x, t)], \end{aligned}\right\} \tag{7.161}$$

with boundary and initial conditions given by (7.157), with r substituted for v and s for w. Using the technique described previously of introducing adjoint variables and replacing again a local constraint (7.159) by a global one, we represent our sensitivity analysis formulas in the form

$$\left.\begin{aligned} \delta J &= \int_0^l \delta h dx = \langle e, \delta h\rangle, \\ \delta J_1 &= \int_0^l \left[\int_0^T \{-(vw + v_x)r_x + \rho r_t v_t + s(w + vv_x) \right. \\ &\quad \left. + \tfrac{1}{4}h^2 s_{xx} w_{xx} - \rho s_t w_t\}dt\right]\delta h dx = \langle e_1, \delta h\rangle. \end{aligned}\right\} \tag{7.162}$$

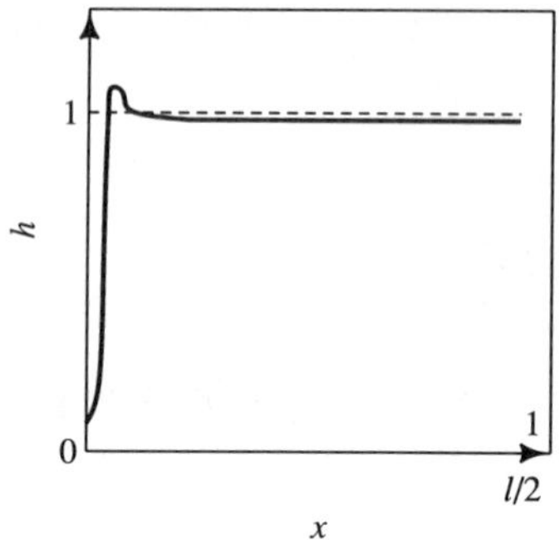

FIGURE 7.16. Optimal thickness of a cylindrical shell.

The two-sided constraints on the design variable h are treated in the same way as before in our discussion of the problem of optimizing dynamically loaded beams. Figure 7.16 illustrates numerically derived results using (7.156), (7.157), and (7.160) to (7.162) and successive optimization algorithm for the thickness function $h(x)$ of a cylindrical shell subjected to the load

$$Q = \begin{cases} q_0 & \text{if } 0 \leq t \leq 0.25 \\ 0 & \text{if } 0.25 \leq t \leq T. \end{cases} \tag{7.163}$$

Noting the symmetry of our problem with respect to the point $x = l/2$, we shall consider in our calculations only the segment $0 \leq x \leq l/2$. In our calculations we take $q_0 = 10^{-7}$, $\rho = 4 \cdot 10^{-3}$, $d = 0.131$, $\nu = 0.3$, $h_{\min} = 10^{-3}$, $h_{\max} = 2 \cdot 10^{-2}$, $l = 4$, and $T = 1$. Comparing the optimal shell, as illustrated in Figure 7.15 with a shell of constant thickness and having the same rigidity in response to a load of the form (7.163), we gain about 3%.

A characteristic property of the thickness distribution, as illustrated in Figure 7.16, is the "edge effect." The thickness is practically constant everywhere except in small regions near the edges.

7.13 Design of Dynamically Loaded Compound Structures

The approach to designing dynamically loaded structural elements pursued in the preceding section used formulas of sensitivity analysis and necessary optimality conditions for systems with distributed parameters. This approach corresponds to continuum modeling of structures (time and space are not discretized). In this model the derivation of the state function and adjoint variables reduces to a problem of solving an initial-boundary-value problem for a system of partial differential equations.

In this section we explain an alternate approach consisting in introducing adjoint variables and then obtaining sensitivity formulas for a discretized description of a structure in terms of the spatial variables (see [1.7]). The state function and the adjoint variables are now functions of time only, and finding their values reduces to the solution of the Cauchy problem for a system of ordinary differential equations. Therefore, in many important practical problems this allows us to compute complex structures whose model is described by a system with a large number of degrees of freedom. This approach also permits us to effectively take care of local constraints on magnitudes of maximal deflections, even in cases when such a maximum is attained at different times and at different spatial locations within the structure.

Let us consider a complex thin-walled structure subjected to dynamic loads during its normal service. To describe the behavior of this structure and its response to external forces we shall use a finite-element computational model. This transition from a continuum representation of a structure to a finite-dimensional model may be justified by a claim that in practical computations and designs of complex mechanical or structural systems, it is a common practice to use discrete models. Let us assume that, after discretization, the state of a structure is characterized by a vector $\mathbf{u}$ of generalized displacements, $\mathbf{u} = \mathbf{u}(t)$, where t denotes time. The response of the structure as a function of time is described by a system of ordinary differential equations

$$A\ddot{\mathbf{u}} + B\dot{\mathbf{u}} + C\mathbf{u} = \mathbf{q}, \tag{7.164}$$

with initial conditions

$$\mathbf{u}(0) = \mathbf{a}, \qquad \dot{\mathbf{u}}(0) = \mathbf{b}, \tag{7.165}$$

where $\mathbf{a}$ and $\mathbf{b}$ are given vectors, A and C are matrices characterizing the inertial and rigidity properties of the system, B is a dissipation matrix, and $\mathbf{q}$ is the vector of external loads applied to the structure. Dots over a variable denote the operation of differentiation with respect to time.

The linearity of the system of differential equations (7.165) is a consequence of assumptions concerning the "smallness" of strains and the linear elastic behavior of the material. The physical interpretation of the vector of generalized displacements and the form of the matrices A, B, and C and the vector $\mathbf{q}(t)$ depend in each specific case on the method of discretization. (For techniques of modeling structures by systems of the form (7.164), see [1.10].)

Let $\mathbf{h}$ denote the vector of design parameters $\mathbf{h} = (h_1, h_2, \ldots, h_n)$, with the mass, rigidity, and geometric properties of the structure chosen as the design parameters. We assume that the elements of the matrices A, B, and C depend smoothly on the parameters h_i, $i = 1, 2, \ldots, n$ (i.e., they are continuously differentiable functions of these parameters).

To estimate rigidity as well as generalized displacements, we need to examine the "real" displacements of certain points P_j within the structure.

Let us introduce a transformation matrix D, transforming the generalized coordinates into "genuine" coordinates

$$\mathbf{U} = D\mathbf{u}, \tag{7.166}$$

where $\mathbf{U}$ is an m-dimensional vector of real displacements of points P_j of the form

$$U^j = (0, 0, \ldots, D^j\mathbf{u}, \ldots, 0), \tag{7.167}$$

where D^j denotes the jth column vector of the transformation matrix D.

Let J^{jk} denote the square of the norm of the displacement of the point P_j at time $t = t_k$ $(0 < t_k < T)$:

$$J^{jk} = (U^j, U^j)_{t=t_k}. \tag{7.168}$$

Here and in the following paragraphs of this section, repeated indices do not imply summation.

We fix the point P^j and the time instant $t = t_k$, and examine the dependence of the quantity J^{jk} on small variations of a parameter h_i. With this in mind, we shall vary the system of equations (7.164) and the initial condition (7.165) without violating (7.166) or (7.167). We compute

$$\delta J^{jk} = 2(D^j\mathbf{u}(t_k))(D^j\delta\mathbf{u}(t_k)), \tag{7.169}$$

$$A\delta\ddot{\mathbf{u}} + B\delta\dot{\mathbf{u}} + C\delta\mathbf{u} + \sum_{i=1}^{n}\left(\frac{\partial A}{\partial h_i}\ddot{\mathbf{u}} + \frac{\partial B}{\partial h_i}\dot{\mathbf{u}} + \frac{\partial C}{\partial h_i}\mathbf{u}\right)\delta h_i = 0, \tag{7.170}$$

$$\delta\mathbf{u}(0) = 0, \qquad \delta\dot{\mathbf{u}}(0) = 0. \tag{7.171}$$

We introduce a vector function of adjoint variables $\mathbf{v} = \mathbf{v}(t)$. We multiply the left-hand side of (7.170) by the adjoint variable and integrate this product between $t = 0$ and $t = t_k$. After integration by parts and substituting conditions (7.171), we obtain an expression for the variation δJ^{jk}:

$$\begin{aligned}\delta J^{jk} = \int_0^{t_k} &\left\{\delta\mathbf{u}[A^*\ddot{\mathbf{v}} - B^*\dot{\mathbf{v}} + C^*\mathbf{v}]\right.\\ &\left.+ \mathbf{v}\sum_{i=1}^{n}\left(\frac{\partial A}{\partial h_i}\ddot{\mathbf{u}} + \frac{\partial B}{\partial h_i}\dot{\mathbf{u}} + \frac{\partial C}{\partial h_i}\mathbf{u}\right)\delta h_i\right\}dt\\ &+ (\mathbf{v}, A\delta\dot{\mathbf{u}})_{t=t_k} + [(\mathbf{v}B - \dot{\mathbf{v}}A + 2(D^j\mathbf{u})D^j)\delta\mathbf{u}]_{t=t_k}.\end{aligned} \tag{7.172}$$

(The asterisk denotes the transpose.) The adjoint variable is defined as a solution of the Cauchy problem for a system of ordinary differential equations

$$A^*\ddot{\mathbf{v}} - B^*\dot{\mathbf{v}} + C\mathbf{v} = 0, \tag{7.173}$$

with initial conditions

$$\mathbf{v}(t_k) = 0, \qquad A^*\dot{\mathbf{v}}(t_k) = 2(D^j\mathbf{u}(t_k))D^j. \tag{7.174}$$

In solving the Cauchy problem (7.173) and (7.174) for the adjoint variables, the integration of equations is carried out "backward" from the time instant t_k to $t = 0$. For fixed values of the parameters h_i, the relation between problem (7.173) and (7.174) for $\mathbf{v}$ and the Cauchy problem (7.164) and (7.165), which governs the dynamic behavior of our structure, is explicitly given by the second equation in (7.174).

Solving for the adjoint variable satisfying (7.173) and (7.174) permits us to eliminate from (7.172) the expressions containing the variation of the state function and to obtain a direct dependence of δJ^{jk} on δh_i $(i = 1, 2, \ldots, n)$:

$$\delta J^{jk} = \sum_{i=1}^{n} \left\{ \int_0^{t_k} \mathbf{v} \left(\frac{\partial A}{\partial h_i} \ddot{\mathbf{u}} + \frac{\partial B}{\partial h_i} \dot{\mathbf{u}} + \frac{\partial C}{\partial h_i} \mathbf{u} \right) \mathrm{d}t \right\} \delta h_i. \tag{7.175}$$

Equation (7.175) permits us to derive formulas for improvement in the design by varying the design parameters, assuring a monotone decrease in the value of the optimized functional. However, (7.175) has been obtained by assuming that the value of the optimized functional is given at any instant of time t_k and that it is the magnitude of the displacement at a fixed point. If the maximal deflection within the structure occurs at some point P_j at the time instant t_k, then (7.175) allows us to construct changes in the design that improve the magnitude of that maximal deflection.

Let us now consider an optimization problem consisting in finding the values of the design parameters h_i $(i = 1, 2, \ldots, n)$ such that the minimum is attained for the cost functional which is given by the square of the maximum deflection

$$J = \max_{j,k} (J^{jk}) \to \min_{\mathbf{h}}, \tag{7.176}$$

with the constraints $f_l(\mathbf{h}) \leq c_l$ $(l = 1, 2, \ldots, m)$, where f_l are given functions and c_l are known constants. Using slack variables μ_l^2, we reduce these inequality constraints to the equalities

$$f_l(\mathbf{h}) + \mu_l^2 - c_l = 0. \tag{7.177}$$

The problem of conditional optimization defined by (7.176) and (7.177) can be solved by iterative techniques. First we seek a solution to the systems of equations (7.176) on the interval $[0, T]$ for given values of the parameters h_i. The actual values of the displacements of certain points within the structure and their squares are computed from (7.166) and (7.168), respectively. First we choose the largest value from the set of values of J^{jk}. Now we fix the index j and the time instant t_k for which this maximum was attained. Then we integrate the system of equations (7.173) for the adjoint variables with initial conditions (7.174). We can compute δh_i from formulas for improvements in design by varying the design parameters, and utilizing some version of the gradient method, based on (7.175) and (7.177) (see [5]). Having found the new values of the design parameters h_i, we can perform the next iteration. This iteration is terminated if the error of the necessary optimality conditions and constraints is judged to be sufficiently small.

In this process of varying the design the values of the maximal deflection may be attained at different points within the structure and at various times. Therefore, the sensitivity formulas derived above should be generalized to take care of the multiple maxima of the rigidity functional. In this case $J = J^{\alpha\beta}$ represents the maximal values of the deflections at times t_β and at specific locations $P^{\alpha\beta}$. For each fixed value of β, the index α assumes values that enumerate points having maximal deflection $\alpha = 1, 2, \ldots, r_\beta$. The total number of points attaining maximal deflection at different instants of time is equal to $\sum r_\beta = r$ (with $\beta = 1, 2, \ldots, s$). Note that with our assumptions the rigidity functional does not have to be differentiable. Therefore, in solving this type of optimization problem, during the process of making design improvements we can only use directional derivatives of our functional and must generalize relations that define the adjoint variables.

Here we consider r adjoint variable vector functions $\mathbf{v}^{\alpha\beta}$. To find their values we formulate the Cauchy problem

$$A^*\ddot{\mathbf{v}}^{\alpha\beta} - B^*\dot{\mathbf{v}}^{\alpha\beta} + C\mathbf{v}^{\alpha\beta} = 0, \tag{7.178}$$

$$\mathbf{v}^{\alpha\beta}(t_\beta) = 0, \qquad A^*\dot{\mathbf{v}}^{\alpha\beta}(t_\beta) = 2(D^{\alpha\beta}u(t_\beta))D^{\alpha\beta}. \tag{7.179}$$

In this case the variation of the rigidity functional is given by the formula:

$$\delta J = \sum_{i=1}^{n} \left\{ \max_{\alpha,\beta} \int_0^{t_\beta} \mathbf{v}^{\alpha\beta} \left(\frac{\partial A}{\partial h_i} \ddot{\mathbf{u}} + \frac{\partial B}{\partial h_i} \dot{\mathbf{u}} + \frac{\partial C}{\partial h_i} \mathbf{u} \right) \mathrm{d}t \right\} \delta h_i. \tag{7.180}$$

The presence of multiple maxima of the deflection function makes the process of solving an optimization problem far more complex. Thus, after identifying the set of points $P^{\alpha\beta}$ at which a maximum is attained, it is necessary to solve r problems posed by (7.178) and (7.179) to determine the values of the adjoint variables. The formula for the variation of the cost functional is given by the expression

$$\delta J = \sum_{i=1}^{n} \left\{ \sum_{\alpha,\beta} \lambda^{\alpha\beta} \int_0^{t_\beta} \mathbf{v}^{\alpha\beta} \left(\frac{\partial A}{\partial h_i} \ddot{\mathbf{u}} + \frac{\partial B}{\partial h_i} \dot{\mathbf{u}} + \frac{\partial C}{\partial h_i} \mathbf{u} \right) \mathrm{d}t \right\} \delta h_i, \tag{7.181}$$

where $\lambda^{\alpha\beta}$ are real numbers satisfying the conditions

$$\lambda^{\alpha\beta} \geq 0, \qquad \sum_{\alpha,\beta} \lambda^{\alpha\beta} = 1. \tag{7.182}$$

Recall that the maximum of the square of a deflection in the structure may be attained over an entire time interval $[t_1, t_2]$ (where $0 < t_1 < t_2 \leq T$). In this case, to compute the variations of the cost functional, it is necessary to solve the adjoint problem (7.173). The solution is a function of $t_\beta \in [t_1, t_2]$, which may be regarded as a continuously varying parameter. Therefore, we need to determine the solution of the adjoint problem as a function of two variables $\mathbf{v} = \mathbf{v}(t, t_\beta)$. In the formula (7.180) for the variation of the cost functional, the maximum should be taken with respect to t_β. However, in carrying out the computations and using (7.181) and (7.182), we usually discretize the time variable.

As an example, we shall consider a solution to the problem of optimizing the rigidity of a thin-walled structure, with an ideal gas flowing around it, and impact-type dynamic loads acting on it. A trapezoidally shaped plate is discretized into a mesh of finite elements of piecewise constant thickness. Discretized masses are assigned to the plate elements. The response of the discretized structure to the flow of gas is described by a system of ordinary differential equations

$$\left.\begin{aligned}&A\ddot{\mathbf{u}} + (B + B_\alpha)\dot{\mathbf{u}} + (C + C_\alpha)\mathbf{u} = \mathbf{q},\\&\mathbf{u}(0) = 0, \qquad \dot{\mathbf{u}}(0) = 0,\end{aligned}\right\} \tag{7.183}$$

where A is the inertia matrix, B_α and C_α are aerodynamic matrices, B is a damping matrix, and C is the rigidity matrix for the structure. To speed up the computations, we transform the system (7.183) into canonical form. This is usually done by considering a free vibration problem $(C - \omega^2 A)w = 0$, where ω is the natural frequency of the vibrations and w is the corresponding eigenvector. By using the transformation $\mathbf{u} = W\mathbf{u}'$ (where W is the matrix of eigenvectors), the system (7.183) is reduced to the form

$$\left.\begin{aligned}&A'\ddot{\mathbf{u}}' + B'\dot{\mathbf{u}}' + C'\mathbf{u}' = \mathbf{q}',\\&A' = W^*AW, \qquad B' = W^*(B + B_\alpha)W, \qquad C' = W^*(C + C_\alpha)W.\end{aligned}\right\} \tag{7.184}$$

The matrices A' and W^*CW in (7.184) are diagonal. We choose the magnitudes of the discrete masses of structural elements as our design parameters, and assign the following constraints

$$\sum_{i=1}^{n} h_i = M, \qquad a_i \le h_i \le b_i, \qquad i = 1, 2, \ldots, n. \tag{7.185}$$

The first constraint implies that the total weight M of the structure remains invariant. The second constraint assigns bounds to the admissible values of the design parameters.

Computations are carried out for four separate cases of dynamic loading applied to the plate. Graphs of loads plotted as functions of time are displayed in Figure 7.17a to d. Figures 7.18 to 7.22 display isoclinic lines representing graphs of dimensionless gradients for two types of design variables, which are discrete mass distribution (a) and plate thickness (b). Figures 7.18 to 7.22 correspond to the first three load variants (i.e., to Figure 7.17a to c). The graphs representing the distribution of gradients of the cost functional (with discrete masses) for given loads are quantitatively similar. A specific property of these gradient functions is the presence of negative gradients with a rather large area in the central part of the plate and in the lower right-hand corner. In Figure 7.20a the regions of negative gradients merge into a single region. In all three cases of minimizing the maximal deflection for a structure, the minimum is attained by reducing the mass in the middle and in the lower right-hand corner of the plate. The mass is increased during the minimization process in the upper left corner. If we choose as design variables the (discrete) thicknesses of the plate segments, the region of negative gradients for the functional, which is absent in Figure 17b, reappears in the upper left-hand corner for the second loading variant (as shown in Figure 7.17b) and becomes larger in Figure 7.20b. The region of maximal positive gradient is displaced, correspondingly, from the lower right-hand corner (see Figure 7.18b) into the central part of the plate (Figures 7.19 and 7.20b).

Figures 7.21 and 7.22 relate to the last case of a sign-changing load (see Figure 7.17d). For this type of load, the maximal deflection occurs at two

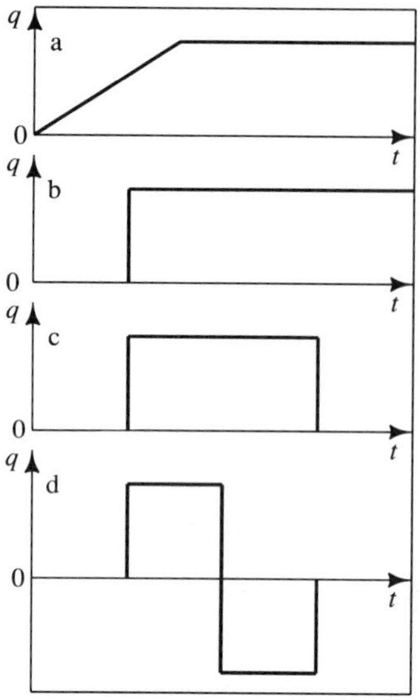

FIGURE 7.17. Types of time-dependent loads.

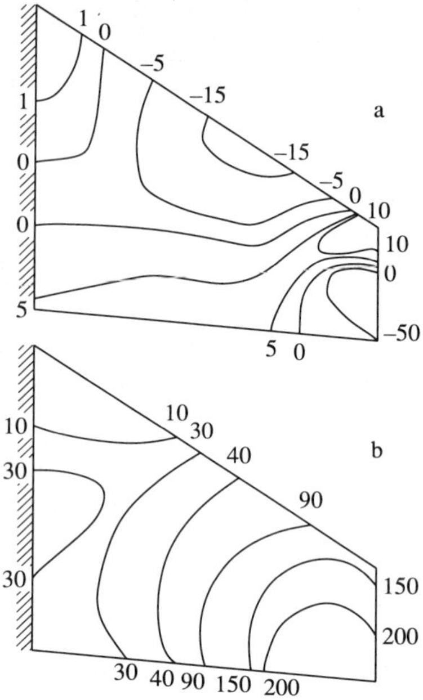

FIGURE 7.18. Isoclinic gradient lines for (a) mass distribution and (b) plate thickness.

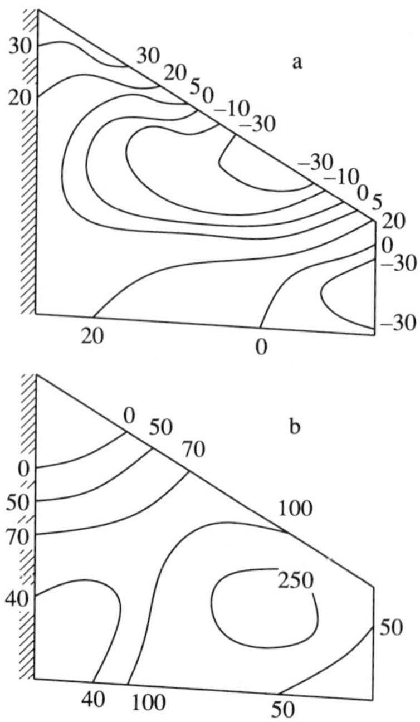

FIGURE 7.19. Corresponding gradients of the cost functional.

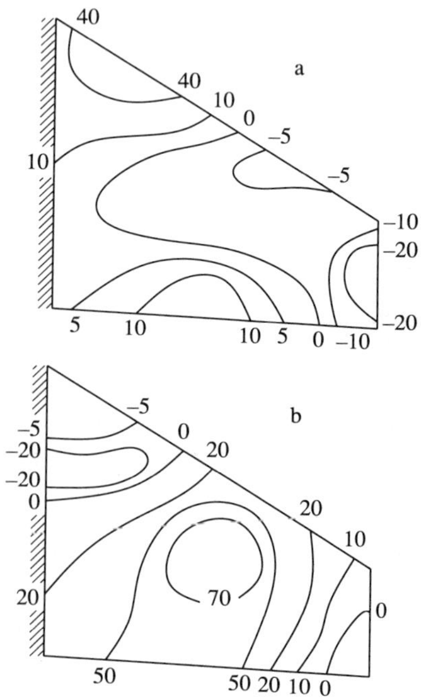

FIGURE 7.20. Corresponding gradients of the cost functional.

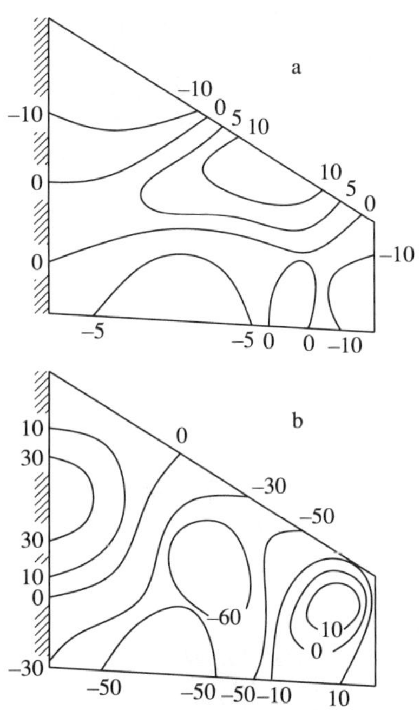

FIGURE 7.21. Gradients for load in Figure 7.17d corresponding to different time and space maxima.

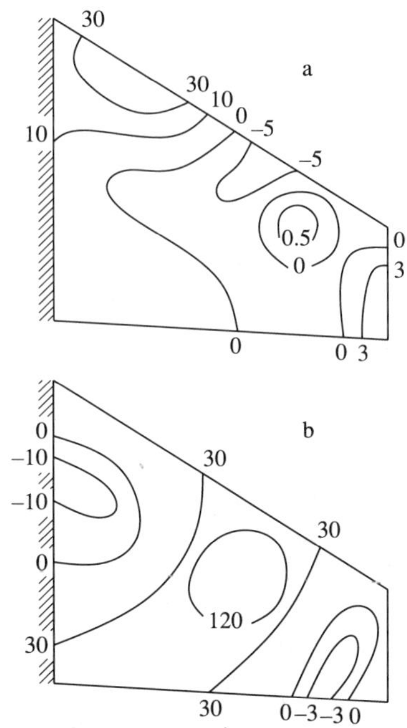

FIGURE 7.22. Gradients for load in Figure 7.17d corresponding to different time and space maxima.

different locations within the structure and at different instants of time. To each maximum there corresponds a different distribution of gradients as shown in Figures 7.21a and b and 7.22a and b. We should mention that for each load, essentially different gradient distributions are shown, and there is a relation between regions with positive and negative gradients, respectively in Figures 7.21 and 7.22. A common characteristic of the designs revealed by such computations is the increase in the values of the natural frequencies accompanying a decrease of the maximal deflection in the dynamic response of a plate. This technique allows us to decrease effectively the value of the structure's cost functional. In the example illustrated above, the cost functional was decreased by about 60% to 70% after only 20 to 30 iterations.

References for Chapter 7

7.1. Adamovich, I.S., and Ricards, R.B., Weight optimization of an othotropic cylindrical plate with variable properties and constraints on the natural frequencies of vibrations, *Izv. Akad. Nauk S.S.S.R., MTT,* 1977, No. 2, pp. 120–125.

7.2. Annin, B.D., Optimal design of elastic, anisotropic inhomogenious bodies, *Third National Congress on Theoretical and Applied Mechanics*, Varna, Bulgaria, 1977, pp. 275–280.

7.3. Armand, J.L., Minimum mass design for a plate-like structure for a specified fundamental frequency, *AIAA J.*, 1971, *9*, No. 9, pp. 1739–1745.

7.4. Arora, J.S., and Haug, E.J., Optimum structural design with dynamic constraints, J. Structural Division, *Proc. Amer. Soc. Civil Engr.*, 1977, *103*, NST 10, pp. 2071–2074.

7.5. Artiunian, N.H., and Abramian, B.L., *Torsion of Elastic Bodies*, Moscow, Fizmatgiz, 1963, 686 pp.

7.6. Banichuk, N.V., Determination of the optimal shape for elastic curved beams, *Izv. Akad. Nauk SSSR, MTT*, 1975, No. 6, pp. 124–133.

7.7. Banichuk, N.V., On a game-theoretic problem in optimization of elastic bodies, *Dokl. Akad. Nauk SSSR*, 1976, *226*, No. 3, pp. 497–499.

7.8. Banichuk, N.V., On a two-dimensional optimization problem in the theory of torsion of elastic beams, *Izv. Akad. Nauk SSSR, MTT*, 1976, No. 5, pp. 45–52.

7.9. Banichuk, N.V., On optimal anisotropy for torsioned shafts, *Izv. Akad. Nauk SSSR, MTT*, 1978, No. 4, pp. 73–79.

7.10. Banichuk, N.V., On an extremization problem for a system with distributed parameters, and on determination of the optimal properties of an elastic medium, *Dokl. Akad. Nauk SSSR*, 1978, *242*, No. 5, pp. 1042–1045.

7.11. Banichuk, N.V., Optimization problems for elastic anisotropic bodies, *Archiwum Mechaniki Stosowanej*, 1981, *33*, No. 3, pp. 347–363.

7.12. Banichuk, N.V., and Kartvelishvili, V.M., Design of plates accounting for effects of constraints on changes in plates, *Izv. Akad. Nauk SSSR, MTT*, 1983, No. 6, pp. 130–136.

7.13. Banichuk, N.V., Kartvelishvili, V.M., and Litvinov, V.G., On optimal solutions for problems in design of elastic plates, *Prikl. Mekhan.*, 1984, *20*, No. 11, pp. 82–85.

7.14. Banichuk, N.V., and Larichev, A.D., Optimization of initially bent plates. In: *Proceedings of the XII-th All-Union Conference on Plates and Shells*, Erevan, Erevan University, 1980, Vol. 1, pp. 126–130.

7.15. Banichuk, N.V., and Larichev, A.D., Optimization problem for reinforced composite beams subjected to compression and bending, *Stroit. Mechanika i Rashchet Sooruzhenii*, 1981, No. 6, pp. 9–12.

7.16. Banichuk, N.V., and Larichev, A.D., Maximization of rigidity for torsioned elastic shafts made of composite materials, *Izv. Akad. Nauk Arm. SSR, Mekhan.*, 1983, *36*, No. 6, pp. 31–38.

7.17. Banichuk, N.V., and Larichev, A.D., Optimal design problems for curvilinear shallow elements of structures, *Optim. Control Applied Methods*, 1984, *5*, No. 3, pp. 197–205.

7.18. Banichuk, N.V., and Mironov, A.A., Optimization of the frequency of vibrations for an elastic plate in an ideal liquid, *Prikl. Mat. Mekh.*, 1975, *39*, No. 5, pp. 889–899.

7.19. Bliss, G.A., Lectures on Calculus of Variations, University of Chicago Press, Chicago, 1946 (reprinted by Dover Co).

7.20. Brach, R.M., Minimum dynamic response for a class of simply supported beam shapes, *Intern. J. Mech. Sci.*, 1968, *10*, pp. 429–439.

7.21. Brach, R.M., Optimum design of beams for sudden loading, J. Eng. Mech. Division, *Proc. Amer. Soc. Civ. Engr.*, 1968, *96*, N-EM 6, pp. 1395–1407.

7.22. Bratus, A.S., Design of circular cylindrical shells of minimum weight with a fixed frequency of natural vibrations, *Prikl. Mat. Mekh.* 1983, *47*, No. 5, pp. 805–814.

7.23. Cassis, J.H., and Schmit, L.A., Optimum design with dynamic constraints, J. Structural Division, *Proc. Amer. Soc. Civ. Engr.*, 1976, *102*, N-ST 10, pp. 2053–2071.

7.24. Chern, J.M., and Prager, W., Optimal design of rotating disc for given radial displacement of edge, *J. Optimization Theory Appl.*, 1970, *6*, No. 2, pp. 161–170.

7.25. Cinquini, C., Optimal elastic design for prescribed maximum deflection, *J. Structural Mechanics*, 1979, 7, No. 1, pp. 21–34.

7.26. Fox, R.L., an Kapoor, M.P., Structural optimization in the dynamic response regime-A computational approach, *AIAA J.*, 1970, *8*, No. 10, pp. 1798–1804.

7.27. Gol'dshtein, Iu.B., and Entov, V.M., *Some Qualitative Techniques in Failure Mechanics*, Moscow, Institute for Problems in Mechanics, 1976, 53 pp.

7.28. Gol'dshtein, Iu.B., and Solomeshch, M.A., On optimal design of beams with dynamic loads, *Stroit. Mekhanika i Rashchet Sooruzhenii*, 1968, No. 4, pp. 27–30.

7.29. Grinev, V.B., and Vasil'chenko, V.F., Optimization of beams for nonstationary loads, *Problemy Machinostroienia*, 1981, *14*, pp. 9–18.

7.30. Grinev, V.B., and Filippov, A.P., Optimal design of structures with given natural frequencies, *Prikl. Mekhanika*, 1971, 7, No. 10, pp. 19–25.

7.31. Gura, N.M., Shell of revolution with maximal rigidity, subjected to torsion, *Izv. Akad. Nauk SSSR, MTT*, 1971, No. 1, pp. 138–144.

7.32. Gyunter, N.M., *Course in Variational Calculus*, Moscow and Leningrad, OGIZ, GITL, 1942, 308 pp.

7.33. Haug, E.J., and Arora, J.S., Distributed parameter structural optimization for dynamic response. In: *Optimization of Distributed Parameter Structures*, E.J. Haug and J. Cea (eds.), Alphen aan den Rijn, Sijthoff and Noordhoff, 1981, pp. 474–515.

7.34. Haug, E.J., Arora, J.S., and Feng, T.T., Sensitivity analysis and optimization of structures for dynamic response, *ASME Journal Mech. Des.* 1978, *100*, pp. 311–318.

7.35. Haug, E.J., and Feng, T.T., Optimization of distributed parameter structures under dynamic loads, *Control and Dynamic Systems*, 1977, *13*, pp. 207–246.

7.36. Haug, E.J., and Feng, T.T., Optimal design of dynamically loaded continuous structures, *Intern. J. Numer. Meth. Eng.*, 1978, *12*, pp. 299–307.

7.37. Hegemeier, G.A., and Tang, H.T., A variational principle, the finite element method and optimal structural design for a given deflection. In: *Optimization in Structural Design*. A. Sawczuk and Z. Mróz (eds.). Springer Verlag, Berlin, 1975, p. 464–483.

7.38. Huang, N.C., Optimal design of elastic structures for maximum stiffness, *Intern. J. Solids Struct.* 1968, *4*, pp. 689–700.

7.39. Huang, N.C., Optimal design of elastic beams for minimum-maximum deflection, *J. Appl. Mech.*, 1971, *38*, No. 4, pp. 1078–1081.

7.40. Huang, N.C., and Sheu, C.Y., Optimal design of elastic circular sandwich beams for minimum complience, *J. Appl. Mech.*, 1970, *37*, p. 569.

7.41. Huang, N.C., and Tang, H.T., Minimum weight design of eastic sandwich beams with deflection constraints, *J. Optimization Theory Appl.*, 1969, *4*, No. 4, pp. 277–298.

7.42. Hutchinson, J.W., and Niordson, F.I., Designing vibrating membranes. In: *Mechanics of Continuous Media and Related Problems of Analysis*, Moscow, 1972, pp. 581–590.

7.43. Jamakawa, H., Optimum structural designs for dynamic response. In: *New Directions in Optimum Structural Design*, E. Atrek et al. (eds.), New York, Wiley and Sons, 1984, pp. 249–268.
7.44. Kamat, M.P., Effect of shear deformation and rotary inertia on optimum beam frequencies, *Intern. J. Numer. Meth. Eng.*, 1975, *9*, pp. 51–62.
7.45. Karihaloo, B.L., and Niordson, F.I., Optimum design of vibrating cantilevers, *J. Optimization Theory Appl.*, 1973, *11*, No. 6, pp. 638–654.
7.46. Kartvelishvili, V.M., and Kobelev, V.V., Rational schemes for reinforcing layered plates made of composite materials, *Prikl. Mat. Mekh.*, 1984, No. 1, pp. 68–80.
7.47. Kartvelishvili, V.M., Mironov, A.A., and Samsonov, A.M., A numerical technique for solving optimization problems for reinforced structures, *Izv. Akad. Nauk SSSR, MTT*, 1981, No. 2, pp. 93–103.
7.48. Krein, M.G., On some problems of maxima and minima for eigenvalues, and Liapunov zones of stability, *Prikl. Mat. Mekh.*, 1951, *15*, No. 3, pp. 323–348.
7.49. Kurshin, L.M., and Rastorguyev, G.I., On the optimal shape of cross-section of a torsioned shaft, *Izv. Akad. Nauk Arm. SSR, Mekh.*, 1979, *22*, No. 6, pp. 17–19.
7.50. Kurshin, L.M., On the problem of determining the cross-section of a shaft having maximal torsional stiffness, *Dokl. Akad. Nauk SSSR*, 1975, *223*, No. 3, pp. 585–588.
7.51. Kurshin, L.M., and Onoprienko, P.N., Determination of the shape of a doubly connected cross-section for a shaft with maximal torsional stiffness, *Prikl. Mat. Mekh.*, 1976, *40*, No. 6, pp. 1078–1084.
7.52. Lepik, U., Optimal design of beams with minimum complience, *Intern. J. Nonlinear Mechanics*, 1978, *13*, pp. 33–42.
7.53. Lipin, E.K., ad Groshev, G.P., Design of structures with minimal volume with constraints on generalized rigidity and minimum thickness, *Ucheb. Zapiski TsAGI*, 1979, *10*, No. 2, pp. 143–148.
7.54. Litvinov, B.G., Bending problem for plates with variable thickness, *Prikl. Mekhan.*, 1975, *11*, No. 5, pp. 54–61.
7.55. Litvinov, B.G., Optimal eigenvalue problems, *Ukr. Mat. Zh.*, 1981, *33*, No. 5, pp. 610–614.
7.56. Litvinov, B.G., Optimal control of coefficients in elliptic systems, *Differen. Uravn.*, 1982, *18*, No. 6, pp. 1036–1047.
7.57. Lur'ie, A.I., *Theory of Elasticity*, Moscow, Nauka, 1970, 940 pp.
7.58. Lur'ie, K.A., and Cherkaev, A.V., On an application of Prager's theorem to optimal design of thin plates, *Izv. Akad. Nauk SSSR, MTT*, 1976, No. 6, pp. 157–159.
7.59. Lavrov, N.A., Lur'ie, K.A., and Cherkaev, A.V., Inhomogeneous shaft with maximal torsional rigidity, *Izv. Akad. Nauk SSSR, MTT*, 1980, No. 6, pp. 82–92.
7.60. Marchuk, G.I., *Computational Techniques of Mathematics*, Moscow, Nauka, 1980, 535 pp.
7.61. Martin, J.B., The optimal design of beams and frames with complience, *Int. J. Solids Struct.*, 1971, *7*, pp. 63–81.
7.62. McCart, B.R., Haug, E.J., and Streeter, T.D., Optimal design of structure with constraints on natural frequency, *AIAA J.*, 1970, *8*, No. 6, pp. 1012–1019.
7.63. Miele, A., Mangiavacchi, A., Mohanty, V.P., and Wu. A.K., Numerical determination of minimum mass structures with specified natural frequencies, *Intern. J. Numer. Meth. Eng.*, 1978, *13*, No. 2, pp. 203–223.

7.64. Niordson, F.I., On the optimal design of a vibrating beam, *Quart. Appl. Math.* 1965, *23*, No. 1, pp. 47–53.
7.65. Niordson, F.I., Some results concerning optimal design of vibrating plates. In: *Optimization Methods in Structural Design, H. Eschenauer and N. Olhoff* (eds.), Mannheim Bibl. Inst. 1983, pp. 380–386.
7.66. Niordson, F.I., Optimal design of elastic plates with constraint on the slope of thickness function, *Intern. J. Solids Struct.*, 1983, *19*, No. 2, pp. 141–151.
7.67. Olhoff, N., Optimal design of vibrating circular plates, *Int. J. Solids Struct.*, 1970, *6*, pp. 139–156.
7.68. Olhoff, N., Optimal design of vibrating rectangular plates, *Int. J. Solids Struct.*, 1974, *10*, No. 1, pp. 93–109.
7.69. Olhoff, N., Optimization of vibrating beams with respect to higher order natural frequencies, *J. Struct. Mech.*, 1976, *4*, pp. 87–122.
7.70. Olhoff, N., Maximizing higher order eigenfrequencies of beam with constraints on the design geometry, *J. Struct. Mech.*, 1977, *5*, pp. 107–134.
7.71. Olhoff, N., Optimization of transversely vibrating beams and rotating shafts. In: Optimization of distributed parameter structures, E.J. Haug and J. Cea (eds.), Alphen aan den Rijn, Sijthof and Noordhoff, 1981, pp. 177–199.
7.72. Pister, K.S., Optimal design of structures under dynamic loading. In: Optimization of distributed parameter structures, E.J. Haug and J. Cea (eds.), Alphen aan den Rijn, Sijthof and Noordhoff, 1981, pp. 569–585.
7.73. Pierson, B.L., An optimal control approach to minimum weight vibrating beam design, *J. Struct. Mech.*, 1977, *5*, pp. 147–148.
7.74. Plaut, R.H., Optimal structural design for given deflection under periodic loading, *Quart. Appl. Math.*, 1971, *29*, pp. 315–318.
7.75. Plaut, R.H., On minimizing the response of structures to dynamic loading, *ZAMP*, 1970, *21*, pp. 1004–1010.
7.76. Polya, G., Torsional rigidity, principal frequency, electrostatic capacity and symmetrization, *Quart. Appl. Math.*, 1948, *6*, No. 3.
7.77. Polya, G., Legt die Stelle der grössten Beanspruchung der Oberfläche? *Zeitschr. angew. Math. Mech.*,1930, *10*, No. 4, pp. 353–360.
7.78. Prager, W., Optimal design of statically determinate beams for given deflection, *Intern. J. Mech. Sci.*, 1971, *13*, p. 893.
7.79. Prager, W., Optimal thermoelastic design for given deflection, *Intern. J. Mech. Sci.*, 1970, *12*, pp. 705–509.
7.80. Rao, S.S., Structural optimization under shock and vibration environment, *Shock and Vibration Digest*, 1979, *11*, No. 2, pp. 3–12.
7.81. Reiss, R., Optimal compliance criterion for axisymmetric solid plates, *Intern. J. Solids Struct.*, 1976, *12*, pp. 319–329.
7.82. Samsonov, A.M., Necessary conditions for an optimal distribution of stiffness of a rib in an elastic plate, *Izv. Akad. Nauk SSSR, MTT*, 1980, No. 1, pp. 136–144.
7.83. Samsonov, A.M., Optimal location for a thin elastic rib in an elastic plate, *Izv. Akad. Nauk SSSR, MTT*, 1978, No. 1, pp. 132–138.
7.84. Samsonov, A.M., Optimal design of rib-reinforced plates. In: *Optimization Methods in Structural Design*. H. Eschenauer and N. Olhoff (eds.), Mannheim Bibl. Inst., 1983, pp. 387–393.
7.85. Seyranian, A.P., Optimal design of beams with constraints on deflections, *Izv. Akad. Nauk Arm. SSR, Mekh.*, 1975, No. 6, pp. 24–33.

7.86. Seyranian, A.P., Study of an extremum in an optimality problem concerning vibrations of a circular plate, *Izv. Akad. Nauk SSSR*, No. 6, pp. 113–118.

7.87. Sheu, C.Y., Elastic minimum weight design for specified fundamental frequency, *Intern. J. Solids Struct.*, 1968, *4*, No. 10, pp. 953–958.

7.88. Shield, R.T., and Prager, W., Optimal structural design for given deflection, *J. Appl. Math. Phys.*, 1970, *21*, pp. 513–523.

7.89. Taylor, J.E., Minimum mass bar for axial vibration at specified natural frequency, *AIAA J.*, 1967, *5*, No. 10, pp. 1911–1913.

7.90. Taylor, J.E., Optimal design of a vibrating bar with specified minimal cross-section, *AIAA J.*, 1968, *6*, No. 7, pp. 1379–1381.

7.91. Thermann, K., Optimal design criteria of dynamically loaded elastic structures. In: *Optimization in Structural Design*, A. Sawczuk and Z. Mróz (eds.), 1975, Springer Verlag, New York, pp. 152–167.

7.92. Timoshenko, S.P., and Goodier, G., *Theory of Elasticity*, McGraw-Hill, New York, 1956.

7.93. Timoshenko, S.P., *Computation of Elastic Shells* (in Russian), Gostekhizdat, 1933, 124 pp.

7.94. Timoshenko, S.P., *Strength and Vibrations of Structural Elements*, (in Russian), Moscow, Nauka, 1975, 704 pp.

7.95. Troitskii, V.A., Optimization of elastic beams for free vibrations, *Izv. Akad. Nauk SSSR, MTT*, 1976, No. 3, pp. 145–152.

7.96. Turner, M.J., Design of minimum mass structure with specified natural frequencies, *AIAA J.*, 1966, *5*, No. 3, pp. 406–412.

7.97. Ukraintsev, G.V., and Frolov, V.M., Technique for optimizing the stiffness of the load-bearing wing structure, with varying the distribution of relative thickness and the shape, *Uch. Zap. TsAGI*, 1972, *3*, No. 4.

7.98. Venkayya, V.B., and Khot, N.S., Design of optimum structures to impulse type loading, *AIAA J.*, 1975, *13*, No. 8, pp. 989–994.

7.99. Wasiutyński, Z., On the congruency of forming according to the minimum potential energy with that according to the equal strength, *Bull. Acad. Pol. Sci., Series Sci. Techn.* 1960, *8*, No. 6, pp. 259–268.

7.100. Wasiutyński, Z., On the criterion of minimum deformability design of elastic structures; effect of own material, *Bull. Acad. Pol. Sci., Series Sci. Techn.* 1966, *14*, No. 9, pp. 875–878.

7.101. Wasiutyński, Z., On the equivalence of design principles: constant volume and minimum volume-constant potential, *Bull. Acad. Pol. Sci., Series Sci. Techn.* 1966, *14*, No. 9, pp. 883–885.

7.102. Weisshaar, T.A., Optimization of simple structures with higher mode frequency constraints, *AIAA J.*, 1972, *10*, pp. 691–693.

8
Application of the Stability and Weight Criteria

8.1 Criteria of Elastic Stability

Stability constitutes one of the basic demands that must be satisfied in designing elastic structures. It is particularly important in designs of slender structures or structures made of high-strength materials.

The concept of elastic stability is closely related to the response of a structure to applied forces. A study of the change in geometry as a function of time is the primary purpose of any investigation of structural stability. Therefore, a dynamic approach to problems of stability is most general. The idea of a static approach in studies of problems of elastic stability goes back to Euler (see [3.8]). We shall offer a brief outline of some approaches and of criteria that may be applied to evaluate the elastic stability of structural elements. (A more detailed exposition may be found in [8.4], [8.5], [8.9], [8.19] to [8.21], [8.23], [8.29], [8.38], [8.64], [8.70] to [8.72], [8.80], and [8.94], [8.95], [8.97], and [8.98].)

Let the structure be subjected to the action of static loads that are proportional to a parameter p and that perform small elastic vibrations around an equilibrium point. The function $w(x, t)$ describing the displacement of the structure with respect to the statically loaded but nonbuckled state satisfies the differential equation

$$Aw_{tt} + Bw_t + Cw - pKw = 0. \tag{8.1}$$

Here A, B, C, and K are linear differential operators, p is the load parameter, and the subscript t denotes partial differentiation with respect to time. The inertial operator A is represented by a symmetric matrix of mass coefficients. The stiffness operator C characterizes the static response of the elastic structure whose application to (vector) w reduces to differentiation of that function with respect to space variables. The operator K contains operators of differentiation with respect to both time and space variables. Let us represent the state function $w(x, t)$ in the form

$$w(x, t) = u(x)e^{i\omega t} \tag{8.2}$$

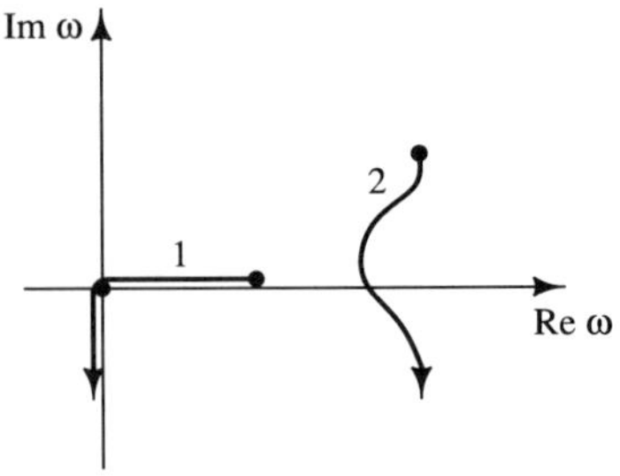

FIGURE 8.1.

and substitute (8.2) in (8.1). As a result, we obtain an equation for the amplitude function $u(x)$

$$-\omega^2 Au + i\omega Bu + Cu - pKu = 0. \tag{8.3}$$

The stability properties of the structure are determined by the magnitude of the eigenvalues $\omega = \omega(p)$ of eq. (8.3). The structure will not become unstable if all eigenvalues $\omega(p)$ have positive imaginary parts or, in the limiting case, if they are all real. A loss of stability occurs if at least one eigenvalue has a negative imaginary part. Curves 1 and 2 in Figure 8.1 represent the static and dynamic loss of stability, respectively. The eigenvalues $\omega(p)$ represented by curve 1 lie in the lower half-plane as the value of the parameter p increases and the curve passes through the origin Re $\omega = 0$, Im $\omega = 0$). The loss of stability occurs due to a monotone increase in the amplitude (of the disturbance) and has a basically static character. Curve 2 corresponds to the dynamic loss of stability and is characterized by initiation of vibrations with increasing amplitudes.

Let us assume that the dissipative terms are absent in our system (i.e., $B = 0$) and that external forces can be derived from a potential. The operator K (as shown in [8.20]) is independent of ω, and eq. (8.3) reduces to

$$-\omega^2 Au + Cu + pKu = 0. \tag{8.4}$$

The eigenvalue ω^2 is a function of the parameter p, that is, $\omega^2 = \omega^2(p)$. Equation (8.4) with appropriate boundary conditions constitutes a self-adjoint boundary-value problem. Therefore, the eigenvalues ω^2 are real. A loss of stability occurs when ω^2 changes from positive to negative real values. The critical value of the parameter p corresponds to the equality $\omega^2 = 0$. A transition into the unstable regime which occurs when the square of the natural frequency vanishes, turns out to be a consequence of the continuous dependence of this value on the load parameter.

Consequently, for a conservative system, the loss of stability must occur when $\omega^2 = 0$. That is, we have a static type of instability. For this reason our research into the elastic stability of conservative systems will generally follow static techniques originated by Leonhard Euler. Let us set $\omega^2 = 0$. Then

$$Cu - pKu = 0. \tag{8.5}$$

The problem of elastic stability now consists in finding the minimal value of the parameter p (i.e., the first eigenvalue) and the corresponding deflection function (the eigenfunction) that solves the boundary-value problem. As we have stated above, the self-adjoint property of a boundary-value operator is a sufficient condition for all eigenvalues p to have real values.

Static techniques for investigating stability can be reduced to a formulation and solution of a bifurcation problem. From a mathematical point of view, this depends on the self-adjoint and positive-definite properties of the eigenvalue problem. This means that for fairly arbitrary functions u and v,

$$\int_\Omega uCv\mathrm{d}\Omega = \int_\Omega vCu\mathrm{d}\Omega, \qquad \int_\Omega uCu\,\mathrm{d}\Omega > 0,$$

$$\int_\Omega uKv\mathrm{d}\Omega = \int_\Omega vKu\mathrm{d}\Omega, \qquad \int_\Omega uKu\mathrm{d}\Omega > 0.$$

We note that a self-adjoint positive-definite operator (for a boundary-value problem of classical elasticity) that does not contain a parameter p always has an infinite spectrum comprised of positive eigenvalues (see [8.4] and [8.57]).

In structural stability problems it generally suffices to find the smallest one, which determines the critical load.

For a self-adjoint, positive-definite eigenvalue problem, the theorem assigning a minimum to the Rayleigh quotient is true (see [1.11], [1.27], and [8.1]). If one uses energy methods that are based on the validity of this theorem, the stability problem reduces to a variational problem of minimizing the nonadditive functional

$$p = \min_u \frac{\int_\Omega uCu\mathrm{d}\Omega}{\int_\Omega uKu\mathrm{d}\Omega}. \tag{8.6}$$

The minimum with respect to u in (8.6) is sought in the class of sufficiently smooth functions that satisfy the assigned boundary conditions.

To this class of "static" research techniques in elastic stability belongs also the "nonideal" approach. This technique consists in introducing into the homogeneous equations of equilibrium (e.g., equation of longitudinal compression and bending for beams and plates, or of buckling for shells) a small additional term that makes these equations inhomogeneous. These additional terms may, for example, represent a small eccentricity in the application of the compressive load to a column, the action of a small transverse force in the compression of a plate by forces acting in the plane of the plate, the initial curvature of panels working in compression, or an initial inhomogeneity in the distribution of the load-bearing material in a shell. Taking into account these "nonideal" corrections described by a function ζ, the basic equilibrium equations assume the form

$$Cu - pKu + \zeta = 0. \tag{8.7}$$

Unlike the bifurcation equation (8.5), eq. (8.7) has a unique solution u (with $\|u\| \neq 0$) even for an arbitrarily small value of the load parameter p. As the value of p increases, the maximal value (amplitude) of the deflection function increases. This maximum approaches infinity as the load parameter approaches a critical value. This value coincides with Euler's value for the critical load, which can be found by computing the first eigenvalue for the bifurcation problem (8.5). Therefore, the study of elastic stability may be carried out using this "nonideal" approach, which leads to the problem of finding the magnitude of a load (i.e., of the parameter p) for which the deflections of the "nonideal" system become infinitely large.

8.2 Formulation of Certain Problems of Optimal Design

Problems in optimization of stability for elastic structural elements are closely related to many classical problems of the theory of design optimization. In previous research (see [16], [26], [8.2], [8.10] to [8.18], [8.22], [8.24] to [8.28], [8.30] to [8.56], [8.58] to [8.63], [8.67] to [8.69], [8.71], [8.73] to [8.84], [8.86] to [8.93], [8.96], [8.99], and [8.100], it was shown that during the optimization process one discovers very significant phenomena, which then provide a new perspective for future research pointing in the same direction. We need to note that the majority of this research and accepted techniques relate to optimization of stability for elastic systems that are described by self-adjoint boundary-value problems. Problems of optimization of nonconservative systems and, in particular, of structures with follower-type loads have been researched to a much lesser degree (see [8.16], [8.53], [8.65], [8.73], [8.75], [8.77], [8.78], and [8.90]).

We shall consider problems of optimal design of thin-walled structures using a general dynamic approach. The problem of optimal design consists in minimizing the weight functional

$$J = \int_{\Omega} h(x) d\Omega \to \min_{h \in R_h} \tag{8.8}$$

with conditions that p is given and the frequencies ω determined as the eigenvalues of the boundary-value problem for eq. (8.3) satisfy the inequality

$$\operatorname{Im} \omega(p) > 0. \tag{8.9}$$

The dual problem consists in maximizing the critical parameter determining the loss of stability

$$p \to \max_{h \in R_h} \tag{8.10}$$

with a constraint that the weight is given and the frequencies ω satisfy inequality (8.9).

The problem of weight minimization for given values of critical loads and the problem of maximizing the critical load that causes a loss of stability for a given weight are closely related to several difficult nonlinear problems of optimal design. Theories for such problems and certain effective algorithms that explore specific characteristics of stability are being intensively pursued at this time.

In applying a static approach to problems of optimal design, it is possible to restate them as max-min problems for the eigenvalue p of eq. (8.5) with a constraint on the weight of the structure, or else as a problem of maximal weight reduction for a given first eigenvalue.

Problems of optimization of elastic stability can also be restated by using the "nonideal" approach. In that case we maximize the load parameter p in (8.7), which causes structural displacements obtained by solving the inhomogeneous boundary-value problem to become infinitely large. Using this "nonideal" technique, we could consider also the dual problem of minimizing the weight under a constraint on the magnitude of the load that causes a loss of stability. The construction of an optimal solution based on this approach reduces to finding the state variables and design variables from conditions for a minimum, which depend on the magnitude of the compressive loads and the magnitude of the greatest deflection $u_0 (u \leq u_0)$. Then we let u_0 approach infinity.

In analyzing the stability and optimal design of elastic structural elements using the maximality of the critical loads that cause a loss of stability, we may encounter well-known difficulties that are caused by the appearance of multiple eigenvalues (see [1.5], [8.15], [8.25], [8.35], [8.47], [8.52], [8.54], [8.55], [8.62], and [8.76]). In such cases a genuine simplification can be achieved by decomposing the spectrum into a sum of auxiliary spectra, that do not contain multiple eigenvalues.

Here we shall describe a technique of such spectral decomposition that uses symmetry properties. We decompose the formulas for loss of stability into symmetric and antisymmetric parts (see [1.5] and [8.15]). A corresponding classification of eigenfunctions according to the type of symmetry also produces a classification of eigenvalues. At the same time, we also separate them according to the specific properties of such multiple loads, and the reformulated optimization problem is now reduced to a classical optimization problem with simple eigenvalues that can be solved by applying some of the previously discussed algorithms (see [5], [20], [25], [28], [32], [24], [50], and [51]).

Let us now offer some details of this approach. We will study the stability of a conservative elastic system following Euler's static approach, and arrive at a homogeneous eigenvalue problem for the boundary-value system

$$Cu - pKu = 0, \tag{8.11}$$

$$(Nu)_{|x|=l} = 0, \tag{8.12}$$

where $x \in [-l, l]$, $u(x)$ is a vector function determining the equilibrium state for an elastic structural element, $h = h(x)$ is the control vector function, p is an eigenvalue (load parameter), and $C(x, h(x))$, $K(x, h(x))$, and $N(x, h(x))$ are differential operators, with differentiation carried out with respect to the variable x. The operators C, K, and N are linear, and the coefficients of these operators depend on the control function h. The linear operators C and K with assigned boundary conditions (8.12) are self-adjoint and positive-definite. Also, we assume that the control function $h(x)$ is symmetric with respect to the point $x = 0$, that is, $h(x) = h(-x)$, and that for a symmetric function h the differential operators in (8.11) with boundary conditions (8.12) are invariant under the transformation $x \to -x$.

The smallest eigenvalue p_1 of the boundary-value problem (8.11) and (8.12) determines the critical load that causes the loss of stability. To compute p_1 we use Rayleigh's variational principle

$$p_1 = \min_u \Phi(u, h), \tag{8.13}$$

$$\Phi(u, h) = (Cu, u)/(Ku, u). \tag{8.14}$$

The parentheses in (8.14) denote a scalar product for the indicated elements, and (8.13) is minimized in the class of functions that satisfy the boundary conditions (8.12).

The optimization problem under consideration consists in maximizing the magnitude of the critical load, that is, the load causing the loss of stability, by finding a best (in this sense) control variable function h, such that

$$p^* = \max p_1, \tag{8.15}$$

$$h \in R_h. \tag{8.16}$$

The maximum in (8.15) is found in the class of symmetric functions $h(x)$ belonging to the admissible set R_h.

Recalling that for a symmetric function $h(x)$, the differential operators C, K, and N are invariant under the reflection transformation $x \to -x$, we obtain the following conclusion. For simple eigenvalues (without multiplicities) the corresponding eigenfunctions u are either symmetric or antisymmetric with respect to the origin $x = 0$. At first let us assume that the eigenvalues are simple. We denote the symmetric eigenfunctions by u^s and the antisymmetric ones by u^a. We can use the properties $u^s(x) = u^s(-x)$ and $u^a(x) = -u^a(-x)$ in (8.13) to arrive at variational problems that can be used to compute the first eigenvalues for the symmetric and antisymmetric solutions. The variational problem for finding the symmetric solutions is of the form

$$\left.\begin{aligned} p_1^s &= \min_{u^s} \Phi(u^s, h), \\ (Nu^s)_{x=l} &= 0, \end{aligned}\right\} \tag{8.17}$$

$$\left(\frac{du^s}{dx}\right)_{x=0} = \left(\frac{d^3}{dx^3}u^s\right)_{x=0} = \cdots = \left(\frac{d^{2n-1}}{dx^{2n-1}}u^s\right)_{x=0} = 0, \tag{8.18}$$

where $2n$ is the order of the differential equation (8.11).

The variational problem for the antisymmetric case differs only in the boundary conditions assigned to $x = 0$:

$$\left.\begin{aligned} p_1^a &= \min_{u^a} \Phi(u^a, h), \\ (Nu^a)_{x=l} &= 0, \end{aligned}\right\} \tag{8.19}$$

$$u^a(0) = \left(\frac{d^2}{dx^2} u^a\right)_{x=0} = \cdots = \left(\frac{d^{2n-2}}{dx^{2n-2}} u^a\right)_{x=0} = 0. \tag{8.20}$$

These symmetry properties are used in computing Rayleigh relationships for (8.17) and (8.18). Here the scalar products are integrals over the interval $[0, l]$. With the help of these relations, the original optimization problem is reduced to the equivalent formulation

$$p_* = \max_h \min_i \{p_1^i\}, \qquad i = s, a, \tag{8.21}$$

where the computation of the maximum with respect to h takes into account (8.16). Now we admit the possibility that multiple critical loads (i.e., multiple eigenvalues) may occur. In the presence of multiple critical loads, the eigenfunction shapes corresponding to the loss of stability may be not only symmetric or antisymmetric, but could be any linear combination of both. However, even in this case the solution of our problem is still reduced to the study of symmetric and antisymmetric forms of a loss of stability and related eigenvalues p_1^s and p_1^a. These forms constitute the basis.

8.3 Optimal Design of Compressed Rods with Pin-Jointed Supports

We shall consider the problem of optimizing the stability of a pin-jointed rod compressed by forces p applied to its ends. For some critical value of this parameter, the rod buckles. Buckling takes place in the xOz plane. The origin of the coordinate system coincides with one end of the rod, and the x axis passes through the point of the rod's support. We restrict our analysis to the case of small deflections and study the rod's equilibrium within the framework of the linear theory of elasticity. Let $u(x)$ denote the magnitude of the deflection of the bent axis of the rod from the line of action of forces compressing the rod and l its length. First we write the basic relations for the problem of maximizing the load that causes a loss of stability, and eventually we find the best (in this sense) distribution of the cross-sectional area $S = S(x)$ along the length of the rod.

$$\left.\begin{aligned} &EIu_{xx} + pu = 0, \qquad EI = A_\alpha S^\alpha, \qquad \alpha = 1, 2, 3; \\ &u(0) = u(l) = 0, \qquad \int_0^l S(x)dx = V, \; p \to \max_S. \end{aligned}\right\} \tag{8.22}$$

The volume of the rod is assumed to be known. For the sake of convenience, we shall introduce the dimensionless variables $x' = x/l$, $S' = lS/N$, and $p' = p/A_\alpha E$ (primes will be omitted). The basic relations now assume the form

$$\left.\begin{array}{ll} S^\alpha u_{xx} + pu = 0, & u(0) = u(1) = 0, \\ \int_0 S(x)\mathrm{d}x = 1, & p \to \max\limits_{S}. \end{array}\right\} \tag{8.23}$$

The optimality conditions for this problem can be derived by standard arguments. We construct the generalized Lagrangian functional

$$\left.\begin{array}{ll} J^\lambda = J - \lambda\left(\left(\int_0^l S(x)\mathrm{d}x - 1\right)\right), & J = J_1/J_2, \\ J_1 = \int_0^l (u_x)^2\mathrm{d}x, & J_2 = \int_0^l (u^2 S^{-\alpha})\mathrm{d}x \end{array}\right\} \tag{8.24}$$

and write the expression for the first variation of this functional by varying the design variable $S(x)$ and making use of eq. (8.23). We derive

$$\delta J^\lambda = \int_0^l [(\alpha J_1 u^2)/(J_2^2 S^{\alpha+1}) - \lambda]\delta S(x)\mathrm{d}x. \tag{8.25}$$

We apply the extremality condition $\delta J^\lambda = 0$ to (8.25) and normalize the function u to obtain the necessary optimality condition

$$u^2(x) = S^{\alpha+1}(x). \tag{8.26}$$

By manipulating the optimality condition (8.26), we arrive at a well-posed problem and solve for the design variable $S(x)$ and also find the state function u and the value of the maximized functional $p = J$. The derivation of the solution has been accomplished in several ways (see [8.24], [8.26], [8.40], [8.41], and [8.60]). We shall offer a technique for solving this problem offered by Barsuk. Let us differentiate the necessary optimality condition (8.26)

$$uu_x = ((\alpha + 1)/2)S^\alpha S_x, \tag{8.27}$$

then multiply eq. (8.23) by u_x and substitute from (8.27) for the term uu_x. After integrating the resultant expression, we have

$$u_x^2 \to p(\alpha + 1)S = c_0, \tag{8.28}$$

where c_0 is an unknown constant. Deriving an expression for u_x^2 from (8.26) and (8.27) and substituting it in (8.28), we arrive at a differential equation in the design variable S

$$S_x^2 = (4/(\alpha + 1))((c - pS)/S^{\alpha-1}). \tag{8.29}$$

Here and in what follows, the optimal distribution of the surface area is sought in the class of functions symmetric with respect to the center of the rod($x = \frac{1}{2}$), that is, we assume that this function has the property

$$S(x) = S(1 - x). \tag{8.30}$$

It follows directly from the symmetry of the design function that $S_x(x) = 0$ when $x = \frac{1}{2}$.

It is easy to show that the eigenvalues of the boundary-value problem (8.23) are simple. Let us assume to the contrary that for some eigenvalue p there exist two linearly independent solutions $u^1(x)$ and $u^2(x)$, and therefore,

$$u^i_{xx} + p(u^i/S^\alpha) = 0, \qquad u^i(0) = u^i(1) = 0, \qquad i = 1, 2.$$

We conclude that $u^1_{xx}u^2 - u^2_{xx}u^1 = 0$. Integrating this equality we have $u^1_x u^2 - u^1 u^2_x = \kappa$, where κ is a constant of integration. From the boundary conditions for u, we find that $\kappa = 0$. Another integration gives us $u^1(x) = \mu u^2(x)$, where μ is a constant. This means that u^1 and u^2 are multiples of each other. Hence, all the eigenvalues of our boundary-value problem are simple. By applying this property of the spectrum of eigenvalues, we are able to determine that the second derivative of the eigenfunction corresponding to the first eigenvalue retains its sign over the entire interval (0, 1) and that the first derivative u_x is a monotone function of x. This conclusion and eq. (8.27) imply that $S(x)$ attains its maximum at $x = \frac{1}{2}$. Let us denote this maximum value by a, that is, $S(\frac{1}{2}) = a$. From $S_x(\frac{1}{2}) = 0$ and eq. (8.29), we find a relation between the parameter a and the constant c: $c = pa$. Introducing a function $s(x)$ defined by $S(x) = as(x)$, we formulate a differential equation for $s(x)$ on the interval $0 \leq x \leq \frac{1}{2}$, as follows:

$$s_x = 2\{\sqrt{p_0}/(\sqrt{\alpha + 1})\}\{(\sqrt{1 - s(x)})/s^{(\alpha-1)/2}\}, \tag{8.31}$$

where $p_0 = pq^{-\alpha}$. Integrating the differential equation (8.31), we represent the solution by the quadrature formula

$$\int_0^s \frac{\xi^{(\alpha-1)/2}\,\mathrm{d}\xi}{\sqrt{(1 - \xi)}} = 2[\sqrt{p_0}/\sqrt{(\alpha + 1)}]x. \tag{8.32}$$

The condition $s(\frac{1}{2}) = 1$ is equivalent to the relation

$$\int_0^1 \frac{\xi^{(\alpha-1)/2}\,\mathrm{d}\xi}{\sqrt{(1 - \xi)}} = \sqrt{p_0}/\sqrt{(\alpha + 1)}. \tag{8.33}$$

The constant volume constraint of the rod leads to the additional relation

$$1 = 2\int_0^{1/2} S\mathrm{d}x = 2\int_0^{1/2} \frac{S\mathrm{d}s}{(\mathrm{d}S/\mathrm{d}x)} = \frac{a\sqrt{(\alpha + 1)}}{\sqrt{p_0}}\int_0^1 \frac{\xi^{(\alpha+1)/2}\,\mathrm{d}\xi}{\sqrt{(1 - \xi)}}. \tag{8.34}$$

We introduce the notation:

$$I_1(\alpha, \beta, s) = \int_\beta^s \frac{\xi^{(\alpha-1)/2}\,\mathrm{d}\xi}{\sqrt{(1 - \xi)}}, \qquad I_2(\alpha, \beta) = \int_\beta^1 \frac{\xi^{(\alpha+1)/2}\,\mathrm{d}\xi}{\sqrt{(1 - \xi)}}. \tag{8.35}$$

Then the solution given by (8.32) to (8.34) for our optimization problem can be displayed in the form

$$\left.\begin{aligned} a &= \frac{I_1(\alpha, 0, 1)}{I_2(\alpha, 0)}, \qquad p_0 = (\alpha + 1)I_1^2(\alpha, 0, 1), \\ I_1(\alpha, 0, s(x)) &= 2I_1(\alpha, 0, 1)x, \qquad S(x) = as(x), \qquad p = p_0 a^\alpha. \end{aligned}\right\} \quad (8.36)$$

For specific values $\alpha = 1, 2, 3$ of the parameter α in (8.36), the integrals I_1 and I_2 can be expressed in terms of elementary functions. Let us just quote the results of our computations, and for the sake of comparison with previous results, we note that the loss of stability for a rod with constant cross-section (i.e., $S(x) \equiv 1$) occurs when $p = \pi^2$.

In the case $\alpha = 1$ (rectangular rods with variable width and constant height, or trilayered rods with variable thickness of reinforcing layers)

$$I_1(1, 0, 1) = 2, \qquad I_2(1, 0) = 4/3, \qquad I_1(1, 0, s) = 2(1 - \sqrt{1 - s}),$$
$$p_0 = 8, \qquad a = 3/2, \qquad p = 12, \qquad S(x) = 6x(1 - x). \quad (8.37)$$

In the case $\alpha = 2$ (rods with similar cross-sections),

$$I_1(2, 0, 1) = \pi/2, \qquad I_2(2, 0) = 3\pi/8,$$
$$I_1(2, 0, s) = \arcsin(\sqrt{s} - \sqrt{s(x)(1 - s(x))}), \qquad p_0 = 3\pi^2/4, \qquad a = 4/3,$$
$$p = 4\pi^2/3, \qquad S(x) = as(x), \quad \text{and} \quad \arcsin\sqrt{s(x)} - \sqrt{s(x)(1 - s(x)} = \pi x. \quad (8.38)$$

In the case $\alpha = 3$ (rods with rectangular cross-section having variable height and constant width),

$$I_1(3, 0, 1) = 4/3, \qquad I_2(3, 0) = 16/15,$$
$$I_1(3, 0, s) = 4/3[1 - (1 + s/2)\sqrt{(1 - s)}], \qquad p_0 = 64/9, \qquad a = 5/4,$$
$$p = 125/9, \qquad S(x) = as(x), \quad \text{and} \quad 1 - (1 + s/2)\sqrt{(1 - s)} = 2x. \quad (8.39)$$

The gain for the optimal solutions of (8.36) to (8.39) in comparison with the rod having constant cross-section estimated for the magnitude of the critical load causing the loss of stability is 21.6%, 33.3%, and 40.7% for values of α equal to 1, 2, and 3, respectively. For all types of cross-section, the optimal function $S(x)$ vanishes at the supported ends of the rod, and consequently, the mass of the rod is concentrated in its central portion. We have the following assymptotic behavior of the optimal solution in the vicinity of the endpoints (on the interval $0 \le x \le 1$): $S(x) \simeq 6x$ for $\alpha = 1$, $S(x) \simeq (16\pi^2/3)^{1/3}x^{2/3}$ for $\alpha = 2$, and $S(x) \simeq (5/\sqrt{3})x^{1/2}$ for $\alpha = 3$.

Analytic solutions for the problem of optimizing the stability of an elastic rod may be obtained with constraints on the minimal size of the cross-sectional area (or on thickness):

$$S(x) \ge S_{\min}. \quad (8.40)$$

The magnitude of the constant $S_{\min}$ is given. It is a consequence of the isoperimetric condition (8.23) that inequality $S_{\min} < 1$ must be obeyed. Earlier we developed a necessary optimality condition in the absence of the constraint (8.40). The changes that must be made to account for this constraint consist in requiring that the optimality condition (8.26) be satisfied for $x_* \le x \le \frac{1}{2}$, with $S > S_{\min}$, while for $0 \le x \le x_*$ the thickness is constant: $S(x) = S_{\min}$. The representation of the optimal solution by quadratures is similar to the one displayed above for the case when the constraint (8.40) is absent. We have

$$I_1(\alpha, \beta, s) = \int_\beta^s \frac{\xi^{(\alpha-1)/2} d\xi}{\sqrt{(1-\xi)}} = \frac{2\sqrt{p_0}}{\sqrt{(\alpha+1)}}(x - x_*), \tag{8.41}$$

where $\beta = S_{\min}/a$.

The isoperimetric condition (8.23) of constant volume, the condition of continuous differentiability of the displacement function at the point $x = x_*$, and condition $s(\frac{1}{2}) = 1$ can be written, respectively, in the form

$$I_2(\alpha, \beta) = [2\sqrt{p_0}/\sqrt{(\alpha+1)}](\tfrac{1}{2} - S_{\min}\, x_*), \tag{8.42}$$

$$\tan[(p_0/\beta^\alpha)^{1/2}]x_* = [1/\sqrt{(\alpha+1)}](\beta/(1-\beta))^{1/2}, \tag{8.43}$$

$$I_1(\alpha, \beta, 1) = 2(\sqrt{p_0}/\sqrt{(\alpha+1)})(\tfrac{1}{2} - x_*). \tag{8.44}$$

Equations (8.42) to (8.44) represent a system of three equations that may be solved to find the values of p_0, x_*, and β. The solution of this system of equations may be accomplished in the following manner. First eqs. (8.42) to (8.44) are solved for p_0, x_*, and $S_{\min}$ (regarding β as a known quantity):

$$\begin{aligned} S_{\min} &= \frac{2\beta^{\alpha/2}\psi(\beta) + \sqrt{(\alpha+1)}I_1(\alpha, \beta, 1)}{2\beta^{(\alpha+2)/2}\psi(\beta) + (\alpha+1)^{1/2} I_2(\alpha, \beta)}, \\ \psi(\beta) &\equiv \arctan\left(\frac{\beta}{(\alpha+1)(1-\beta)}\right)^{1/2}, \\ p_0 &= (\alpha+1)\frac{[I_2(\alpha, \beta) - \beta I_1(\alpha, \beta, 1)]^2}{(1-S_{\min})^2}\left(\frac{S_{\min}}{\beta}\right)^2, \\ x_* &= \frac{S_{\min}\, I_2(\alpha, \beta) - \beta I_1(\alpha, \beta, 1)}{2S_{\min}(I_2(\alpha, \beta) - \beta I_1(\alpha, \beta, 1))}. \end{aligned} \tag{8.45}$$

Since the values of β are restricted to the interval $0 \le \beta \le 1$, for each value of $\beta \in [0, 1]$ we can compute the corresponding value of $S_{\min}$, as given by the formula in the first line of (8.45). For such corresponding values of β and $S_{\min}$, by using (8.45) (third and fourth line) we can compute p_0 and x_*, and then compute $s(x)$ making use of (8.41). Consequently, we can find the parameters a and p and the shape of the design function: $a = S_{\min}/\beta$, $p = p_0 a^\alpha$, and $S(x) = as(x)$.

For specific values $\alpha = 1, 2, 3$, we introduce formulas for I_1 and I_2, and describe the solutions of our optimization problems in terms of these functions:

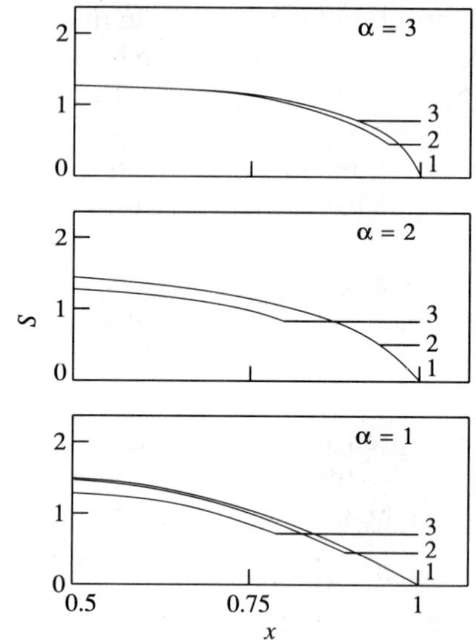

FIGURE 8.2. Distribution of rod area $S(x)$ $\alpha = 1, 2, 3$.

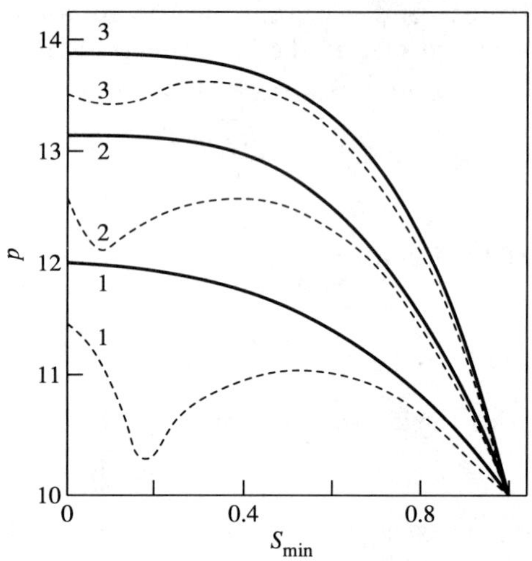

FIGURE 8.3. Critical loads as functions of minimal values of thickness.

$$\left.\begin{aligned}
I_1(1, \beta, s) &= 2(\sqrt{(1-\beta)} - \sqrt{(1-s)}, \qquad I_2(1, \beta) = 2/3(2+\beta)\sqrt{(1-\beta)}\\
I_1(2, \beta, s) &= \arcsin\sqrt{s} - \arcsin\sqrt{\beta} - \sqrt{s(1-s)} + \sqrt{\beta(1-\beta)},\\
I_2(2, \beta) &= \tfrac{1}{4}(3\arcsin\sqrt{(1-\beta)} + (3+2\beta)\sqrt{\beta(1-\beta)}),\\
I_1(3, \beta, s) &= 2/3\{(2+\beta)\sqrt{(1-\beta)} - (2+s)\sqrt{(1-s)}\},\\
I_2(3, \beta) &= 2/15(8 + 4\beta + 3\beta^2)\sqrt{1-\beta}.
\end{aligned}\right\} \tag{8.46}$$

Figure 8.2 illustrates the distribution of the area function $S(x)$ for the values $\alpha = 1, 2, 3$. The curves 1, 2, 3 correspond to the values $S_{\min} = 0.1, 0.3$, and 0.5, respectively. The values of $S(x)$ are graphed for $\frac{1}{2} \le x \le 1$. It is clear from Figure 8.2 that the minimal values of the thickness are attained near the ends of the rod and that the length of such segments increases as the value of the parameter $S_{\min}$ increases. For a rod with freely supported ends, the principal mass of material is concentrated in its middle part. The graph of the values of the critical loads causing the loss of stability and their dependence on $S_{\min}$ is shown in Figure 8.3. Curves 1, 2, and 3 correspond to the values $\alpha = 1, 2, 3$, respectively.

8.4 Design of Compressed Rods with Built-In Ends

We shall consider the problem of optimizing the stability of a compressed elastic rod with rigidly clamped ends. In terms of the dimensionless variables introduced in Section 8.3, the basic relations can be written as

$$\left.\begin{aligned}
&(S^\alpha u_{xx})_{xx} + pu_{xx} = 0, \qquad u(0) = u_x(0) = u(1) = u_x(1) = 0,\\
&\int_0^1 S(x)dx = 1, \qquad S \ge S_{\min}, \qquad p \to \max_S.
\end{aligned}\right\} \tag{8.47}$$

This problem has been investigated in many articles (see [17], [1.5], [8.15], and [8.62]). The problem in which the bound on the minimum admissible thickness is absent has been researched in great detail (see [8.52] and [8.76]). For the case when $S_{\min} \ne 0$, the area distribution function $S = S(x)$ will be found numerically.

Assuming that $S_{\min} \ne 0$ and that the critical load is simple (i.e., without multiplicities) we shall describe a technique that reduces the solution of this problem for a rod with rigidly held ends to the problem for a rod with simple supports (the ends of the rod are free to rotate). The subsequent discussion of multiple eigenvalues and construction of bimodal solutions shall be carried in Sections 8.8 to 8.10. We seek an optimal solution in the class of functions representing a symmetric distribution of the cross-sectional areas, that is, $S(x) = S(1-x)$. For such symmetric distribution of the design function, the solutions may be classified as either odd or even functions. Even functions satisfy the condition $u^s(x) \equiv u^s(1-x)$, while for the odd (i.e., asymmetric) functions, $u^a(x) \equiv -u^a(1-x)$. The eigenvalues p^s and p^a correspond to the displacement functions u^s and u^a, respectively.

We introduce a function $M = S^\alpha u_{xx}$ to describe the elastic moment distribution. The equilibrium equation for bending of a compressed rod may be written in terms of the variable M,

$$M_{xx} + pM/S^\alpha = 0. \tag{8.48}$$

The boundary values for (8.48) are introduced in the following manner: We write the bending (equilibrium) equation in the form $M_{xx} + pu_{xx} = 0$ and integrate both sides twice. We obtain

$$M(x) = -pu(x) + c_1 x + c_2. \tag{8.49}$$

Taking into account the assigned boundary conditions on displacements results in boundary conditions for the bending moment,

$$M(0) + M_x(0) = M(1), \qquad M_x(0) = M_x(1). \tag{8.50}$$

The functions $M(x)$ can also be classified into symmetric and assymmetric moment distribution $M^s(x)$ and $M^a(x)$. The boundary values are obtained by first substituting $(M^s(\frac{1}{2}), u^s(\frac{1}{2}))$ and $(M^a(\frac{1}{2}), u^a(\frac{1}{2}))$ in eq. (8.49). We arrive at the conditions

$$M_x^s(0) = 0, \qquad M_x^s(1) = 0, \tag{8.51}$$

$$M^a(0) = -\tfrac{1}{2}M_x^a(0), \qquad M^a(\tfrac{1}{2}) = 0. \tag{8.52}$$

We note that for a rod with constant cross-section, the minimal eigenvalue corresponds to a symmetric displacement function (symmetric eigenfunction). Therefore, in the case when critical loads are simple (and in the optimization process, we take $p^s \neq p^a$), the resulting optimization problem is equivalent to the one given below:

$$\left.\begin{aligned} &M_{xx}^s + p(M^s/S^\alpha) = 0, \qquad M_x^s(0) = 0, \qquad M_x^s(\tfrac{1}{2}) = 0,\\ &2\int_0^{1/2} S(x)dx = 1, \qquad S(x) \geq S_{\min}, \qquad p_* = \max_S p(S). \end{aligned}\right\} \tag{8.53}$$

We remark that problem (8.53) is solved on the interval $0 \leq x \leq \frac{1}{2}$, in which all the conditions in (8.53) are symmetric with respect to the point $x = \frac{1}{4}$. Let us assume that $S(x) = S(\frac{1}{2} - x)$ and decompose again the spectrum of the boundary-value problem that defines the stability of the rod relative to symmetry properties. The eigenfunctions M^s that are symmetric or antisymmetric on $0 \leq x \leq \frac{1}{2}$ are classified again and are denoted by $M^{ss}(x)$ and $M^{sa}(x)$. That is, $M^{ss}(x) = M^{ss}(\frac{1}{2} - x)$ and $M^{sa}(x) = -M^{sa}(\frac{1}{2} - x)$.

The eigenfunctions $M^{ss}(x)$ and $M^{sa}(x)$ correspond to the eigenvalues p^{ss} and p^{sa}, respectively. For rods of constant cross-section, the minimal eigenvalue of (8.53) corresponds to the antisymmetric solution. Recalling this property and the fact that the eigenvalues (critical loads) are simple, we can reduce the problem (8.53) to an equivalent problem defined on the interval $[0, \frac{1}{4}]$:

$$\left.\begin{array}{lll} M_{xx} + p(M/S^{\alpha}) = 0, & M_x(0) = 0, & M_x(\tfrac{1}{4}) = 0, \\ 4\displaystyle\int_0^{1/4} S(x)\mathrm{d}x = 1, & S(x) \geq S_{\min}, & p_* = \max\limits_{s} p(S). \end{array}\right\} \quad (8.54)$$

Problem (8.54) is equivalent to the problem of the loss of stability for a freely supported rod having length $l = \frac{1}{2}$ with a symmetric distribution of its displacements. In this manner the optimization problem for the loss of stability of a rod with rigidly built-in ends and simple eigenvalues has been derived from a solution of exactly the same problem but with free support at each end. The computation of the eigenvalue relies on the formula $p = 4p_0$, where p_0 is the critical load for a freely supported rod. In that case the distribution of its thickness (or, equivalently, its cross-sectional area) for a rod of length $l = 1$ with rigidly built-in ends on the interval (0, 0.25) is identical with the distribution of thickness for a freely supported rod of length $l = .5$ on the interval (0.25, 0.5). Among the various articles on the optimal design of compressed rods (or columns), we should mention [8.87] dealing with eccentrically loaded compressed rods and [1.23] and [8.65] dealing with beam-columns. Problems dealing with these types of designs for structural elements made of inhomogeneous materials have been studied in [7.22], [8.17], and [8.2].

8.5 Optimization of Stability for Elastic Beams Subjected to Thermal Loads

We shall consider the problem of the stability of a linear version of equilibrium for a heated rod. We assume that a rod having a variable thickness lies along the x axis and is freely supported at the endpoints $x = 0$ and $x = l$. Initially in the unheated state its length is equal to l. Before it is heated axial stresses are absent. Let ϑ denote the increase in temperature caused by heating the rod. Since the positions of the ends are fixed, compressive reaction forces p will be generated at the supports caused by thermal expansion of the rod. The magnitude of the reaction force p is related to the increase in temperature by the formula

$$p = \beta\vartheta El\left(\int_0^l \frac{\mathrm{d}x}{S}\right)^{-1}, \quad (8.55)$$

where β and E are, respectively, the coefficient of linear expansion and Young's modulus, while $S = S(x)$ is the function describing the distribution of the cross-sectional areas.

With the increase in temperature ϑ rises the magnitude of the compressive reaction force p, and for some critical value of temperature ϑ, the compressive force reaches a value that causes a loss of stability and departure from a straight-line equilibrium shape, and the rod buckles. According to Euler's theory, the magnitude of the critical load p is determined as the smallest eigenvalue for the homogeneous boundary-value problem

$$EIu_{xx} + pu = 0, \tag{8.56}$$

$$u(0) = u(l) = 0. \tag{8.57}$$

Here, as in the preceding sections, we consider only designs of rods for which the moment of inertia of the cross-sectional area I is a function of the cross-sectional area that may be represented by the formula $EI = A_\alpha S^\alpha$, $i = 1, 2, 3$. For a fixed type of cross-section and for a given function $S = S(x)$, we can use the solution of the boundary-value problem (8.56) and (8.57), and relation (8.55) to find the value of the critical temperature ϑ that causes the loss of stability and a change in the corresponding deflection function.

The magnitude of the reaction force p and the rod's rigidity depend on the distribution of cross-sectional areas (thicknesses) $S(x)$ (see (8.55)). A variation of $S(x)$ produces changes in these characteristic properties and produces a change in the value of the critical temperature. Thus, the problem of finding the best distribution of thickness that maximizes the critical temperature is of some interest. Let us introduce a rigorous formulation of this optimization problem (see [8.3]).

We need to find a function describing the distribution of cross-sectional area that satisfies the bounds

$$S_{\min} \le S(x) \le S_{\max}, \tag{8.58}$$

with the volume constraint

$$\int_0^l S(x)dx = V, \tag{8.59}$$

which assigns the maximum value to the critical temperature

$$\vartheta_* = \max_S \vartheta, \tag{8.60}$$

where ϑ is computed from (8.55) to (8.57). $S_{\min}$, $S_{\max}$, and V are positive constants satisfying the relation $S_{\min}\, l \le V \le S_{\max}\, l$. In the formulation of our problem (8.55) to (8.60), the distribution of the area function $S(x)$ plays the role of the control function while eq. (8.56) is a differential connection. We shall offer some remarks concerning the formulation of problem (8.55) to (8.60).

In the stability problem that commences with a straight-line equilibrium configuration of a heated rod, we assume that the position of each end of the rod is fixed on the x axis. As a loss of stability occurs, the length of the rod increases, and reaction loads at the ends decrease. The magnitude of the critical load and of the actual compressive force that produces an equilibrium when the shape of the rod is curved but close to a straight line may be entirely different. Therefore, this problems differs from the classical problem of buckling of an elastic rod compressed by "dead" loads, and some additional analysis needs to be carried out with regard to the applicability of Euler's classical theory. This problem has been examined in [8.47], where it was shown that the difference in the rod's displacement is a quantity of higher order of smallness (relative to the magnitude of a characteristic displacement), and can be ignored in computations of stability.

If in this opttimization problem we do not assign a lower bound to the rod's thickness (i.e., if we substitute in (8.58) $S_{\min} = 0$), then it is possible to show that there exists a distribution of thickness for which $p = 0$ for an arbitrary value of ϑ, and for which the rod does not buckle. Indeed, let the shape of the rod finish with a sharp spike at $x = 0$ and the function $S(x)$ obey the equation $S(x) = cx^m$, with $m \geq 1$ in the neighborhood $0 \leq x \leq \varepsilon < l$ of the endpoint $x = 0$. Then for an arbitrary choice of cross-sectional areas on the segment $\varepsilon \leq x \leq l$, which asymptotically approaches the function assigned to $0 \leq x \leq \varepsilon$, the magnitude of the reaction p (computed from (8.55)) approaches 0.

To prove the necessary optimality conditions, it is convenient to transform this problem and to eliminate the differential connection (8.56).

To do this we note that finding the critical temperature and the corresponding function $u(x)$ may be reduced to an equivalent variational problem of finding a minimum for the nonadditive functional

$$\vartheta = \min_u \frac{A_\alpha}{\beta l E}\left(\int_0^l \frac{dx}{S}\right)\left(\int_0^l (u_x)^2\right)\left(\int_0^l \frac{u^2 dx}{S^\alpha}\right)^{-1}, \tag{8.61}$$

where the minimum in (8.61) is sought in the class of functions satisfying given boundary conditions. Making use of condition (8.61), we reduce the original problem (8.55) to (8.60) to a minimax optimization problem containing no differential connections:

$$\vartheta_* = \max_S \min_u \frac{A_\alpha}{\beta l E}\left(\int_0^l \frac{dx}{S}\right)\left(\int_0^l (u_x)^2\right)\left(\int_0^l \frac{u^2 dx}{S^\alpha}\right)^{-1}. \tag{8.62}$$

The minimum with respect to S is taken in the class of functions $S = S(x)$ that satisfy conditions (8.58) and (8.59). The two-sided inequality (8.58) assigned to the design variable $S(x)$ can be "automatically" satisfied if we introduce an auxiliary variable φ related to S by the equalities

$$S = \mu_1 + \mu_2 \sin\varphi, \qquad \mu_1 = \tfrac{1}{2}(S_{\min} + S_{\max}), \qquad \mu_2 = \tfrac{1}{2}(S_{\max} - S_{\min}).$$

The only restriction on the values of φ comes from the isoperimetric condition. It can be deduced from this condition and the formula for $S(x)$ that

$$\int_0^l \sin\varphi \, dx = \frac{V - \mu_1 l}{\mu_2}.$$

We can now write the formula for the first variation of the optimized functional derived from the first variation of the auxiliary control function.

$$\delta\vartheta = \int_0^l \Lambda \cos\varphi \, \delta\varphi \, dx, \qquad \Lambda = \frac{B_1 u^2}{(\mu_1 + \mu_2 \sin\varphi)^{(\alpha+1)}} - \frac{B_2}{(\mu_1 + \mu_2 \sin\varphi)^2},$$

$$B_1 = \alpha\mu_2\left(\int_0^l \frac{dx}{(\mu_1 + \mu_2 \sin \varphi)}\right)\left(\int_0^l (u_x)^2 \, dx\right)\left(\int_0^l \frac{u^2 \, dx}{(\mu_1 + \mu_2 \sin \varphi)^\alpha}\right)^{-2},$$

$$B_2 = \mu_2\left(\int_0^l (u_x)^2 \, dx\right)\left(\int_0^l \frac{u^2 \, dx}{(\mu_1 + \mu_2 \sin \varphi)^\alpha}\right)^{-1}. \tag{8.63}$$

The isoperimetric condition assigned to all possible variations $\delta\varphi$ is given by the constraint

$$\int_0^l \cos \varphi \, \delta\varphi \, dx = 0. \tag{8.64}$$

Substituting (8.63) and (8.64), we obtain a necessary optimality condition $\delta\vartheta = 0$ in the form

$$(\Lambda - \lambda) \cos \varphi = 0, \tag{8.65}$$

where λ is a Lagrangian multiplier.

For the sake of simplicity, we shall consider the case when the upper bound is inactive, and therefore, the thickness function $S(x)$ and the value of the critical reaction p are functions only of the lower bound parameter $S_{\min}$.

The optimal distribution of thickness will contain intervals on which $S(x) = S_{\min}$, and the region in which $S(x) > S_{\min}$ will change smoothly in accord with the optimality condition

$$\left.\begin{aligned} &u^2 = \gamma_1 S^{\alpha-1} + \gamma_2 S^{\alpha+1}, \\ &\text{where } \gamma_1 = B_2/B_1, \qquad \gamma_2 = \lambda/B_1. \end{aligned}\right\} \tag{8.66}$$

For the sake of convenience, we shall use the following dimensionless variables in the subsequent discussion:

$$\left.\begin{aligned} &x' = x/l, \qquad S' = lS/V, \qquad S'_{\min} = lS_{\min}/V, \\ &\vartheta' = \beta l^{\alpha+1}\vartheta E/(A_\alpha V^{\alpha-1}), \qquad p' = l^{\alpha+2}p/(A_\alpha V^\alpha). \end{aligned}\right\} \tag{8.67}$$

In these dimensionless variables (with primes omitted), the solution of the optimization problem will be derived on the interval $0 \le x \le 1$ for a rod with unit volume. For $\alpha = 2, 3$, an analytic solution was derived by Barsuk (who assumed that the eigenvalues are simple). The case $\alpha = 1$, as was already indicated, is exceptional and we shall construct a numerical solution for that case.

The optimal solution is sought in the class of symmetric distributions of thickness; that is, we assume that $S(x) = S(1 - x)$. The corresponding deflection function shall be classified according to the odd or even property $y(x) = \pm y(1 - x)$, with the sign "+" corresponding to symmetric deflections and "−" to antisymmetric deflections. These relations allow us to restrict our analysis to the half-interval $(0, \frac{1}{2})$ in deriving the solution. The derivation of an analytical solution is carried out exactly as was done before in Section 8.3. Without repeating the identical arguments, let us quote the final results. The optimal distribution of thickness is given by the quadratures

$$\int_\beta^s \frac{d\tau}{\Phi(\tau)} = \frac{2\sqrt{p_0}}{\sqrt{(\alpha-1)}}(x - x_*),$$

$$\Phi(s) \equiv \frac{\sqrt{\gamma_0 + (\alpha+1)s}\sqrt{\gamma_0 + (\alpha-1)s^2}}{(\gamma_0 + (\alpha+1)s^2)s^{(\alpha-1)/2}}\sqrt{1-s},$$

$$\gamma_0 = \int_0^1 s(x)dx \Big/ \int_0^1 \frac{dx}{s(x)}, \qquad p = a^\alpha p_0, \qquad S(x) = aS(x),$$

$$\beta = S_{\min}/a. \tag{8.68}$$

Here a stands for the maximum value attained by the function $S(x)$ (in this context $a = S(\frac{1}{2})$), and x_* is the coordinate of the point at which the graph of $S(x)$ attains its minimal value, that is, $S(x_*) = S_{\min}$.

The system of equations used to derive the values of the constants p_0, β, and x_* is

$$I_1(\alpha, \gamma_0, \beta) = \frac{2\sqrt{p_0}}{\sqrt{(\alpha-1)}}(\tfrac{1}{2} - x_*), \qquad \tan[(p_0/\beta^\alpha)^{1/2}]x_* = \psi(\beta), \tag{8.69}$$

$$S_{\min} I_2(\alpha, \gamma_0, \beta) = \frac{2\sqrt{p_0}}{\sqrt{(\alpha-1)}}(\tfrac{1}{2} - S_{\min} x_*)\beta.$$

Here we use the notation

$$\left.\begin{aligned} I_1(\alpha, \gamma_0, \beta) &= \int_\beta^1 \frac{d\tau}{\Phi(\tau)}, \qquad I_2(\alpha, \gamma_0, \beta) = \int_\beta^1 \frac{\tau d\tau}{\Phi(\tau)}, \\ I_3(\alpha, \gamma_0, \beta) &= \int_\beta^1 \frac{d\tau}{\tau\Phi(\tau)}, \qquad \psi(\beta) = \frac{\sqrt{\gamma_0 + (\alpha-1)\beta^2}}{\sqrt{(\alpha-1)(1-\beta)(\gamma_0 + (\alpha+1)\beta)}}. \end{aligned}\right\} \tag{8.70}$$

The magnitude of γ_0 can be found by solving the equation

$$\gamma_0 = \left[\beta x_* + \tfrac{1}{2}\left(\frac{\alpha-1}{p_0}\right)^{1/2} I_2(\alpha, \gamma_0, \beta)\right] \Big/ \left[\frac{x_*}{\beta} + \tfrac{1}{2}\left(\frac{\alpha-1}{p_0}\right)^{1/2} I_3(\alpha, \gamma_0, \beta)\right]. \tag{8.71}$$

The curves shown in Figures 8.4 and 8.5 illustrate (8.68) to (8.71) for $\alpha = 2, 3$. For the case $\alpha = 1$, these relations were established numerically using an iterative technique.

Figure 8.4 illustrated the dependence of the critical temperature ϑ_* on the value of the parameter $S_{\min}$ for optimal rods. Curves 1, 2, and 3 correspond to the values of $\alpha = 1, 2, 3$, respectively. As $S_{\min}$ increases, the value of the critical temperature decrease in all the cases. This may be explained by pointing out that as $S_{\min}$ increases, the amount of material that can be redistributed decreases and the available choices for optimizing are reduced.

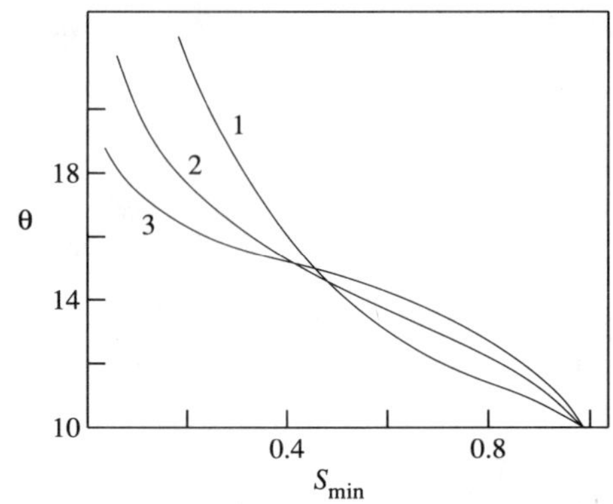

FIGURE 8.4. Dependence of critical temperature on S_{min}.

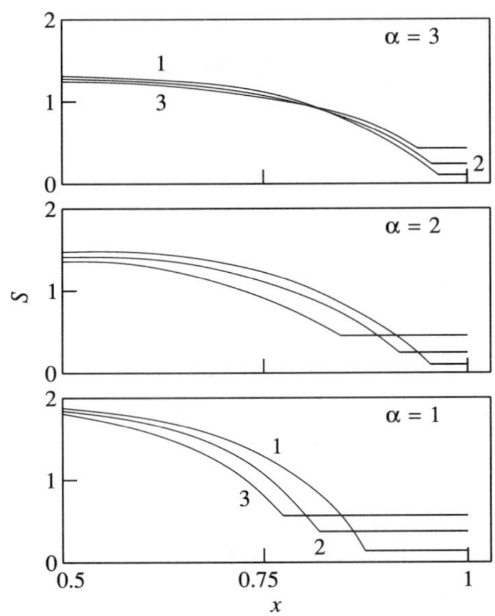

FIGURE 8.5. Optimal distribution of thickness.

For $S_{min} = 1$, there is only one choice of design for the thickness function consistent with conditions (8.58) and (8.59), which is $S(x) \equiv 1$. Here all values of the optimized functionals coincide.

The optimal thickness distribution functions for $\alpha = 1, 2, 3$ are shown in Figure 8.5. Curves 1, 2, and 3 correspond to the values $S_{min} = 0.1, 0.3, 0.5$, respectively. Because of symmetry with respect to the point $x = \frac{1}{2}$, the thickness functions $S(x)$ are drawn on the interval $\frac{1}{2} < x \leq 1$ only. All of the derived solutions indicate that, for a freely supported optimal rod, most of its mass is concentrated in its middle part, and near the ends the thickness assumes the minimum permitted value. As $S_{\min}$ is increased, the lengths of the segments of the optimal rod where the thickness is constant and equal to the minimal allowed value also increase.

Aside from the critical temperature, another important property of our problem is the magnitude of the reaction p related to ϑ by (8.55). Broken lines 1, 2, and 3 (corresponding to values of $\alpha = 1, 2, 3$, respectively) in Figure 8.3 illustrate the dependence of p on $S_{\min}$ for an optimal shape of the rod.

We remark that the solution of the problem of optimizing the critical temperature for a rod with rigidly built-in ends (with only simple eigenvalues) can be easily recomputed from a known solution for a rod with free end supports. The procedure is identical with the one discussed in the preceding section.

8.6 Stability and Design of Torsioned Rods

Let a straight elastic rod of length l be positioned along the x axis of a Cartesian cordinate system xyz. The ends of the rod are rigidity held at the points $x = 0$ and $x = l$. The rod is torsioned by a couple M applied to its end (see Figure 8.6). It is assumed that the rod has constant rigidity in various cross-sectional planes, and therefore that $EI_y = EI_z = a$, where E is Young's modulus for the material, and I_y and I_z are the moments of inertia of the transverse cross-sectional area with respect to lines passing through a point on the neutral axis of bending and parallel to the axes y and z, respectively.

In studying the stability of the rod and the critical values of the applied torque, we shall follow Euler's static approach. Let $y = y(x)$ and $z = z(x)$ be fundtions describing the shape of the axis of the buckled rod. We write down the corresponding equations of equilibrium and the boundary values

$$\left.\begin{aligned} (ay_{xx})_{xx} &= Mz_{xxx}, & (az_{xx})_{xx} &= My_{xxx}, \\ y(0) = y_x(0) = z(0) &= z_x(0) = 0, & y(l) = y_x(l) = z(l) &= z_x(l) = 0. \end{aligned}\right\} \tag{8.72}$$

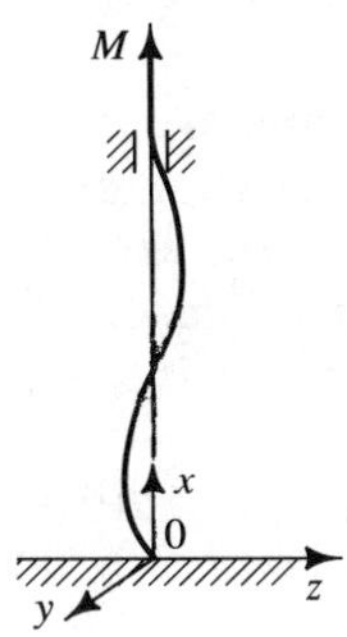

FIGURE 8.6.

We note that eqs. (8.72) are true only if we assume that the deflections are "small." The functions $y(x) = z(x) \equiv 0$ describe the equation of the unbuckled axis satisfying the equilibrium equations and the boundary conditions (8.72).

According to Euler's theory, the magnitude of the critical load and the loss of stability are determined by the smallest eigenvalue and the corresponding eigenfunctions $y(x) \not\equiv 0$ and $z(x) \not\equiv 0$ for the boundary-value problem (8.72).

For the particular case when torsional rigidity is uniformly distributed, that is, $a \equiv \text{const}$, finding the torque that causes a loss of stability is known to reduce to the computation of the smallest root for the (transcendental) equation $\tan(Ml/(2a)) = Ml/(2a)$. The magnitude of the critical torque is equal to $\pm 8.988\, a/l$. Following [8.13], let us derive an analog of this formula for the more general case when the distribution of torsional rigidity is a fairly arbitrary function of the variable x. To accomplish it, we integrate twice both sides of the equations (8.72). Then we multiply the second equation in its integrated form by i (where i is the imaginary unit) and add these equations side by side. After introduction of the complex variable $w(x) = y(x) + iz(x)$, the derived equations and boundary conditions become

$$aw_{xx} = -iMw_x + c_1 x + c_2, \tag{8.73}$$

where c_1 and c_2 are complex constants of integration. By integrating the linear differential equation (8.73) twice, and taking care of the boundary conditions, we derive the formula

$$w = \int_0^x \left[e^{-iM\varphi(t)} \int_0^t ((c_1\xi + c_2)/a(\xi)) e^{iM\varphi(\xi)} \mathrm{d}\xi \right] \mathrm{d}t, \tag{8.74}$$

and relations that determine the values of the constants

$$c_1 \int_0^l \frac{x}{a(x)} e^{iM\varphi(x)} \mathrm{d}x + c_2 \int_0^l \frac{1}{a(x)} e^{iM\varphi(x)} \mathrm{d}x = 0.$$

$$\left(\varphi(x) \equiv \int_0^x \frac{\mathrm{d}t}{a(t)} \right),$$

$$c_1 \int_0^l e^{-iM\varphi(x)} \int_0^x \frac{t}{a(t)} e^{iM\varphi(t)} \mathrm{d}t\mathrm{d}x + c_2 \int_0^l e^{-iM\varphi(x)} \int_0^x \frac{e^{iM\varphi(t)}}{a(t)} \mathrm{d}t\mathrm{d}x = 0. \tag{8.75}$$

The relations (8.75) represent a system of two homogeneous linear equations that may serve the purpose of determining values of unknown constants c_1 and c_2. A necessary condition for the existence of the nontrivial solutions of (8.74) (i.e., when the determinant of the coefficients of (8.75) vanishes) can be reduced to an equation that is to be solved for determining the critical torque

$$\begin{aligned} f(M) \equiv & \int_0^l \frac{e^{iM\varphi(x)}}{a(x)} \mathrm{d}x \int_0^l e^{-iM\varphi(x)} \int_0^x \frac{\xi e^{iM\varphi(\xi)}}{a(\xi)} \mathrm{d}\xi\mathrm{d}x \\ & - \int_0^l \frac{x e^{iM\varphi(x)}}{a(x)} \mathrm{d}x \int_0^l e^{-iM\varphi(x)} \int_0^x \frac{e^{iM\varphi(\xi)}}{a(\xi)} \mathrm{d}\xi\mathrm{d}x = 0. \end{aligned} \tag{8.76}$$

When rigidity is uniformly distributed along the length of the rod, that is, if $a(x) = \text{const}$ in (8.76), we arrive at the well-known formula $\tan(Ml/(2a)) = Ml/(2a)$. We note that the complex equation (8.76) can be written as a system of two real equations $\mathrm{Re}(f(M)) = 0$ and $\mathrm{Im}(f(M)) = 0$. In this manner a study of the stability of torsioned rods with variable rigidity may be reduced to a computation of quadratures for a given function $a(x)$ and the solution of the transcendental equation (8.76) for M.

We shall consider separately the case of the symmetric distribution of bending rigidity when $a(x) = a(l - x)$. In this case $\varphi(x) = \varphi(l/2) + \psi(x)$, where $\psi(x)$ is an antisymmetric function, that is, $\psi(x) = -\psi(l - x)$. Using these properties of the functions $a(x)$ and $\psi(x)$ and the rules of integration on symmetric domains of symmetric or antisymmetric functions, eq. (8.76) is transformed into the form

$$\begin{aligned} f(M) &= \int_0^{l/2} \frac{\cos(M\psi(x))}{a(x)}\mathrm{d}x \int_0^{l/2} [\chi_1 \cos(M\psi(x))\mathrm{d}x + \chi_2 \sin(M\psi(x))]\mathrm{d}x \\ &\quad + \int_0^{l/2} \frac{(x - l/2)}{a(x)} \sin(M\psi(x))\mathrm{d}x \int_0^{l/2} [\chi_3 \cos(M\psi(x))\mathrm{d}x \\ &\quad - \chi_4 \sin(M\psi(x))]\mathrm{d}x = 0. \end{aligned} \tag{8.77}$$

Here

$$\chi_1(x) = \int_0^x \frac{(\xi - l/2)}{a(\xi)} \cos(M\psi(\xi))\mathrm{d}\xi, \qquad \chi_2(x) = \int_0^x \frac{(\xi - l/2)}{a(\xi)} \sin(M\psi(\xi))\mathrm{d}\xi,$$

$$\chi_3(x) = \int_0^x \frac{\sin(M\psi(\xi))}{a(\xi)}\mathrm{d}\xi, \qquad \chi_4(x) = \int_0^x \frac{\cos(M\psi(\xi))}{a(\xi)}\mathrm{d}\xi.$$

The essential difference between (8.76) and (8.77) is that (8.77) is a single equation in the unknown M instead of a system of equations.

Now let us consider a torsioned rod freely supported at the points $x = 0$ and $x = l$. Finding a static condition for the loss of stability (Greenhill's problem) leads to the boundary-value problem

$$\left.\begin{aligned} ay_{xx} &= Mz_x, \qquad az_{xx} = -My_x, \\ y(0) &= y(l) = z(0) = z(l) = 0. \end{aligned}\right\} \tag{8.78}$$

We multiply both sides of the second equation in (8.78) by i and add them side by side. We introduce a function $w(x) = y(x) + iz(x)$ and transform the boundary-value problem (8.78) into the form

$$aw_{xx} = -iMw_x, \qquad w(0) = w(l) = 0. \tag{8.79}$$

Integrating eq. (8.79) twice and taking care of the assigned boundary conditions, we obtain a formula for w and a system of two simultaneous linear, homogeneous equations that can be solved to determine the constants of integration. The necessary condition on the existence of nontrivial solutions (determinant of coefficients for this system vanishing) reduces to a single complex equation in M, which may be written as a system of the two real equations

$$\int_0^l \cos(M\varphi(x))\mathrm{d}x = 0, \qquad \int_0^l \sin(M\varphi(x))\mathrm{d}x = 0. \tag{8.80}$$

The smallest common root of these equations defines a critical divergence torque. Therefore, finding critical torques amounts to finding a simultaneous solution for the system (8.80).

We shall show that for a symmetric distribution of rigidity, the system (8.80) may be replaced by a single equation. Let $a(x)$ be a quite arbitrary distribution of bending rigidity. We write $1/a(x)$ as a sum of the two terms $1/a(x) = \Omega(x) + \Psi(x)$, where $\Omega(x)$ and $\Psi(x)$ are, respectively, symmetric and antisymmetric functions with respect to the midpoint of the interval $[0, l]$. The function $\varphi(x)$ can also be represented as a sum of a symmetric and antisymmetric function $\omega(x)$ and $\psi(x)$, respectively

$$\left.\begin{aligned} &\varphi(x) = \omega(x) + \psi(x), \qquad \omega(x) = \int_0^{l/2} \Omega(t)\mathrm{d}t + \int_0^x \Psi(t)\mathrm{d}t, \\ &\psi(x) = \int_0^x \Omega(t)\mathrm{d}t - \int_0^{l/2} \Omega(t)\mathrm{d}t. \end{aligned}\right\} \tag{8.81}$$

Note the properties of integrals in symmetric domains of symmetric and antisymmetric functions and use (8.81) to transform eqs. (8.80). We obtain

$$\int_0^l \cos(M\psi(x))\cos(M\omega(x))\mathrm{d}x = 0, \qquad \int_0^l \cos(M\psi(x))\sin(M\omega(x))\mathrm{d}x = 0. \tag{8.82}$$

For an asymmetric distribution of rigidity (when $\Psi(x) \equiv 0$, $1/a = \Omega(x)$, and $\omega(x) \equiv \varphi(l/2)$), eqs. (8.82) assume the form

$$\cos(M\varphi(l/2))\int_0^l \cos(M\psi(x))\mathrm{d}x = 0, \quad \sin(M\varphi(l/2))\int_0^l \cos(M\psi(x))\mathrm{d}x = 0.$$

It follows directly that the system (8.80) can be reduced in this case to the single equation

$$\int_0^l \cos(M\psi(x))\mathrm{d}x = 0, \qquad \psi(x) = \varphi(x) - \varphi(l/2). \tag{8.83}$$

For a constant distribution of rigidity, eq. (8.83) predicts the well-known value of the critical torque $M = 2\pi EI/l$. The solution of eq. (8.83) can be found very effectively using one of several analytic or numerical techniques, in particular, Newton's method. Thus, eq. (8.83) results in a real simplification, particularly in solving problems such as the optimization of the rigidity function in which it is necessary to recompute the critical value of M several times for various approximate values of $a(x)$.

Let us examine some of the simplest problems of optimizing stability. Let a rod have similarly shaped cross-sections, and consequently $a(x) = k[S(x)]^2$, where $S(x)$ is the function defining the distribution of magnitude of the area. We formulate the following problem in terms of dimensionless variables: We need to find a function of the cross-sectional area $S(x)$ satisfying $S_{\min} < S(x)$, and also a constant volume condition, and assign a maximum to the critical value of the torque:

$$M_* = \max_S M, \tag{8.84}$$

where M is computed from (8.77) and (8.83), and $S_{\min}$ is a given dimensionless quantity.

For the indicated constraints on the volume of the rod, and the allowable values of its thickness problem (8.84) was solved numerically in [8.13] using an iterative technique algorithm. This algorithm applied to our optimization problem consists in finding an approximate value of $a(x)$ using Newton's method (in the process of iterating critical values of M) from the current value of $a(x)$, and then improving the variation of the control function $a(x)$ by the gradient projection technique. The computation of the critical value of M from (8.77) and (8.83) was carried out to within 10^{-5} accuracy. The computation terminates when the magnitude of the gradient of the optimized functional (i.e., the error in satisfying the optimality conditions) was smaller than 10^{-3}.

The optimal functions $S(x)$ were found for several values of the parameter $S_{\min}$.

Figure 8.7 illustrates the function $S(x)$ found as a result of these numerical computations for a rod freely supported at both ends. Curves 1, 2, and 3 correspond to the values of the parameter $S_{\min} = 0.98$, 0.92, and 0.88, respectively. For these values of the parameter, the magnitude of the critical torque M is 6.56, 7.21, and 7.80, respectively. The relative gain in comparison with a rod having a constant rigidity function and a unit volume amounts to 4.45, 14.9, and 24.2%, respectively.

Numerical solutions were also carried out for rods with built-in ends. The gain for the gradient in the stability optimization problem was carried out with the help of the following representation for the eigenvalue of the boundary-value problem (8.72)

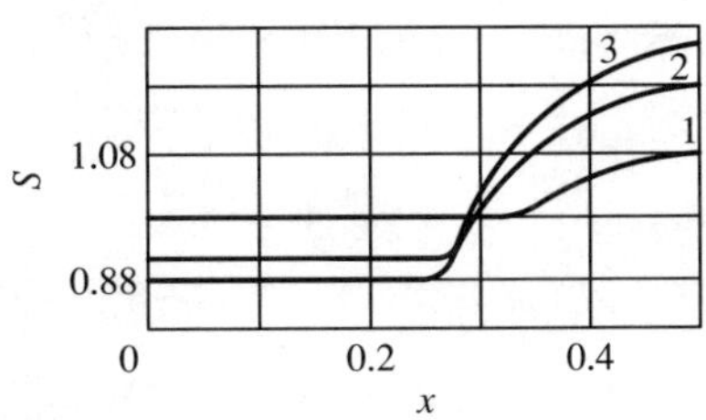

FIGURE 8.7. Optimal area distribution for a freely supported rod.

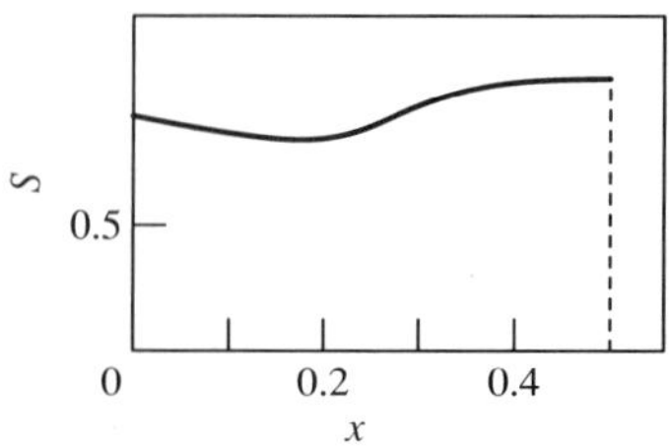

FIGURE 8.8. Optimal distribution of cross-sectional area.

$$M = -\int_0^1 S^2 \operatorname{Re}(w_{xx} w_{xx}^*) dx \Big/ \int_0^1 \operatorname{Im}(w_x w_{xx}^*) dx,$$

where $w(x)$ is computed from (8.74) and (8.75), and w^* is the complex conjugate function. Numerical computation indicated that for $S_{\min} < 0.865$ the constraint $S_{\min} \leq S$ becomes inactive. Figure 8.8 illustrates the optimal distribution function of the cross-sectional area for $S_{\min} < 0.865$. The critical value of the torque M for an optimal rod is 9.2789. For the sake of comparison, the first four eigenvalues M for an optimal rod are 9.2789, 14.9503, 21.149, and 27.274, and the first four eigenvalues for a rod with constant cross-sectional area ($S(x) \equiv 1$) are 8.98688, 15.3506, 21.808, and 28.132.

8.7 Design of Elastic Rods for Simultaneous Compression and Torsion

Let a straight elastic rod be rigidly built in at its endpoints $x = 0$ and $x = l$ in a Cartesian coordinate system xyz. A torsion couple M and a compressive force P are applied to the ends of the rod. We assume that the rod has an identical bending rigidity in the different planes (i.e., $EI_y = EI_z = a$), and we shall represent the dependence of the rigidity a on the magnitude of the cross-sectional area S by the formula $a(x) = kS^2(x)$, where the constant k is determined by the shape of the cross-sectional area. In examining the stability of the rod and in computing the magnitude of the critical torque, we shall use the assumption that this problem is conservative, and apply Euler's static approach (see [8.20], [8.48] and [8.91]). The equilibrium equations and the boundary conditions for the torsioned rod can be written as follows:

$$\begin{aligned}
(az_{xx})_{xx} &= -Pz_{xx} - My_{xxx}, \\
(ay_{xx})_{xx} &= -Py_{xx} + Mz_{xxx}, \\
y(0) = y_x(0) &= z(0) = z_x(0) = 0, \\
y(l) = y_x(l) &= z(l) = z_x(l) = 0. \qquad (8.85)
\end{aligned}$$

After some multiplication and addition of the right sides and left-hand sides of (8.85) side by side, then integration twice from $x = 0$ to $x = l$, we obtain a formula for the magnitude of the critical moment M corresponding to the loss of stability for a given compression load P:

$$M = \frac{\int_0^l (a(x)(y_{xx}^2 + z_{xx}^2) - P(y_x^2 + z_x^2))\mathrm{d}x}{\int_0^l (y_{xx}z_x - y_x z_{xx})\mathrm{d}x}. \tag{8.86}$$

The optimization problem consists in finding the distribution of cross-sectional areas that assigns the largest value to the critical moment causing a loss of stability (as given by (8.86)) and such that the constraint on the volume of material

$$\int_0^l S(x)\mathrm{d}x = v, \tag{8.87}$$

and the constraints on the admissible thickness of the rod are satisfied.

Problem (8.85) to (8.87) and the other problems discussed in this chapter belong to the class of self-adjoint optimization problems and it is not necessary to introduce adjoint variables.

A numerical solution of the last problem can be found in [8.14]. It makes use of an iterative optimization algorithm. An expression for the improvement of variation δS is derived using a gradient projection technique. It has the form

$$\left.\begin{aligned} &\delta S = \tau\psi, && \psi = \Lambda - \frac{1}{l}\int_0^l \Lambda\mathrm{d}x, \\ &\Lambda = \frac{2k}{T}(y_{xx}^2 + z_{xx}^2), && T = \int_0^l (y_{xx}z_x - y_x z_{xx})\mathrm{d}x \end{aligned}\right\} \tag{8.88}$$

where $\tau > 0$ is a given constant. Relations (8.88) ensure that the variation δM is positive and that the variational constraint (8.87) is obeyed. The solution of the direct problems of finding functions $y(x)$ and $z(x)$ and critical moments M is based on ideas of [8.14] and the application of the successive iterations technique.

For the sake of convenience, we introduce a complex function $w = y(x) + iz(x)$ so that formulas for the iterative technique of solving the boundary-value problem (8.85) for the eigenvalues P may be written in the form

$$\left.\begin{aligned} &(a(x)w_{xx}^j)_{xx} + Pw_{xx}^j = -iw_{xxx}^{j-1}, \\ &w^j(0) = w_x^j(0) = w^j(l) = w_x^j(l). \end{aligned}\right\} \tag{8.89}$$

Here i stands for the complex unity, and j is the iteration counter ($j = 1, 2, 3, \ldots$). To compute the value of our functional, we use the formula

$$M_j(S) = \frac{\mathrm{Re}\int_0^l (a(x)w_{xx}^j w_{xx}^{j*} - Pw_x^j w_x^{j*})\mathrm{d}x}{\mathrm{Im}\int_0^l (w_x^j w_{xx}^{j*})\mathrm{d}x}. \tag{8.90}$$

Relations (8.89) become, after integrating twice,

$$a(x)w_{xx}^j + Pw^j = -iw_x^{j-1} + C_1 x + C_2, \tag{8.91}$$

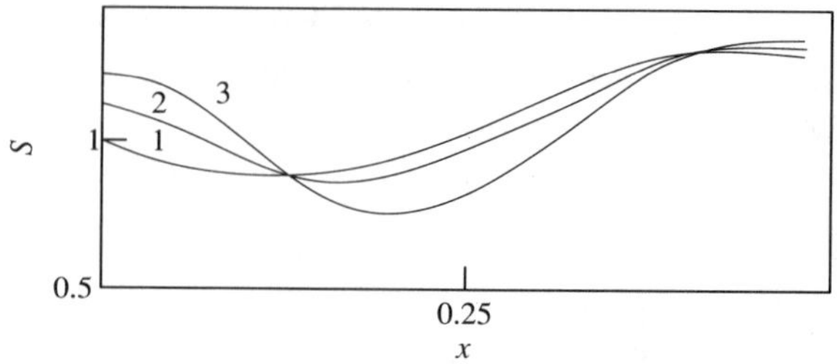

FIGURE 8.9. Optimal mass distribution.

where C_1 and C_2 are arbitrary complex constants. Futhermore, we introduce functions $w_1(x)$, $w_2(x)$, and $w_3(x)$ that satisfy the boundary conditions $w_i(0) = w_i(l) = 0$, (for $i = 1, 2, 3$) and solve the differential equation (8.91) with its right-hand side replaced by $-iw_x^{j-1}$, x, and 1, respectively. Because of the linearity of the differential equation (8.91), its general solution may be written as

$$w_j(x) = w_1^j(x) + C_1 w_2^j(x) + C_2 w_3^j(x). \tag{8.92}$$

The unknown constants C_1 and C_2 that appear in (8.91) and (8.92) may be determined from the conditions $w_x^j(0) = w_x^j(l) = 0$. After deriving the functions $w_i^j(x)$ and constants C_1 and C_2, we see that the function $w^j(x)$ is a solution to the boundary-value problem (8.89). Therefore, the process of solving for the eigenvalues of the boundary-value system (8.89) has been reduced by means of this argument to the solution of a system of three simultaneous auxiliary boundary-value problems with lower order (for the functions $w_i(x)$), which may be computed by a successive elimination technique.

As a result of our computations, the derived distribution of area (in dimensionless variables) is illustrated in Figure 8.9. Because of symmetry with respect to the middle point of the rod ($x = \frac{1}{2}$), the function $S(x)$ is shown only on $0 \leq x \leq \frac{1}{2}$. Curves 1, 2, and 3 correspond to the values $P = 0$, 15.8, and 31.6, respectively. These solutions indicate that for an optimal rod built-in at the ends, most of its mass is concentrated in the middle and at the ends. This effect becomes more pronounced as we increase the value of P. The dependence of the critical torque M^* on the values of the parameter P for an optimally designed rod can be regarded for practical applications as linear. For the computed variants, the gain due to optimization is calculated from the formula $((M_* - M_0)/M_0) \cdot 100\%$, and varies between 3.7% and 28%.

8.8 Multiplicity of Critical Loads in Problems of Optimal Stability

In many problems of optimal design, it has been observed that the points of discrete spectrum frequently come close to each other and multiple critical loads appear. This phenomenon creates real difficulties in designing structures with optimal stability properties. A large number of research works on the subject of structural optimization has as its objective a study of different aspects of cases with multiple eigenvalues.

Let us first discuss the possibility of the occurrence of double eigenvalues. We formulate a general optimization problem depending on a single parameter, and study the behavior of the eigenvalues as we change this parameter. Let H_α denote the set of admissible design parameters, that is, $\mathbf{h} \in H_\alpha$. The subscript signifies the dependence of admissible values on the parameter α. For example, in many problems of optimizing stability, one takes as the design variable the thickness function $h(x)$ for a structural part, which must satisfy a constant volume condition and constraints on the minimal values of the thickness:

$$\int_\Omega h d\Omega = 1, \qquad h \geq h_{\min} = \alpha.$$

Here we have selected the quantity $h_{\min}$ as the parameter defining the admissible set. (We use dimensionless variables!)

Let us examine the following optimization problem. We wish to find a function $h(x)$ that assigns a maximum to the first eigenvalue

$$\left.\begin{aligned} p_1 &= \max_h p_1(h), \qquad &(8.93)\\ h &\in H_\alpha &(8.94)\end{aligned}\right\}$$

for the boundary-value problem

$$L(h)u = pu. \tag{8.95}$$

We assume that the optimization problem (8.93) to (8.95) can be solved for values of α filling some interval and that the exact dependence of the design function $h(x, \alpha)$ on the parameter α is known. The corresponding eigenfunctions and eigenvalues of the boundary-value problem (8.95) also depend on the parameter α and on the parameters $u_i(x, \alpha)$, $p_i(\alpha)$. The behavior of the eigenvalues that depend on a design parameter is of special interest in the theory of optimal design. This problem has been widely discussed in relation to other important problems (see [8.7] and [8.8]). For example, for systems that have only a finite number of degrees of freedom and do not have any symmetry (i.e., systems in general position), a general result given in [8.8] is widely known. According to this result, (in general) a change in the value of a single parameter will not make it possible to merge two natural frequencies. An analogous assertion can be made (see [8.8]) for vibrations of a continuous medium.

Let $h^0(x)$ be a solution of the optimal design problem (8.93) to (8.95) for some fixed value α^0 of the parameter. For a given design variable $h^0(x)$, there exists a corresponding spectrum of eigenvalues and eigenfunctions for the boundary-value problem (8.95). Let us look at the first two eigenvalues p_1 and p_2 of the problem (8.95). Aside from the fixed value α^0, we consider small perturbations of the parameter $\alpha = \alpha^0 + \delta\alpha$, where we assume that α is "small." To this (perturbed) value of α corresponds an optimal solution $h(x) = h^0(x) + \delta h$. For this changed value of the parameter the eigenvalue problem may be written as

$$\left.\begin{aligned} &(L^0 + \delta L)u = pu, \\ &L^0 = L(h^0(x)), \qquad \delta L = L(h^0(x) + \delta h(x)) - L(h^0(x)). \end{aligned}\right\} \tag{8.96}$$

We determine the magnitude of the perturbations for which the eigenvalues are equal to $p_1^0 = p_1(\alpha^0)$ and $p_2^0 = p_2(\alpha^0)$ as we vary the parameter. The solution of problem (8.96) is written as a sum

$$u = c_1 u_1^0(x) + c_2 u_2^0(x),$$

where $u_i^0(x)$ (with $i = 1, 2$) are eigenfunctions corresponding to the eigenvalues p_i^0. After some routine substitutions, we derive the formulas

$$\left.\begin{aligned} &p_{1,2} = \tfrac{1}{2}(p_1^0 + p_2^0 + A_{11} + A_{22}) \pm \sqrt{\tfrac{1}{4}(p_1^0 - p_2^0 + A_{11} - A_{22})^2 + A_{12}^2}, \\ &A_{ij} = \int_0^l u_i^0(x)\delta L u_j^0(x)\mathrm{d}x. \end{aligned}\right\} \tag{8.97}$$

It is clear from (8.97) that merging the first two eigenvalues is possible only if the expression under the square root vanishes. Since the radical is a sum of two squares, it can be equal to 0 only if the two conditions

$$p_1 - p_2 + A_{11} - A_{22} = 0, \tag{8.98}$$

$$A_{12} = 0 \tag{8.99}$$

are simultaneously obeyed. The quantities appearing on the left-hand side of (8.98) and (8.99) are functions of α. Therefore, in general, it is impossible to effect merging of the eigenvalues by altering the value of α. (The curves $p_1 = p_1(\alpha)$ and $p_2 = p_2(\alpha)$ do not, in general, intersect each other.) An exception to this rule is a rare case, the probability of which is equal to the probability of finding equal roots of independent algebraic equations. However, for problems with symmetries such an occurrence is possible, since the condition $A_{12} = 0$ is an identity that is satisfied for any choice of α. Thus, for example, if $u_1^0(x)$ and δA are both symmetric with respect to the middle point of the interval $(0, l)$ and $u_2^0(x)$ is antisymmetric with respect to $x = \frac{1}{2}l$, then $A_{12} \equiv 0$. This condition is realized in studying the optimization of the stability of compressed columns that have symmetric support conditions at both ends.

The oscillation theorem of Sturm is applicable to design problems for compressed columns, where it can be used to make some important arguments.

Let us consider the self-adjoint eigenvalue problem

$$(\varphi(x)u_x)_x + [\lambda\psi(x) + \chi(x)]u = 0, \tag{8.100}$$

where $\varphi(x) > 0$, $\psi(x) > 0$, and $\chi(x)$ are continuous and $\varphi(x)$ is a continuously differentiable function on the interval $(0, l)$. General self-adjoint boundary conditions are written in the form

$$u(l) = au(0) + bu_x(0), \qquad u_x(l) = cu(0) + du_x(0). \tag{8.101}$$

Here $ad - bc = \varphi(0)/\varphi(l)$. The set of all eigenfunctions of the boundary-value problem (8.100) and (8.101) may be labeled in such manner that the corresponding eigenvalues form two unbounded increasing sequences

$$\lambda_1 < \lambda_2 < \cdots, \qquad \lambda_1^* < \lambda_2^* < \cdots, \tag{8.102}$$

with

$$\lambda_n < \lambda_n^* < \lambda_{n+1} \cdots. \tag{8.103}$$

It follows from (8.102) and (8.103) that the multiplicity of the eigenvalues for the problem (8.100) and (8.101) is at most 2. For the "Sturm boundary conditions",

$$\left.\begin{aligned} au(0) + bu_x(0) = 0, \qquad cu(l) + du_x(l) = 0, \\ |a| + |b| > 0, \qquad |c| + |d| > 0, \end{aligned}\right\} \tag{8.104}$$

the spectrum is real and the eigenvalues form an unbounded monotone increasing sequence. The multiplicity is equal to 1 for all eigenvalues. The eigenfunction corresponding to the eigenvalue λ_n has exactly n zeros (knots) in the interval $(0, l)$. It is a direct consequence of this theorem that for columns with freely supported ends, no matter what thickness function is adopted (see [8.33]), all critical loads are simple.

The behavior of columns with rigidly built-in ends comes under boundary conditions of the type (8.101). A corresponding oscillation theorem indicates that there is a possibility that a singularity corresponding to a double multiplicity of eigenvalues may occur. However, if no symmetry is exhibited, the occurrence of such singularity is highly improbable.

The boundary-value problem for the stability of a column with built-in ends and symmetric distribution of thickness (as was indicated in Section 8.4), can be decomposed into two separate problems for eq. (8.48) with boundary conditions (8.51) and (8.52), which result in symmetric (even) and antisymmetric (odd) solutions. For a column of constant cross-section, the spectrum of a full problem consists of interlaced eigenvalues corresponding to the symmetric and antisymmetric eigenfunctions, respectively. We have $p_1^s = 4\pi^2$, $p_1^a = 80.77$, and $p_2^s = 16\pi^2, \ldots$, so that $p_n^s \leq p_n^a < p_{n+1}^s$.

Therefore, in solving optimal problems concerning stability within a class of symmetric thickness functions, multiple eigenvalues must correspond to eigenfunctions with different odd-even properties.

8.9 Bimodal Solutions for Built-In Compressed Rods

We shall consider a problem of optimal design of an elastic straight rod with built-in ends at the points $x = -l$ and $x = l$ subjected to a compressive load p. Many research papers on this problem (see [1.5], [8.15], [8.52], [8.62], and

[8.76]) have provided both analytic and numerical solutions. Following [8.13], we introduce in this section a solution based on symmetry properties and the decomposition of the spectrum consisting of eigenvalues for the boundary-value problem of elastic stability.

At the same time, we eliminate properties connected with the multiplicities of the critical loads, so that the original optimization problem reduces to the classical problem of maximizing a simple eigenvalue.

Let us write down the basic formulas for maximizing a load that causes a loss of stability by finding the best possible distribution of the cross-sectional areas:

$$p = \max_S \min_u \Phi(u, S), \tag{8.105}$$

$$\left.\begin{aligned} \Phi(u, S) &= \int_{-l}^{l} EIu_{xx}^2 \mathrm{d}x \Big/ \int_{-l}^{l} u_x^2 \mathrm{d}x, \\ u(l) &= u_x(l) = u(-l) = u_x(-l) = 0, \end{aligned}\right\} \tag{8.106}$$

$$\int_{-l}^{l} S(x)\mathrm{d}x = V, \qquad S(x) \geq S_{\min}, \tag{8.107}$$

where $I = A_\alpha S^\alpha$, $\alpha = 1, 2, 3$. We seek the maximum in (8.105) in the class of symmetric smooth functions (denoting the distribution of the cross-sectional area) that satisfy a constraint on the minimum admissible value of thickness and a constraint concerning the constant total volume of the material. We observe that for a symmetric distribution of rigidity $EI(x)$, the Euler–Lagrange differential equation for a minimum of the functional (8.105), that is, the equation $(EIu_{xx})_{xx} + pu_{xx} = 0$, with homogeneous boundary conditions, remains invariant under the inversion transformation $x \to -x$. This implies that our arguments of Section 8.2 are applicable. Using (8.17) to (8.21), we arrive at the following formulas for computing the first eigenvalues for symmetric and antisymmetric solutions:

$$p_1 = \min_{u^i} \Phi(u^i, S), \qquad i = s, a, \tag{8.108}$$

$$u_x(0) = u_{xxx}^s(0) = 0, \qquad u^s(l) = u_x^s(l) = 0, \tag{8.109}$$

$$u_x^a(0) = u_{xxx}^a(0) = 0, \qquad u^a(l) = u_x^a(l) = 0. \tag{8.110}$$

In computing scalar products in (8.105) for the functional Φ, integration is carried out on the interval $[0, l]$. The solution of the optimization problem depends in a natural manner on results of numerical computations. In carrying out such computations, we use the dimensionless variables $x' = x/(2l)$, $S' = 2lS/V$, $p' = (2l)^{\alpha+2}p/(A_\alpha EV^\alpha)$. In dimensionless variables (with primes omitted), the search for an optimal thickness function is carried out on the interval $0 \leq x \leq \frac{1}{2}$, while the total volume is equal to 1.

We shall investigate the auxiliary problem of finding the maximum of the first eigenvalue $\max_S p_1^s$ with constraints (8.107) to (8.110), which corresponds

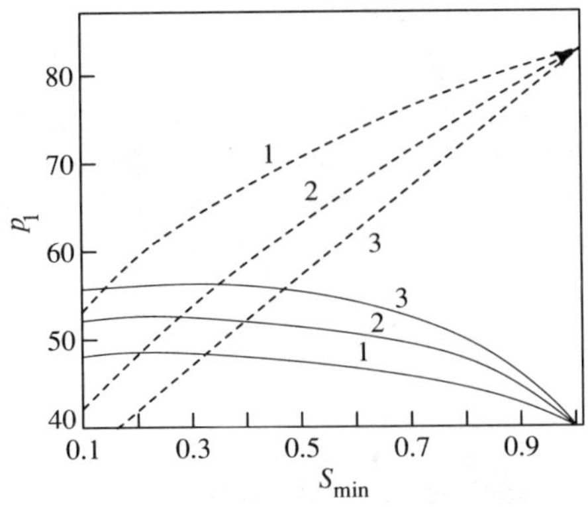

FIGURE 8.10. The magnitude of the first eigenvalue as a function of S_{min}.

to a symmetric shape $u^s(x)$. The solution to this optimization problem was derived by a numerical technique in [8.13] using iterative optimization for different values of the parameter S_{min} ($0 < S_{min} < 1$) and for different types of cross-sections ($\alpha = 1, 2, 3$). Aside from $p_1^s(S_0^s)$, computed for the derived thickness function S_0^s, the values of $p_1^a(S_0^s)$, $p_2^s(S_0^s)$, and $p_2^a(S_0^s)$ were also computed. In Figure 8.10 solid lines 1, 2, and 3 indicate cases corresponding to values of $\alpha = 1$, 2, and 3, respectively. The magnitude of $p_1^a(S_0^s)$ as a function of the parameter S_{min} is shown in Figure 8.10 by the broken lines 1, 2, and 3 again corresponding to the cases $\alpha = 1$, 2, and 3, respectively. Looking at the curves displayed in Figure 8.10, it should become clear that there exist intervals in which the parameter S_{min} varies between $S_{min}^\alpha \le S_{min} \le 1$, on which it is true that

$$p_1^s(S_0^s) < p_1^a(S_0^s). \tag{8.111}$$

Making use of inequality (8.111), it is easy to show that the solution of the auxiliary problem $\max_S p_1$ of maximizing the first eigenvalue p_1^a corresponding to an antisymmetric case with constraints (8.107), (8.108), and (8.110) must satisfy the inequality

$$p_1^s(S_0^a) < p_1^a(S_0^a). \tag{8.112}$$

To prove this inequality, we introduce the chain of inequalities

$$p_1^s(S_0^a) < p_1^s(S_0^s) < p_1^a(S_0^s) < p_1^a(S_0^a). \tag{8.113}$$

The inequality between the first and last item in this chain expresses the statement that for a rod with nonoptimal distribution of thickness, the force causing a loss of stability is smaller than the critical force for an optimal rod. The second inequality in the chain (8.113) is true on any interval $S_{min}^a \le S_{min} \le 1$ (this follows from computations). Therefore, for $S_{min}^a \le S_{min} \le 1$, a solution S_0^s of the auxiliary problem determines the optimal design project for a compressed rod.

For $0 \leq S_{\min} \leq S^a_{\min}$ the optimization problem consists in finding, for each fixed value of the parameter $S_{\min}$, the rod's thickness distribution function $S(x)$, which assigns a maximum to the functional p^s_1, while satisfying constraints (8.107), such that the first eigenvalue corresponding to a loss on the stability of the symmetric shape does not exceed the magnitude of the first eigenvalue corresponding to an antisymmetric shape; that is,

$$p^s_1 \to \max_S \tag{8.114}$$

$$p^a_1 \geq p^s_1. \tag{8.115}$$

It is possible to show that equality is realized in condition (8.115) (i.e., multiple eigenvalues occur). Indeed, let us assume that for some $S_{\min}$ in the interval $0 < S_{\min} < S^a_{\min}$ in (8.115) the strict inequality $p^a_1 > p^s_1$ is true. This does contradict numerical results in solving the problem $p^s_1 \to \max_S$ with the constraint (8.107), since for such solutions $p^a_1(S_0) < p^s_1(S_0)$ (see Figure 8.10). Therefore, in solving problem (8.114) and (8.115) on the interval $0 \leq S_{\min} \leq S^a_{\min}$, we need to consider only the case $p^a_1 = p^s_1$.

To obtain necessary optimality conditions and to derive gradient projection

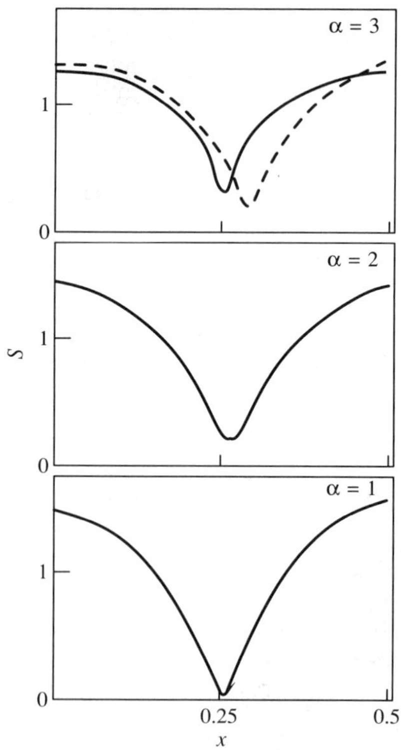

FIGURE 8.11. Optimal distribution of the area $S(x)$.

formulas for the optimized functional, we display the Lagrangian functional for the problem discussed here:

$$\Pi = \min_{u^s} \Phi(u^s, S) - \gamma\left(\min_{u^s} \Phi(u^a, S) - \min_{u^s} \Phi(u^s, S)\right) - \beta \int_0^{.5} S(x)dx, \quad (8.116)$$

where β and γ are unknown constants (Lagrangian multipliers).

The unknown Lagrangian functional Π is varied with respect to S while obeying the constraint $S \geq S_{\min}$.

The orthogonality condition on the full interval between the symmetric and antisymmetric shapes that cause a loss of stability is automatically satisfied, and in the formula for the Lagrangian functional, it is not necessary to introduce sums of terms in order to satisfy this condition. This is one of the advantages of utilizing a decomposition based on classifying shapes by symmetry. We observe that in [8.62], necessary conditions were derived for an extremum using a different representation of the Lagrangian, without making any use of symmetry properties. However, in that case one needs orthogonality condition for the shapes corresponding to a loss of stability, because this condition is not satisfied "automatically."

Optimal functions $S(x)$ for cases corresponding to $\alpha = 1, 2$, and 3 obtained by an iterative optimization algorithm are indicated by the digits 1, 2, and 3 Figure 8.11.

8.10 Design of Elastic Cylindrical Shells

We shall consider the problem of maximizing the critical load for an orthotropic shell with variable thickness. The distribution of the thickness is chosen as the design variable (see [8.54] and [8.55]). The critical load and corresponding shape causing the loss of stability are determined, respectively, as the minimal positive eigenvalue p_1 and the corresponding eigenfunction $\mathbf{u}^1 = (u_1, u_2, u_3)$ of the eigenvalue problem $L(h)\mathbf{u} - \lambda N\mathbf{u} = 0$, where L and N are linear symmetric operators. The operator $L = L(h)$ depends on the distribution of thickness function, which varies in both the longitudinal and circumferential directions, that is, $h = h(z, \vartheta)$.

The optimization problem consists in finding a distribution of thickness that satisfies constraints assigned to the weight and admissible thickness

$$V(h) \leq \beta, \qquad h_{\min} \leq h < h_{\max}, \qquad \|h\|_{W_p^1(\Omega)} \leq c, \qquad (8.117)$$

and such that the critical load assumes, the maximum value

$$p_1^* = \sup_h \inf_u \frac{(L(h)u, u)}{(Nu, u)}, \qquad (8.118)$$

where β, c, $h_{\min}$, and $h_{\max}$ are given constants. To construct a numerical solution to this problem, we replace the design function by a finite-dimensional vector $h_N = \{h_{ij}\}$ having dimension N:

$$h_N(z, \vartheta) = \sum_{i=1}^{n} \sum_{j=0}^{m} h_{ij} s_{ij}(z, \vartheta), \tag{8.119}$$

where $s_{ij}(z, \vartheta)$ are base functions in the space of piecewise continuous functions. Applying finite-dimensional representations for the unknown quantities, we can represent the initial optimization problem in the formalism of nonlinear programming problems:

$$\left.\begin{aligned} &V(h_N^*) \leq \beta, \qquad h_{\min} \leq h \leq h_{\max}, \\ &P_1(h_N^*) = \max_h \inf_u \frac{(L(h)u, u)}{(Nu, u)}. \end{aligned}\right\} \tag{8.120}$$

As an example, we consider an orthotropic cylindrical shell rigidly built-in at its ends with $l = 3$ m, $R = 0.75$, $h = .004$ m, $\nu_1 = \nu_2 = 0.11$, $E_1 = E_2 = 3.7 \times 10^{10}$, and $G = 0.75 \times 10^{10}$. A uniform external pressure is applied to the entire side surface of the shell.

An approximate solution for the stability of this shell can be found in [8.55]. It was computed by the Ritz–Galerkin technique, utilizing tensor products of cubic splines and trigonometric polynomials. As a result, it was found that the critical pressure was 4.3×10^6. The shape corresponding to the loss of stability resembled three waves along the circumferential coordinate.

Further, we considered the maximization of the critical load for this shell. The thickness function for the shell was sought in the class of piecewise linear continuous functions, depending only on the axial coordinate (i.e., $h = h(z)$, symmetric with respect to the center of the shell, with value of the optimization parameter $N = 20$ and with the constraints $h_{\min} = 0.02$ m and $h_{\max} = 0.08$ m.

The nonlinear programming problem using penalty functions was reduced to the following sequence of problems of unconditional minimization:

$$F^i(h) = -p_1(h) + C_p^{(i)} V(h) + C_h^{(i)} \sum_{j=1}^{20} H_j(h),$$

where h_j are the values of the thickness at the nodes along the axial coordinate z. $C_p^{(i)}$ and $C_h^{(i)}$, are penalty coefficients satisfying $C_p^{(i)}, C_h^{(i)} > 0$, $C_p^{(i)} < C_p^{(i+1)}$, $C_h^{(i)} < C_h^{(i+1)}$ (with $i = 0, 1, \ldots$), and $C_p^{(i)}, C_h^{(i)} \to \infty$ as $i \to \infty$. This was solved by a coordinate-wise descent combined with a one-dimensional Fibbonacci search.

In Figure 8.12 the solid line illustrates the optimal distribution of the shell's thickness function, whose weight is the same as the weight of a shell with a uniform thickness (as indicated by a dotted line). After optimization the critical load is increased roughly by a factor of 2. The displayed distribution of thickness indicates that material is concentrated in the vicinity of the rigidly held edge and in the central part, where it resembles a trisymmetric design for a trestle.

In Figure 8.13 we illustrate the dependence of the critical load p_1 on the number of corrugations n along the circumferential coordinate ϑ for a shell of constant thickness (curve 1) and of variable thickness (curve 2).

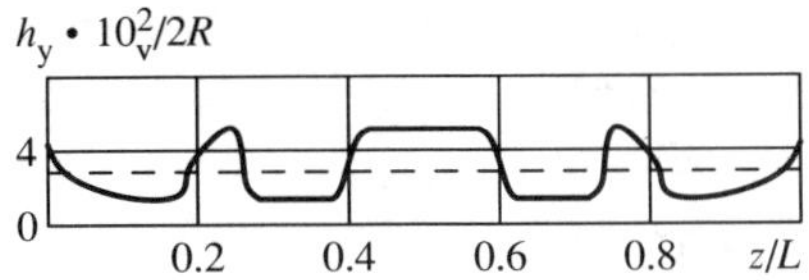

FIGURE 8.12. Thickness of the optimal shell.

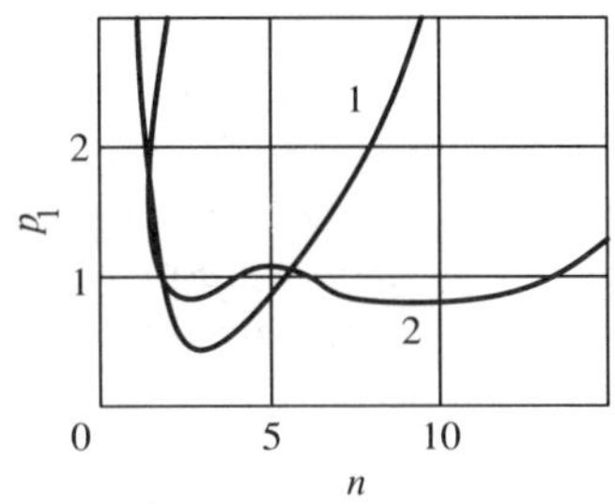

FIGURE 8.13. Dependence of critical load on the number of corrugations.

The minimal values of p_1 are attained for $n = 3, 7, 8, 9$, and 10. In these cases we obtain one common shape for a loss of stability and 11 local shapes. The common shape arises when we have three corrugations along the circumferential coordinate, two local ones for $n = 7$ and $n = 10$, four for $n = 8$, and three for $n = 9$. We do not regard as a local form of the loss of stability any shape such that, for some subsurface $\Omega_1 \in \Omega$, $u \equiv 0$.

Results presented in [8.55] imply that in solving this optimization problem for a shell, the multiplicity of the first eigenvalue increases and the shell becomes more sensitive to larger numbers of possible types of instability as the optimal solution is approached. In turn, this phenomenon may have as a consequence a greater sensitivity to imperfections.

8.11 Optimization in Nonconservative Problems of Elastic Stability

In preceding sections of this chapter, we introduced solutions to problems of optimal structural design in order to investigate stability by static techniques. However, in designs of nonconservative systems it is necessary to conduct stability analysis based on dynamic criteria. A dynamic approach makes optimization problems much more complex, and so far only a limited number of problems has been solved (see [8.16], [8.50], [8.65], [8.66], [8.73], [8.75], [8.77], [8.78], and [8.96]). Since research in this area is in its infancy, we shall limit our discussion to selected results contained in a single paper [8.27].

The study of the stability of linear systems can be based on the analysis of eq. (8.3) for the amplitude function, which after substituting $\lambda = i\omega$ becomes

$$[C + pK + \lambda B + \lambda^2 A]u = 0. \tag{8.121}$$

The natural frequency $\lambda = \lambda_{Re} + i\lambda_{Im}$ corresponds to complexification of the (right and left) eigenfunctions u and v. We assume that the operator K is not self-adjoint. A dynamic loss of stability (flutter) occurs if at least one of the characteristic curves of the system (8.121) drawn in the space spanned by λ_{Re},

λ_{Im}, p intersects the plane $\lambda_{\mathrm{Re}} = 0$ for some values of $p = p_{\mathrm{fl}}, \lambda_{\mathrm{Im}} = (\lambda_{\mathrm{Im}})_{\mathrm{fl}}$. The optimization problem consists in maximizing the critical value of the load parameter p_{fl} by selecting a suitable design variable h for a given total weight of the structure. The operators of (8.121) depend on h.

Sensitivity analysis is conducted as we use the idea of a generalized solution. The equation for the amplitude function can be written in the form of a scalar product

$$(v, [C + pK + \lambda B + \lambda^2 A]u) = 0, \tag{8.122}$$

where we assume differentiability of (8.122) with respect to the variable h. Now we compute the variation of (8.122) caused by varying the design variable (i.e., we replace h by $(h + \delta h)$, where δh represents a "small" variation of the function h). Recalling that this variation is taken for critical values of the parameters, we derive

$$\begin{aligned}(v_{\mathrm{fl}}, [\delta C + p_{\mathrm{fl}}\delta K + i(\lambda_{\mathrm{Im}})_{\mathrm{fl}}\delta B + (\lambda_{\mathrm{Im}})_{\mathrm{fl}}^2\delta A]u_{\mathrm{fl}}) + (v_{\mathrm{fl}}, Ku_{\mathrm{fl}})\delta p_{\mathrm{fl}}&\\ + (v_{\mathrm{fl}}, [B + 2i(\lambda_{\mathrm{Im}})_{\mathrm{fl}}A]u_{\mathrm{fl}})i(\delta\lambda_{\mathrm{Im}})_{\mathrm{fl}} = 0.&\end{aligned} \tag{8.123}$$

We introduce the notation

$$\begin{aligned}\Delta_{\mathrm{Re}} + i\Delta_{\mathrm{Im}} &= (v_{\mathrm{fl}}, [\delta C + p_{\mathrm{fl}}\delta K + i(\lambda_{\mathrm{Im}})_{\mathrm{fl}}\delta B + (\lambda_{\mathrm{Im}})_{\mathrm{fl}}^2\delta A]u_{\mathrm{fl}}),\\ K_{\mathrm{Re}} + iK_{\mathrm{Im}} &= (v_{\mathrm{fl}}, Ku_{\mathrm{fl}}),\\ F_{\mathrm{Re}} + iF_{\mathrm{Im}} &= (v_{\mathrm{fl}}, [D + 2i\omega_{\mathrm{fl}}A]u_{\mathrm{fl}}),\end{aligned}$$

permitting us to write this variational equation in compact form:

$$\Delta_{\mathrm{Re}} + i\Delta_{\mathrm{Im}} + (K_{\mathrm{Re}} + iK_{\mathrm{Im}})\delta p_{\mathrm{fl}} + (F_{\mathrm{Re}} + iF_{\mathrm{Im}})i(\delta\lambda_{\mathrm{Im}})_{\mathrm{fl}} = 0. \tag{8.124}$$

As a consequence of eq. (8.124), we can derive formulas for the variations δp_{fl} and $(\delta\lambda_{\mathrm{Im}})_{\mathrm{fl}}$, which determine the variation of the design variable δh:

$$\left.\begin{aligned}\delta p_{\mathrm{fl}} &= -(F_{\mathrm{Re}}\Delta_{\mathrm{Re}} + F_{\mathrm{Im}}\Delta_{\mathrm{Im}})/(K_{\mathrm{Re}}F_{\mathrm{Re}} + K_{\mathrm{Im}}F_{\mathrm{Im}}),\\ (\delta\lambda_{\mathrm{Im}})_{\mathrm{fl}} &= (K_{\mathrm{Im}}\Delta_{\mathrm{Re}} + K_{\mathrm{Re}}\Delta_{\mathrm{Im}})/(K_{\mathrm{Re}}F_{\mathrm{Re}} + K_{\mathrm{Im}}F_{\mathrm{Im}}).\end{aligned}\right\} \tag{8.125}$$

Formulas (8.125) allow us to apply the gradient projection techniques to maximize the critical value for the stability parameter (see [8.59]).

As an example, we consider a numerical solution for the problem of a column compressed by a follower force p (as shown in Figure 8.14). If we assume that vibrations are "small," the state equation in terms of the amplitude of bending for a column is of the form

$$(EIu_{xx})_{xx} + pu_{xx} + \lambda\eta(EIu_{xx})_{xx} + \lambda^2 mu = 0, \tag{8.126}$$

where EI is the bending rigidity, η is the coefficient of viscosity, and m is the mass of the follower column.

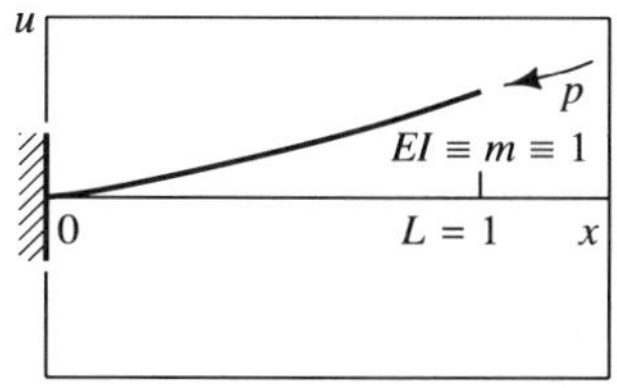

FIGURE 8.14. Follower column.

Internal damping in equation (8.126) obeys the Kelvin–Voigt law. In this case (8.122) becomes

$$\int_0^l (v^*_{xx} EIu_{xx}) + v^* pu_{xx} + \lambda\eta v^*_{xx} EIu_{xx} + \lambda^2 v^* mu)\mathrm{d}x = 0. \quad (8.127)$$

The star superscript in (8.127) denotes the complex conjugate. The best distribution of material, that is, distribution of the mass $m(x)$, is derived for the unit length $l = 1$, for the coefficient $\eta = 0.01$. We assume, for the sake of convenience, that $EI = m^2$. The design improving iterations are constructed by following the gradient-projection technique, while observing the constant mass constraint of the column ($m(x) \equiv 1$). Numerical results related to the initial approximation and to the improved design of the column (after 18 iterations) are displayed, respectively, in Figures 8.15 and 8.16. The distribution of mass for the redesigned column is displayed in Figure 8.17.

In this project the magnitude of the force that caused the loss of stability increased by more than 250% in comparison with that for the initial shape of a column with constant thickness.

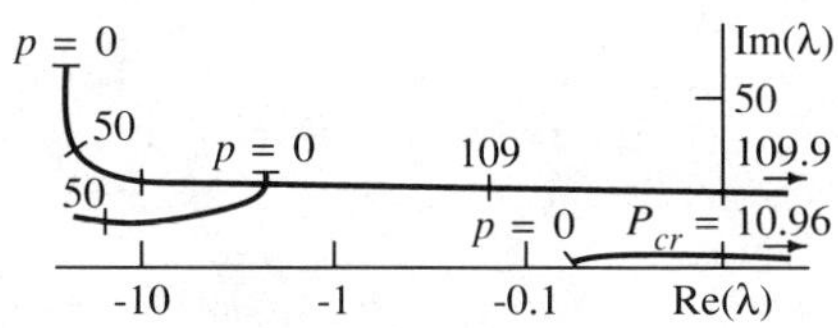

FIGURE 8.15. Initial approximation.

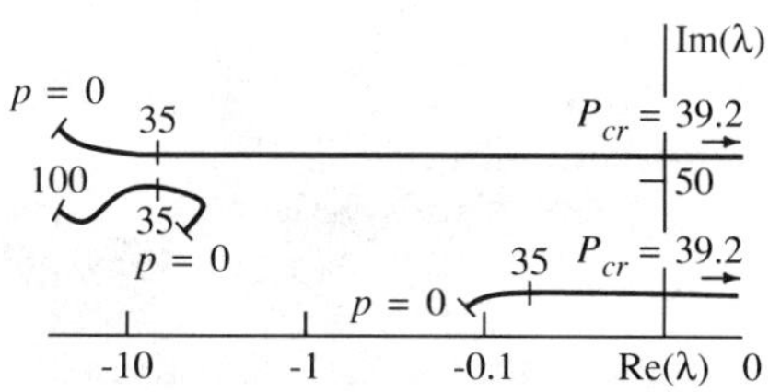

FIGURE 8.16. Improved column design.

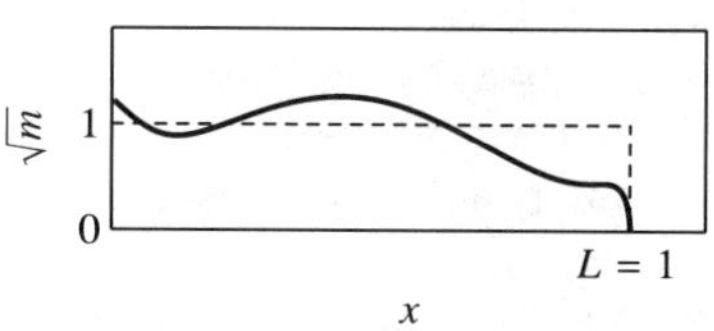

FIGURE 8.17. Final shape of redesigned column.

References for Chapter 8

8.1. Abramov, IU.Sh., *Variational Methods in the Theory of Operator Bundles. Spectral Optimization*, Leningrad, Leningrad State University, 1983, 180 pp.

8.2. Adali, S., Optimal shape and nonhomogenuity of nonuniformly compressed column, *Intern. J. Solids Struct.*, 1979, *15*, pp. 935–949.

8.3. Albul, A.V., Banichuk, N.V., and Barsuk, A.A., Optimization of stability for elastic rods with thermal loads, *Izv. Akad. Nauk S.S.S.R., M.T.T.*, 1980, No. 3, pp. 127–133.

8.4. Alfumov, N.A., *Foundations for Computation of Stability of Elastic Systems*, Moscow, Mashinostroienie, 1978, 312 pp.

8.5. Alfumov, N.A., and Balabukh, L.I., Energy criteria for stability of elastic bodies which do not require determination of the initial stressed state, *Prikl. Mat. Mekh.*, 1972, *32*, No. 1, pp. 703–770.

8.6. Andreev, L.V., Mossakovskii, V.I., and Obodan, N.I., On optimal thickness of a cylindrical shell loaded by external pressure, *Prikl. Mat. Mekh.*, 1972, *36*, No. 4, pp. 717–726.

8.7. Arnol'd, V.I., Modes and quasi-modes, *Functional Analysis and Its Applications*, 1972, *6*, No. 2, pp. 12–20.

8.8. Arnol'd, V.I., *Mathematical Methods of Classical Mechanics* (translated from Russian by K. Vogtman and A. Weinstein), 1978, Springer-Verlag, Berlin and Heidelberg, 464 pp.

8.9. Artunian, N.Kh., and Kolmanovskii, V.B., *Theory of Creep for Inhomogeneous Bodies*, Moscow, Nauka, 1983, 336 pp.

8.10. Ashley, H., and McIntosh, S.C., Application of aeroelastic constraints in Structural optimization. In: *Proceedings 12th International Congress Applied Mechanics*, Stanford University, 1969. Springer Verlag, Berlin, pp. 100–113.

8.11. Banichuk, N.V., Optimization of stability for elastically embedded rod, *Izv. Akad. Nauk S.S.S.R., M.T.T.*, 1974, No. 4, pp. 150–154.

8.12. Banichuk, N.V., Minimization of the weight of a wing with bounded speed of divergence, *Nauchn. Zap. TsAGI*, 1978, *9*, No. 5.

8.13. Banichuk, N.V., and Barsuk, A.A., On stability of torsioned elastic rods, *Izv. Akad. Nauk S.S.S.R., M.T.T.*, 1982, No. 6, pp. 148–154.

8.14. Banichuk, N.V., and Barsuk, A.A., Optimization of stability of elastic beams against simultaneous compression and torsion. In: *Applied Problems of Strength and Plasticity. Automation and algorithms for Solving Problems in Elastoplastic State* (collection). Gorkii, Gorkii State University, 1982, pp. 122–126.

8.15. Banichuk, N.V., and Barsuk, A.A., Application of decomposition of the spectrum of eigenvalues to problems of optimal design. In: *Problems of Stability and Critical Load Carrying Capacity of Structures*, Leningrad, L.I.S.I., 1983, pp. 17–24.

8.16. Banichuk, N.V., and Gura, N.M., On a dynamic problem of optimal design. In: *Mechanics of Deformable Solid Bodies*, Novosibirsk, 1979, *41*, pp. 20–24.

8.17. Banichuk, N.V., and Kobelev V.V., Optimization of structures made from chaotically reinforced composites, *Mekh. komp. materialov*, 1981, No. 4, pp. 668–676.

8.18. Bochenek, B., and Gajewski, A., Certain problems of unimodal and bimodal optimal design of structures. In: *Proceedings Euromech. In: Optimization Methods in Structural Design*, H. Eschenauer and N. Olhoff (eds.), Mannheim, Bibl. Institute, 1983, pp. 204–209.

8.19. Bolotin, V.V., *Dynamic Stability of Elastic Systems*, Moscow, Gostekhizdat, 1956.

8.20. Bolotin, V.V., *Non-conservative Problems in the Theory of Elastic Stability*, Moscow, Fizmatgiz, 1961, 339 pp.

8.21. Bolotin, V.V., Variational Principles in Theory of Elastic Stability. In: Problems in Mechanics of Solid Deformable Bodies, Moscow, Subostroienie, 1973, pp. 83–88.

8.22. Budianski, B., Frauenthal, J.C., and Hutchinson, J.W., On optimal arches, *J. Applied Mech., Trans. A.S.M.E.*, 1969, *36*, No. 4, pp. 239–240.

8.23. Bushnell, D., Buckling of shells; Pitfall for Designers, *A.I.A.A. J., 1981, 19*, No. 9, pp. 1183–1226. (Russian translation in *Raket. Tekh. Kosmon.*).

8.24. Chentsov, P.G., Minimal weight supports, *Trudy TsAGI*, 1936, *265*, pp. 1–48.

8.25. Choi, K.K., and Haug, E.J., Optimization of structures with repeated eigenvalues. In: *Optimization of Distributed Parameter Structures*, E.J. Haug and J. Cea (eds.), Alphen aan den Rijn, Sijthoff and Noordhoff, 1981, pp. 219–277.

8.26. Clausen, T., Uber die Formarchitektonischen Saulen, *Bull. Phys. Math. Acad. St. Petersburg*, 1851, *9*, pp. 279–294.

8.27. Claudon J.-L., and Sunakawa M., Optimizing distributed structures for maximum flutter load, *AIAA J.*, 1981, *19*, No. 7, pp. 957–959.

8.28. Danielov, E.R., Girder with uniform strength against longitudinal-transverse bending, Tr. Khabar. Politechn. In. 1967, *8*, pp. 26–34.

8.29. Fedoseev, V.I., *Selected Problems and Exercises on Strength of Materials*, Moscow, Nauka, 1973, 400 pp.

8.30. Frauenthal, J.C., Constrained optimal design of circular plates against buckling, *J. Structural Mech.*, 1972, *1*, pp. 159–186.

8.31. Gajewski, A., and Życzkowski, M., Optimal design of elastic columns subject to general conservative behavior of loading, *ZAMP*, 1970, *21*, No. 5, pp. 806–818.

8.32. Geniev, G.A., On the equigradient principle and its applications to problems of optimizing stability of systems of rods, *Stroit. Mekh. i Raschet Sooruzhenii*, 1979, No. 6, pp. 8–13.

8.33. Grinev, V.B., and Filippov, A.P., On optimal shape of beams in problems of stability, *Stroit. Mekh. i Raschet Sooruzhenii*, 1977, No. 2, pp. 21–27.

8.34. Grinev, V.B., and Filippov, A.P., Optimization of circular plates in problems of stability, *Stroit. Mekh. i Raschet Sooruzhenii*, 1977, No. 2, pp. 16–20.

8.35. Haug, E.J., Optimization of distributed parameter structures with repeated eigenvalues. In: *New Approaches to Nonlinear Problems in Dynamics*, P.J. Holmes (eds), Philadelphia, S.I.A.M., 1980, pp. 511–520.

8.36. Haug, E.J., and Rousselet, B., Design sensitivity analysis in structural mechanics II: Eigenvalues variations, *J. Struct. Mech.*, 1980, *8*.

8.37. Hu, K.K., and Kirmser, P.G., Numerical solution of a nonlinear differential-integral equation for the optimal shape of the tallest column, *Intern. J. Eng. Science*, 1980, *18*, pp. 333–339.

8.38. Iasinskii, F.S., *Collected Works on Stability of Compressed Columns*, Moscow and Leningrad, Gostekhizdat, 1952, 427 pp.

8.39. Karihaloo, B.L., and Parberry, R.D., Optimal design of beam-columns subjected to concentrated moments, *Eng. Optim.*, 1980, *5*, pp. 59–66.

8.40. Keller, J.B., The shape of the strongest column, *Arch. Rational Mechanics and Analysis*, 1960, *5*, No. 4, pp. 275–285.

8.41. Keller, J.B., and Niordson, F.I., The tallest column. *J. Math. and Mech.*, 1966, *16*, No. 5, pp. 433–446.

8.42. Komkov, V., Application of invariant principles to the optimal design of a column, *ZAMM*, 1981, *61*, pp. 75–80.
8.43. Komkov, V., and Haug, E.J., On the optimum shape of columns. In: *Optimization of Distributed Parameter Structures*, E.J. Haug and J. Cea (ed.), Alphen aan den Rijn, Sijthoff and Noordhoff, 1981, pp. 219–277.
8.44. Kruželecki, J., Optimization of shells under combined loading via the concept of uniform stability. In: *Optimization of Distributed Parameter Structures*, E.J. Haug and J. Cea (eds.), Alphen aan den Rijn, Sijthoff and Noordhoff, 1981, pp. 929–950.
8.45. Lagrange, J.L., Sur la figure des colonnes, *Miscellanea Taurinensia*, 1770–1773, *5*.
8.46. Kuznetsov, Iu.M., Compression and combined tension and bending of beams with minimal weight, *Dopovidi Akad. Nauk Ukr.S.S.R.*, 1959, No. 14, pp. 372–378.
8.47. Larichev, A.D., Finding a minimum volume for a beam on an elastic foundation, for a given magnitude of a critical load. In: *Applied and Theoretical Research into Building Structures*, Moscow, Kucherenko TsNIINSK, 1981, pp. 19–25.
8.48. Leipholz, H.H.E., On conservative elastic systems of the first and second kind, *Ing. Arch.*, 1974, *43*, No. 5, pp. 255–271.
8.49. Litvinov, V.G., and Medvedev, N.G., Stability problem for shells of revolution and the Ritz–Galerkin techniques, *Prikl. Mekhanika*, 1977, *13*, No. 7, pp. 8–16.
8.50. Masur, E.F., Optimal design of symmetric structures against postbuckling collapse, *Intern. J. Solids Struct.*, 1978, *14*, pp. 319–326.
8.51. Masur, E.F., and Mróz, Z., On nonstationary optimality conditions in structural design, *Intern. J. Solids Struct.*, 1979, *15*, pp. 503–512.
8.52. Masur, E.F., Optimal structural design under multiple eigenvalue constraint, *Intern. J. Solids Struct.*, 1984, *20*, No. 3, pp. 211–231.
8.53. McIntosh, S.C., and Ashley, H., On optimization of discrete structures with aeroelastic constraints, *Comput. and Structures*, 1978, *8*, pp. 411–419.
8.54. Medvedev, N.G., Some spectral properties of optimal stability problems for shells with variable thickness, *Dokl. Akad. Nauk U.S.S.R., Ser. A,* No. 9, pp. 59–63.
8.55. Medvedev, N.G., and Totskii, N.P., Multiplicity of eigenvalues in the spectrum in optimal stability problems for cylindrical shells with variable thickness, *Prikl. Mekhanika*, 1984, *20*, No. 6, pp. 113–116.
8.56. Mikisheva, V.I., Optimal weaving of shells made of glassy plastics, subjected to loss of stability due to external pressure or to axial compression, *Polymer Mechanics*, 1968, No. 5, pp. 864–875.
8.57. Naimark, M.A., *Linear Differential Operators*, Moscow, Nauka, 1969, 526 pp.
8.58. Nemirovskii, Iu.V., and Samsonov, V.I., On rational reinforcement of cylindrical shells compressed by an axial force, *Izv. Akad. Nauk S.S.S.R., M.T.T.*, 1974, No. 1, pp. 103–112.
8.59. Nemirovskii, Iu.V., and Samsonov, V.I., Cylindrical reinforced shells most stable against external pressure, *Polymers Mechanics*, 1974, No. 1, pp. 75–83.
8.60. Nicolai, E.L., Lagrange's problem on the best shapes of columns, *Izv. Peterburg. Polytekh. In.*, 1907, *5*.
8.61. Nicolai, E.L., *Research Works in Mechanics*, Moscow, Gostekhizdat, 1955, 584 pp.
8.62. Olhoff, N., and Rasmussen, S.H., On single and bimodal optimum buckling loads of clamped columns, Intern. J. Solids Struct., 1977, 13, *No.* 7, pp. 605–614.

8.63. Olhoff, N., and Taylor, J.E., Designing continuous columns for minimum total cost of material and interior supports, *J. Struct. Mech.*, 1978, *6*, pp. 367–382.
8.64. Papkovich, P.F., *Studies on Shipbuilding Structural Mechanics*, vol. 4, Leningrad, (Sudostroienie), 1963, 552 pp.
8.65. Pedersen, P., and Seyranian, A.P., Sensitivity analysis for problems of dynamic stability, *Intern. J. Solids Struct.*, 1983, *19*, No. 4, pp. 315–335.
8.66. Pierson, B.L., Pannel flutter optimization by gradient projection, *Intern. J. Numer. Meth Eng.*, 1975, *9*, pp. 271–296.
8.67. Popelar, C.H., Optimal design of beams against buckling; a potential energy approach, *J. Struct. Mech.*, 1976, *4*, pp. 181–196.
8.68. Popelar, C.H., Optimal design of structures against buckling; a complementary energy approach, *J. Struct. Mech.*, 1977, *5*, pp. 45–66.
8.69. Prager, S., and Prager, W., A note on optimal designs of columns, *Intern. J. Mech. Science*, 1979, *21*, pp. 249–251.
8.70. Prasad, S.N., and Herman, G., The usefulness of adjoint system in solving nonconservative problems of elastic continua, *Intern. J. Solids Struct.*, 1969, *5*.
8.71. Roorda, J., and Reiss A.J., Nonlinear interactive buckling; sensitivity and optimality, *J. Struct. Mech.*, 1977, *5*, pp. 207–232.
8.72. Rzhanitsyn, A.P., *Stability of Equilibrium for Elastic Bodies*, Moscow, Gostekhizdat, 1955, 476 pp.
8.73. Sandararaian, C., Optimization of a nonconservative elastic system with stability constraint, *J. Optimiz. Theory Appl.*, 1975, *16*, pp. 355–378.
8.74. Schmit, L.A., Jr., and Ramanathan, R., Multilevel approach to minimum weight design including buckling constraints, *AIAA J.*, 1978, *16*, No. 2, pp. 97–104.
8.75. Seyranian, A.P., Optimization of stability for a plate in a supersonic gas stream, *Izv. Akad. Nauk S.S.S.R., M.T.T.*, 1974, No. 1, pp. 103–112.
8.76. Seyranian, A.P., On a problem of Lagrange, *Izv. Akad. Nauk S.S.S.R, M.T.T.*, 1974, No. 1, pp. 103–112.
8.77. Seyranian, A.P., and Sharaniuk, A.V., Seneitivity analysis for optimization of critical parameters in problems of dynamic stability, *Izv. Akad. Nauk S.S.S.R., M.T.T.*, 1983, No. 5, pp. 173–182.
8.78. Seyranian, A.P., Sensitivity analysis and optimization of aeroelastic stability characteristics, *Intern. J. Solids Struct.*, 1982, *18*, No. 9, pp. 791–808.
8.79. Shanley, F.R., Optimum design for eccentrically loaded columns, *J. Structural Division, Proc. Amer. Soc. Civ. Eng.*, 1967, *93*, pp. 201–226.
8.80. Smirnov, A.F., *Stability and Vibration of Reinforcements*, Moscow, Transzheldorizdat, 1958, 572 pp.
8.81. Smirnov, A.F., Beams and arches of least weight in longitudinal bending, Tr. *Mosk. In. Zheldor. Transp.*, 1950, *74*, pp. 3–40.
8.82. Sofonov, Iu.D., Computation of the least weight of beams, with consideration of stability for a planar shape of deformed shape, *Zap, Kazan'. Aviat. In.*, Kazan', 1974, *168*, pp. 3–43.
8.83. Solodovnikov, V.N., On techniques of optimization of shells for stability of the stressed state. In: *Dynamics of Continuous Media*, Novosibirsk, Institute of Hydrodynamics, 1976, *27*, pp. 135–143.
8.84. Solodovnikov, V.N., Optimization of elastic shells of revolution, *Prikl. Math. Mekh.* 1978, *42*, No. 3, pp. 511–520.
8.85. Stadler, W., Stability of the natural shapes of sinusoidally loaded uniform shallow arches, *Quart. J. Mech. Appl. Math.*, 1983, *33*, pt. 3, pp.

8.86. Świsterski, W., Wróblewski, W., and Życzkowski, M., Geometrically nonlinear excentrically compressed columns of uniform creep strength vs optimal columns, *Intern. J. Nonlinear Mechanics*, 1983, *18*, No. 4, pp. 287–296.
8.87. Tadjbaksh, I., and Keller, J.B., Strongest columns and isoperimetric inequalities for eigenvalues, *J. Appl. Mech.*, 1962, *29*, No. 1, pp. 159–164.
8.88. Taylor, J.E., The strongest column; an energy approach. J. Appl. Mechanics, *Trans. A.S.M.E.*, 1967, *34*, No. 2, pp. 486–487.
8.89. Taylor, J.E., Optimal prestress against buckling; an energy approach, *Intern. J. Solids Struct.*, 1971, 7, pp. 213–223.
8.90. Taylor, J.E., and Liu, C.Y., Optimal design of columns, A.I.A.A. J. 1968, *6*, pp. 1496–1502.
8.91. Thomson, J.M.T., Optimization as a generator of structural instability, *Intern. J. Mech. Science*, 1972, *14*, No. 9, pp. 627–630.
8.92. Thomson, J.M.T., and Hunt, G.W., Dangers of structural optimization. *Eng. Optim.*, 1974, *1*, pp. 99–100.
8.93. Thomson, J.M.T., and Supple, W.D., Erosion of optimal design by compound branching phenomena, *J. Mech. Phys. Solids*, 1972, *21*, No. 3, pp. 135–144.
8.94. Timoshenko, S.P., *Stability of beams plates and shells. Selected Works*, Moscow, Nauka, 1971, 808 pp.
8.95. Vol'mir, A.S., *Stability of Deformed Systems*, Moscow, Nauka, 1967, 984 pp.
8.96. Weisshaar, T.A., and Plaut, R.H., Structural optimization under nonconservative loading. In: *Optimization of Distributed Parameter Structures*, E.J. Haug and J. Cea (eds), Alphen aan den Rijn, Sijthoff and Noordhoff, 1981, pp. 843–864.
8.97. Ziegler, G., On stability of elastic systems. In: *Problems of Mechanics*, Moscow, Foreign Literature, 1959, pp. 116–160.
8.98. Ziegler, G., *Foundations of the Theory of Structural Stability*, Moscow, Mir, 1971, 192 pp.
8.99. Życzkowski, M., Optimal design of shells with respect to their elastic stability. In: *Stability of Elastic Structures*, H.H.E. Leipholz (ed.), Wien, Springer-Verlag, 1978, pp. 268–291.
8.100. Życzkowski, M., and Krużelewski, J., Optimal design of shells with respect to their stability. In: *Optimization in Structural Design*, IUTAM, A. Sawczuk and Z. Mróz (eds.), Springer-Verlag, Berlin and New York, 1975, pp. 229–247.

Bibliography

Related Monographs and Expository Articles on the Theory of Structural Design

1. Armand, J.L., *Applications of Optimal Control Theory for Systems with Distributed Parameters to Problems of Structural Optimization*, Moscow, Mir, 1977, 142 pages.
2. Ashley, H., On making the best aeronautical uses of optimization, *J. Aircraft*, *19*, No. 1, pp. 5–28.
3. Atrek, E., et al. (eds.), New Directions in Optimum Structural Design. Chichester, Wiley, 1984, 727 pp.
4. Bakulenko, L.D., and Mazalov, V.N., *Optimal Design of Structures. Bibliography of Soviet and Foreign Literature 1948–1974*, Novosibirsk, Institute of Hydrodynamics, S.O. Akad. Nauk. S.S.S.R., 1975, Parts I and II, 472 pp.
5. Banichuk, N.V., *Optimization of the Shapes of Elastic Bodies*, Moscow, Nauka, 1980, 256 pages. (Translated into English as *Problems and Methods of Optimal Structural Design* (see [8]).
6. Banichuk, N.V., Optimality conditions and analytical methods of shape optimization. In: *Optimization of Distributed Parameter Structures*, E.J. Haug and J. Cea (eds.), Alphen aan den Rijn, Sijthoff and Noordhoff, 1981, pp. 973–1004.
7. Banichuk, N.V., Modern problems in structural optimization, *Izv. Akad. Nauk S.S.S.R., MTT*, 1982, No. 2, pp. 110–124.
8. Banichuk, N.V., *Problems and Methods of Optimal Structural Design*, Plenum Press, New York, 1983, 313 pp. (with preface by E.J. Haug and V. Komkov, and supplementary bibliography.)
9. Barnett, R.L., Survey of optimum structural design. *Exp. Mech.* 1966, *6*, No. 12, pp. 19–26.
10. Biriuk, V.I., Lipin, E.K., and Frolov, V.M., *Methods of Aircraft Structure Design*, Moscow, Mashinostroienie, 1977, 232 pp.
11. Bogomolov, S.I., and Simpson, E.A., *Optimization of Mechanical Systems in Resonance States*, Kharkov, Vishcha Shkola, 1983, 153 pp.
12. Brousse, P., Gallagher, R.H., and Gellatly, R.A., *Optimum Structural Design*, Paris, Inst. pour la promotion des sciences d l'ingenieur, 1976, 333 pp.
13. Chiras, A.A., *Mathematical Models for the Analysis of Elastoplastic Systems*, Vilnius, Mokslas, 1982, 112 pp.
14. Optymizacja wytrzymałościowa konstrukcji (praca zbiorowa), Dzieniszewski, W., et al. (eds.), Warsaw, Ossolineum, 1983, 522 pp.

15. Optimization methods in structural design: *Intern. Proc. Euromech Colloqium.* H. Eschenauer and N. Olhoff (eds), Mannheim Bibl. Institute, 1983, 452 pages.
16. *Optimum Structural Design; Theory and Applications*, Gallagher R.H. and Zienkiewicz O.C. (eds.), London, 1973, Wiley, 358 pages.
17. Gol'dshtein, Iu.V., and Solomets, M.A., *Variational Problems of Statics for Optimal Systems of Beams*, Leningrad, Leningrad State University, 1980, 208 pp.
18. Goluber I.S., *Analytical Methods for Design of Wing Structures*, Moscow, Mashinostroienie, 1970, 288 pp.
19. Grinev, V.B., and Filippov, A.P., *Optimization of Structural Elements with Respect to Mechanical Properties*, Kiev, Naukova Dumka, 1975, 294 pp.
20. Grinev, V.B., and Filippov, A.P., *Optimization of Beams with Respect to the Spectrum of Eigenvalues*, Kiev, Naukova Dumka, 1979, 212 pp.
21. Haug, E.J., and Arora, J.S., *Applied Optimal Design.* Wiley, New York, 1979, 506 pp.
22. Haug, E.J., and Cea, J. (eds.), Optimization of distributed parameter structures, *Proc. Advanced Study Institute,* NATO, Iowa City, Iowa, 1980. Publ. Sijthoff and Noordhoff, Alphen aan den Rijn, Holland, 1981, in 2 volumes, 1609 pages.
23. Haug, E.J., A review of distributed parameter structural optimization literature, in: *Optimization of Distributed Parameter Structures*, E.J. Haug and J. Cea (eds.), Alphen aan den Rijn, Sijthoff and Noordhoff, 1981, pp. 3–68.
24. Haug, E.J., and Arora, J., *Applied Optimal Design*, Moscow, Mir, 1983, 478 pp. (See [31].)
25. Komkov, V., *Optimal Control Theory for Damping of Vibrations of Simple Elastic Systems*, Springer-Verlag, Lecture Notes in Mathematics, *253*, Berlin, 1972 (Russian translation 1975, Mir, Moscow. Foreword by Selezov).
26. Komkov, V. (eds.), *Sensitivity of Functionals with Applications to Engineering Sciences. Proceedings of a Special Session of American Mathematical Society*, New York, 1983, Springer-Verlag, Berlin, 1984, 130 pp.
27. Lepik, Iu., *Optimal Design of Nonelastic Structures for Dynamic Loads*, Tallin, Valgus, 1982, 160 pp.
28. Malkov, V.P., and Ugodchikov, A.G., *Optimization of Elastic Systems*, Moscow, Nauka, 1981, 288 pp.
29. Mazhid, K.I., *Optimal Design of Structures*, Moscow, Vyssh. Shk. 1979, 237 pp.
30. Niordson, E.I., and Pedersen, P.A., Review of optimal structural design. In: Applied Mechanics, Becker and Mikhailov (eds.), Berlin, Springer Verlag, 1973, pp. 264–278.
31. Obraztsov, I.F., Vasil'ev, V.V., and Bunakov, V.A., *Optimal Reinforcement for Shells of Revolution Made from Composite Materials*, Moscow, Mashinostroenie, 1977, 144 pp.
32. Olhoff, N., *Optimal Design of Structures*, Moscow, Mir, 1977, 109 pp.
33. Olhoff, N., A survey of the optimal design of vibrating structural elements. Part I: Theory; Part II: Applications. Shock and Vibration Digest, 1976, vol. 8, No. 8, pp. 3–10, & No. 9.
34. Olhoff, N., Optimal design with respect to structural eigenvalues, in Proceedings XV-th International Congress on Theoretical and Applied Mechanics, Toronto, Canada, 1980. North Holland, Amsterdam, 1980, pp. 133–149.
35. Pierson, B.L., A survey of optimal structural design under dynamic constraints, *Intern. J. Numer. Meth. Eng.*, 1972, *4*, pp. 491–499.
36. Prager, W., Optimization of structural design, *J. Optimiz. Theory Appl.*, *6*, No. 31, pp. 1–21.

37. Prager, W., *Introduction to Structural Optimization*, Springer-Verlag, Wien, 1972, 80 pp.
38. Prager, W., *Foundations of the Theory of Structural Optimization*, Moscow, Mir, 1977, 109 pp. (See reference [43].)
39. Radtsig, Iu.A., *Statically Indeterminate Trusses of Minimal Weight*, Kazan, Publ. Kazan University, 1969, 287 pp.
40. Reitman, M.I., and Shapiro, G.S., Optimal Design of Deformable Bodies. In: Directions in Science and Technology: Mechanics of Deformable Solid Bodies, Moscow, VINITI, 1978, *12*, pp. 5–90.
41. Rozvany, G.I.N., *Optimal Design of Flexural Systems*, Pergammon Press, Oxford, 1976, 297 pp.
42. Rozvany, G.I.N., and Mróz, Z., Analytical methods in structural optimization, *Appl. Mech. Reviews*, 1977, *30*, No. 11, pp. 1461–1470.
43. Rozvany, G.I.N., *Optimal Design of Flexural Systems*, Moscow, Stroiizdat, 1980, 316 pp. (See [45]).
44. Sawczuk, A., and Mroz, Z. (eds.), *Optimization in Structural Design: Symposium IUTAM*, Berlin, Springer-Verlag, 1975, 585 pp.
45. Sheu, C.Y., and Prager, W., Recent developments in optimal structural design, *Appl. Mech. Reviews*, 1968, *21*, No. 10, pp. 985–992.
46. Stadler, W., A survey of multicriteria optimization, or the vector maximum problem. Part 3: 1776–1960, *J. Optimiz. Theory and Appl.*, 1979, *29*, No. 1, pp. 1–52.
47. Teters, G.A., Rikards, R.B., and Narusberg, V.L., *Optimization of Shells Made of Layered Composites*, Riga, Zinatne, 1978, 240 pp.
48. Troitskii, V.A., and Petukhov, L.V., *Optimization of the Shape of Elastic Bodies*, Moscow, Nauka, 1982, 432 pp.
49. Vasil'ev, V.V., Optimal designs of shells and plates. In: *Proc. VII-th All-Union Conference on Theory of Plates and Shells*, Moscow, Nauka, 1970, pp. 721–735.
50. Venkayya, V.B., Design of optimum structures, *Comput. and Struct.*, 1971, *1*, No. 1–2, pp. 265–309.
51. Venkayya, V.B., Structural optimization: A review and some recommendations, *Intern. J. Numer. Meth. Eng.*, 1978, *13*, pp. 203–228.
52. Wasiutyński, Z., and Brandt, A., The present state of knowledge in the field of optimum design of structures, *Appl. Mech. Reviews*, 1963, *16*, No. 5, pp. 341–350.
53. Arora, J.S., and Wu, C.C., Design sensitivity analysis and optimization of nonlinear structures. In: *Computer-Aided Optimal Design: Structural and Mechanical Systems, Portugal, 1986*, C.A. Mota-Soares, (ed.), NATO-ASI Series, Springer-Verlag, Berlin, 1987, pp. 589–603.
54. Haftka, R.T., and Kamat, M.P., *Elements of Structural Optimization*, Martinus Nijhoff, Dordrecht, Holland, 1985.

Supplementary Selected Bibliography Added in Translation

55. Haug, E.J., Choi, K.K., and Komkov, V., *Design Sensitivity Analysis of Structural Systems*, Academic Press, New York, 1986.
56. Kohn, R.V., and Strang, G., Optimal design and relaxation of variational problems. Parts I, II, and III, *Comm. Pure Appl. Math. 39*, 1986, pp. 113–137, 139–182, 353–377.

57. Komkov, V., *Variational Principles of Continuum Mechanics with Engineering Applications*; Volume II: *Introduction to Optimal Design Theory*, D. Reidel, Dordrecht, Holland, 1988.
58. Komkov, V., Sensitivity techniques for systems with distributed parameters, *J. Math. Analysis Appl.*, *128*, No. 2, 1987, pp. 443–455.
59. Komkov, V., Properties of sensitivity functions, *Arch. Mech.*, *39*, No. 6, 1987, pp. 693–698.
60. Li, H.L., *Design Optimization Strategies with Global and Local Knowledge*, Ph.D. Dissertation, University of Michigan, 1985, Ann Arbor, Michigan.
61. Mota-Soares, C.A. (ed.), *Computer-Aided Optimal Design: Structural and Mechanical Systems. Proceedings of a Meeting Held in Troia, Portugal, July, 1986*, NATO-ASI Series, Springer-Verlag, Berlin, 1987.
62. Murat, F., and Tartar, L.C., Calcul des variations et homogeneisation. In: *Les Méthodes de l'homogeneisation: Théorie et Applications en Physique*, Coll. de la Dir. des Etudes et Recherches de Elect. de France. *57*, Eyrolles (Paris), 1985, pp. 319–369.
63. Olhoff, N., Structural optimization by variational methods. In: *Computer-Aided Optimal Design: Structural and Mechanical Systems*, Troia, Portugal, 1986. C.A. Mota-Soares (ed.), NATO-ASI Series, Springer-Verlag, Berlin, 1987, pp. 87–164.
64. Ozawa, S., Spectra of domains with small spherical Neumann boundary, *J. Fac. Tokyo Univ., Section IA*, *30*, 1983, pp. 259–277.
65. Papalambros, P.Y., Knowledge Based Systems in Optimal Design. In: *Computer-Aided Optimal Design: Structural and Mechanical Systems. Troia, Portugal, 1986.* C.A. Mota-Soares (ed.), NATO-ASI Series, Springer-Verlag, Berlin, 1987, pp. 759–804.
66. Papalambros, P.Y., and Wilde, D.J., *Principles of Optimal Design. Modelling and Computation*, Cambridge University Press, Cambridge, 1988.
67. Pironneau, O., *Optimal Shape Design for Elliptic Systems. Springer Series in Computational Physics*, Springer-Verlag, New York & Berlin, 1984.
68. Polak, E., On mathematical foundations of nondifferentiable optimization in engineering design, *SIAM Review*, *29*, No. 1, 1987, pp. 21–89.
69. Pshenichnyi, B.N., and Danilin, Yu.M., *Numerical Methods in Extremal Problems* (English translation, second edition), Mir, Moscow, 1982.
70. Rousselet, B., and Haug, E.J., Design sensitivity in structural mechanics. Part III; Effects of shape vibration, *J. Struct. Mech. 10*, No. 3, 1982, pp. 273–310.
71. Taylor, J.E., Distributed parameter optimal structural design: Some basic problem formulations and their applications. In: *Computer-Aided Optimal Design: Structural and Mechanical Systems, Troia, Portugal, 1986.* Mota-Soares, C.A. (ed.), NATO-ASI Series, Springer-Verlag, Berlin, 1987, pp. 3–85.
72. Vanderplaats, G.N., *Numerical Optimization Techniques for Engineering Design*, McGraw-Hill, New York, 1984.